MINING ENVIRONMENTAL POLICY

Mining Environmental Policy

Comparing Indonesia and the USA

MICHAEL S. HAMILTON
University of Southern Maine, USA

Routledge
Taylor & Francis Group

LONDON AND NEW YORK

First published 2005 by Ashgate Publishing

Reissued 2018 by Routledge
2 Park Square, Milton Park, Abingdon, Oxon, OX14 4RN
605 Third Avenue, New York, NY 10017

First issued in paperback 2021

Routledge is an imprint of the Taylor & Francis Group, an informa business

A Library of Congress record exists under LC control number: 2005014419

Notice:
Product or corporate names may be trademarks or registered trademarks, and are used only for identification and explanation without intent to infringe.

Publisher's Note
The publisher has gone to great lengths to ensure the quality of this reprint but points out that some imperfections in the original copies may be apparent.

Disclaimer
The publisher has made every effort to trace copyright holders and welcomes correspondence from those they have been unable to contact.

ISBN 13: 978-0-815-39053-4 (hbk)
ISBN 13: 978-1-351-15300-3 (ebk)
ISBN 13: 978-1-138-35841-6 (pbk)

DOI: 10.4324/9781351153003

Contents

List of Figures

List of Tables

Preface

Comparative public policy may be defined as "the study of how, why and to what effect different governments pursue particular courses of action or inaction" (Heidenheimer, Heclo and Adams, 1990, p. 3). To ask how governments pursue particular policies focuses our attention on what goes on inside and at the fringes of the state, requiring some familiarity with the structures and processes of governmental decision making. Answering "why" questions requires some familiarity with history and tradition, political culture, and changes in the social and economic conditions that affect issue saliency and agenda setting in the specific context in which each policy development occurs. Examining the effects of public policy requires evaluation of success or failure to achieve policy goals during policy implementation. Thus it is essential to examine three theoretically and practically important dimensions of environmental policy: policy history, policy process, and policy performance (Desai, 2002, pp. 3–7).

The measure of policy performance lies in its impact on actual people. Both the content of policy statements and practices of actual policy implementation must be studied to understand how a policy actually works on the ground. "Studying comparative public policy, rather than comparative government or political behavior, gives special attention to the effects of government action on people's lives" (Heidenheimer, Heclo and Adams, 1990, p. 4). Policy statements contained in laws, regulations and executive decrees describe goals and policy intentions, but are not enough in themselves to reveal what actually happens, which nearly always includes unexpected events and consequences.

In effect, a policy is not entirely determined until it is implemented. "What matters for purposes of comparative public policy is the string of decisions that add up to a fairly consistent body of behavior on the part of the government" (Heidenheimer, Heclo and Adams, 1990, p. 5). Thus assessing a government's capacity for coping with unintended consequences, and adapting to its social context, requires examination of the results of policy implementation. It is this focus on consequences that distinguishes comparative policy analysis from comparative government and comparative political behavior of individuals (Antal, Dierkes and Weiler, 1987, p. 18).

Comparative policy research is an outgrowth of the comparative study of politics, stimulated in part by Almond and Powell's assertion that broad comparative analysis of politics and political systems "cannot take the place of substantive comparative policy analysis in specific fields such as ... environmental protection ..." (1978, p. 288). They suggested the content and implementation of public policy may be studied as indicators of the performance of political systems, as have others subsequently (Almond and Powell, 1978, p. 283; Antal, Dierkes and Weiler, 1987, p. 17; Heidenheimer, Heclo and Adams 1990, pp. 1–6).

Does comparative policy research produce information that is useful to policy makers or other scholars? Does the research serve the "intelligence needs of policy" or "improve the rationality of the flow of decision" (Lerner and Lasswell, 1951, pp. 3–4)? Comparative policy analysis has been more successful in coping with cross-cultural factors than comparative politics or comparative public administration research (Heady, 1996, p. 49). The problem-orientation and multidisciplinarity of comparative policy research make it more relevant to decision makers and scholars than earlier approaches to the study of comparative politics (Antal, Dierkes and Weiler, 1987, p. 17). Researchers conducting such investigations "often find they learn as much, if not more, about their own political system by studying others" (Antal, Dierkes and Weiler, 1987, p. 15; Heidenheimer, Heclo and Adams, 1990, pp. 1–9).

Using a comparative policy approach, this book explores how contextual factors, legal resources, and organizational variables function to determine regulatory decisions in Indonesia concerning environmental control of coal mining operations. Cross-national comparative analysis of coal development policies and environmental policies in the United States and Indonesia reveals how political culture and institutions affect policy structure and implementation at the convergence of these two related policy areas, described here as "mining environmental policy."

In both the United States and Indonesia, public policies are enacted on a piecemeal, incremental basis, to solve particular problems which are salient at the time they are considered. Incremental policy making produces numerous narrow responses to particular problems or perceived needs. New policies, missions, organizations and government functions accumulate over time at all levels of government. Most environmental policies in both countries were developed in isolation from one another, as if new concerns were separable from existing policies on related aspects of the same subject. Thus while we often aggregate several policies together and refer to them in shorthand as "environmental policy" or "energy policy," their many component parts were enacted at different times to address different problems. No comprehensive strategy or overview was evident during their accumulation, few priorities were established between them, and there is little evidence of any overall logic within each area. Few of these policies were designed to mesh together neatly, and legislative requirements for operational coordination between environmental programs were minimal.

Particular concerns for mineral development and environmental protection were addressed separately, at different times, with few attempts to relate new functions to old, resulting in a patchwork of policies. In both Indonesia and the United States, mineral development policies were established and coal development proceeded for many years before environmental assessment and control policies were created and applied to mining operations. Early environmental legislation in each country was designed to apply to many different industrial activities, and was not tailored specifically to mining operations until much later. Administrators were left to their own devices to figure out how to apply general goals to specific cases.

Fragmentation of regulatory policies and processes may impede assessment of relevant information prior to regulatory decisions. The degree of integration between two policies may determine the success or failure of policy implementation. Consequently, related policies cannot really be understood unless examined in the combinations in which they are implemented. This is especially important when the goals of one policy appear to pull decision making in the direction of environmental degradation, while goals of another policy pull decision making towards environmental protection. In these circumstances, we must expect implementation of each policy to be influenced by goals of the other. This book explores the degree of integration between coal development policies and environmental policies in two countries, with an eye towards explaining their successes and failures during policy implementation. It examines cultural, structural and institutional reasons why Indonesian mining environmental policy implementation suffers from frequent failures to attain policy goals.

Previous policy history and recent efforts to develop institutional capacity, formulate and implement new public policies for control and management of the environmental impacts of surface mining coal in Indonesia during the period 1995–2004 are examined. Although mining environmental policies in the United States were relatively stable during this period, it was a time of great turmoil in Indonesia, as the country moved rapidly away from an authoritarian Presidency of 32 years duration towards a decentralized form of democracy.

This book makes original contributions principally to comparative public policy research, and secondarily to the literatures of international development and institutional capacity building. It should be of special interest to scholars and practitioners in the fields of public policy, regulatory politics, international relations and foreign policy, comparative public policy and public administration. Scholars and practitioners in engineering concerned with policies affecting mineral development should also find the book of interest, especially those in Australia, Indonesia, South Korea, Singapore, Japan and China, and those in European countries having large international development programs or home to multinational firms developing resources in Asia (e.g., Great Britain, Germany, the Netherlands).

Despite the fact Indonesia is now the eighth largest producer of coal in the world, and the largest exporter of coal in the Asia region, its surface mining regulatory policies have received little attention in the literature. They have received only brief treatment in books with a broader focus (Warren and Elston, 1994), and in conference proceedings (Simatupang and Wahju, 1994; UNCTAD, 1994), mostly with Indonesian and Australian publishers. Surface mining regulation in the United States has received somewhat more attention, but not recently, and not in comparative analysis. Harris (1985) provided an economic interpretation of behavioral responses by individual coal mining firms in the United States to the Surface Mining Control and Reclamation Act of 1977 in terms of variables related to size of the firm and geographic location of its operations. Shover, Clelland and Lynzwiler (1986) provided a sociological interpretation of the early years (1977–1982) of program development and implementation by the Office of Surface Mining, applying *neo-Marxist* theories of the state. Neither book provided a cross-national comparison of mining regulatory policy, or addressed the integration of mining and environmental policy. Both books are substantially outdated following many years of administrative policy change, including significant shifts in enforcement strategies from use of regulatory design standards to performance standards, and from national to state agency primacy in policy implementation.

Occasionally U.S. surface mining policy has been treated in journal articles and anthologies (Desai, 1993; Sheberle, 1997; Miller, 1993, 1999), but usually not in a comparative analysis. Galloway and McAteer (1980) provided a comparative examination of surface mining regulatory policy in Germany, Great Britain, Australia and the United States but did not include Indonesia or systematically discuss the degree of integration of mining regulation with environmental impact assessment requirements. Indonesian environmental assessment policies have occasionally been described in a comparative context, but usually without much analysis. Modak and Biswas (1999) described Indonesian environmental assessment in comparison with several other developing countries. Welles (1995) compared environmental impact assessment practices in Indonesia, the Philippines, and Sri Lanka. Dick and Bailey (1992) contributed a case study of Indonesian environmental assessment policies to the international development literature. MacAndrews (1994) described the creation and early development of the Indonesian Environmental Impact Management Agency (BAPEDAL). However all these publications are dated, published before the traumatic events of Presidential succession and enactment of sweeping decentralization laws in Indonesia circa 1999, which produced dramatic changes in mining environmental policy and implementation. Indonesian environmental assessment practices have not previously been systematically compared with those of the United States.

Occasional in-house "country studies" prepared by staff of international development organizations reviewed environmental experience and prospects for sponsor activity in their areas of interest (World Bank, 1994a; McMahon, et al, 2000). These lacked a focus on environmental policy, did not utilize a comparative approach, and found only limited distribution outside sponsoring organizations.

There is great interest and a vast literature analyzing environmental assessment policies of the United States. Most numerous are case books for use in law school classes (Anderson, Mandelker and Tarlock, 1999), anthologies (Clark and Canter, 1997), and how-to-do-it handbooks (Hildebrand, 1993). Some early scholarly studies (Anderson, 1973; Liroff, 1976) are still useful, some excellent studies of litigation as an instrument of U.S. environmental policy change have been published (O'Leary, 1993), and there are numerous journal articles (Bear, 1989), and a few government publications evaluating the performance of U.S. environmental assessment practices (U.S. Council on Environmental Quality, 1976). None of this literature compares environmental assessment policies or practices in the U.S. to those of Indonesia, or attempts to address their degree of integration with mining regulatory policies. This book is the first scholarly examination of the degree of integration between environmental assessment policies and mining regulatory policies in Indonesia and the United States, or any two countries.

A significant contribution is made by this book to continuing development of a relevant knowledge base for both domestic and foreign policy making and implementation. The origins and implementation of bilateral science and technology exchanges and technical assistance agreements between the U.S. and other countries are examined. Gaps in our understanding about how other countries deal with environmental problems are narrowed. The book specifies conditions under which one country can learn from another by comparing structural, institutional and cultural constraints imposed on policy develop-ment and implementation in the U.S. and Indonesia.

The development of mining environmental policy in Indonesia during 1995–2004 is a multifaceted story that illustrates how international development assistance may be used to advance environmental protection goals in industrializing countries. It is also a story about how U.S. foreign policy objectives – such as improved understanding and maintenance of friendly relations with a country having the largest Muslim population in the world, Indonesia – may be successfully pursued by the U.S. State Department through mechanisms other than direct diplomacy, utilizing personnel and expertise of "technical agencies" of American government other than those in that Department or the U.S. Agency for International Development.

Although they are seldom examined and there is virtually no literature about their efficacy, numerous agencies of the U.S. government have for many years maintained science and technology or technical assistance relationships with counterpart agencies in foreign governments, approved by the U.S. State Department. For example, in 2001, the various bureaus, services and offices of the U.S. Department of the Interior alone had over 100 formal agreements with 52 foreign governments and 15 multilateral agencies, authorized by international treaties, program-specific legislation, general provisions of the Foreign Service Act of 1961, and the National Environmental Policy Act of 1969 (U.S. DOI, 2001). This number does not include similar agreements between the U.S. Environmental Protection Agency, the U.S. Department of Agriculture/Forest Service or other agencies with foreign governments or multilateral agencies, which are beyond the scope of this book.

Successes of diplomacy are not always reported in media stories about the signing of major international treaties, but are often unheralded, based on small actions and relationships which build confidence, trust, and stability in bilateral relations. Such relationships are often valued by Embassy staff in turbulent times such as Indonesia experienced during the rapid succession from long-term President Soeharto, to short-term President Habibie, through the truncated term of the first elected and first impeached President Abdurrahman Wahid, to reform President Megawati – all of which transpired in less than four years during 1998–2002. Continuation of regular, scheduled science and technology exchanges and technical assistance activities between American and Indonesian personnel contributed to stable and generally positive bilateral relations during that period of turmoil.

The spread of new ideas may be either spontaneous or planned (Rogers, 1995, p. 7). There is a substantial literature about diffusion of innovations and policy entrepreneurship among the various states in the United States (Walker, 1969; Gray, 1973; Savage, 1985; Berry, 1994; Mintrom, 1997). However there is little literature on the spread of new ideas between countries through science and technology cooperation, technical assistance arrangements, or other small instruments of international relations. This book illustrates how technology diffusion (Rogers, 1995), policy transfer, and reformulation of environmental policy by a foreign government may be stimulated and assisted by the U.S. government through technical training activities, advice on regulatory program development, and assistance in solving environmental problems. This finding constitutes an original and significant extension of the theory of international environmental relations, which previously suggested that for policy transfer to occur, "sustained diplomatic efforts must be initiated or explicit threats communicated" (DeSombre, 2000, Ch. 1). In several instances documented in this book, neither sustained diplomatic efforts of persuasion nor explicit threats were evident, but policy change was facilitated by a more subtle, but deliberate exchange of ideas.

Another facet of this story concerns "entrepreneurial bureaucracy," evident in the initial request for technical assistance by an employee of the Indonesian Ministry of Mines and Energy in 1991, and later during pursuit of funding for a specific project. This book provides evidence suggesting the "politics of getting" (Lasswell, 1958; Wengert, 1979, p. 17) international financial assistance appears to motivate pursuit of development projects by some recipient governments. It suggests provision of financial assistance by The World Bank for environmental programs is based on familiar clientele relationships rather than any recent greening of that institution.

Initiation of these activities also reveals how entrepreneurial bureaucracy in an American agency seeking expansion of its mission engaged in activities beneficial to the agency, but not entirely accepted by its existing organizational culture. The Office of Surface Mining Reclamation and Enforcement (OSM) in the U.S. Department of the Interior was originally a regulatory agency providing financial and technical assistance to states while they developed their own mining regulatory programs. As state governments assumed primacy of regulatory functions over surface coal mining in the 1980s, OSM's role shifted to regulatory overseer, and later in the 1990s increasingly towards provider of technical assistance to state regulatory agencies. As its regulatory activities and number of full-time personnel declined by nearly half, OSM moved haltingly to further develop its domestic and international technical assistance capabilities.

However despite a steady stream of inquiries and visiting delegations from foreign governments, OSM's ability to develop programs of international technical assistance was restrained by an aging and somewhat chauvinistic organizational culture among upper and middle-management. This forced greater use of personnel from state regulatory authorities and less use of OSM personnel than originally planned to deliver technical assistance in Indonesia, producing an unexpected benefit of improved communication and intergovernmental relations between technical personnel in state and national government agencies. Thus a somewhat obstructive organizational culture among career managers in OSM forced project adaptation which resulted in greater professional development opportunities for personnel in state regulatory agencies, and fewer opportunities for OSM professional staff. Thus description here of the different motivations for entrepreneurial bureaucracy apparent in Indonesia and the U.S. provide a significant contribution to the literature of public administration.

Because the executive branch of government in a unitary system like Indonesia generally exercises more initiative and power than the legislative branch, because Indonesia functions as a bureaucratic polity, and because its Ministry of Energy and Mineral Resources has very general, ambiguous legislative mandates which grant broad discretionary powers, administrative decision makers there have great latitude in designing the information search and decision making procedures they use. Thus Ministry decision making practices and procedures specified in legislation, regulations, decrees and guidelines can be viewed as outputs of regulatory politics, which have great potential impact on the nature of subsequent regulatory decisions such as permits to mine coal, inspection of mining operations, and enforcement of regulatory requirements. Because legislative mandates generally assume decision making procedures and information used will allow attainment of policy goals, they may be examined to see whether rational decision making does occur when it appears to be required.

Rather than focusing on the desirability of specific decisions in terms of some set of implicit or explicit values, this research examined how the procedures used may or may not be improved in their capability to provide information adequate to minimize decision making errors which may impair attainment of policy goals (Janis and Mann, 1977). Thus this book contributes to scholarship in public administration and regulatory politics by examining factors which influence the design and reform of regulatory decision making practices.

Overview of the Book

A brief description of the environmental impacts of surface mining and a conceptual framework for understanding policy development and implementation in Indonesia are presented in the first chapter. The role of contextual factors such as political culture and traditional patterns of land use in shaping legal resources and organizational variables during implementation of mining environmental policy are outlined.

The role of mining in economic development and the evolution of a policy framework for regulating exploitation of coal resources in Indonesia are examined in the second chapter. Difficulties encountered during implementation of mining regulatory policies in a period of extremely rapid resource development, political upheaval involving struggles over succession to the Presidency, and decentralization of national government functions in Indonesia are described and evaluated. Although Indonesia's Basic Mining Law of 1967 predates enactment of the Surface Mining Control and

Reclamation Act of 1977 in the United States, evidence of technology diffusion from the U.S. to Indonesia immediately thereafter is examined. In the third chapter, examination of regulatory policy for surface coal mining in the United States provides a basis of comparison with the Basic Mining Law of 1967 and its implementation in Indonesia. Similarities and differences in substantive and procedural elements of policy requirements and implementation in these two countries are analyzed.

The fourth chapter describes and compares environmental assessment policies and practices applied to coal mining development proposals in the United States and Indonesia. Although the resulting policies display some similarities in policy goals and language, it is evident Indonesian policy makers created a regulatory program to identify and control environmental impacts of development proposals, while American policy makers attempted a broader reform of administrative decision making in national government agencies. The results of policy implementation and degree of integration between mining regulatory policies and environmental assessment requirements in each country are examined.

In the fifth chapter, some of the consequences of failure during mining environmental policy implementation in Indonesia are illustrated. Specific instances of degradation of environmental quality by coal mining operations are described, and substantial losses of government revenues and company profits are estimated, based on empirical observation of environmental conditions at Indonesian coal mines and application of best practices used routinely in the United States.

The sixth chapter describes recent efforts to improve mining environmental policy in Indonesia. Establishment of a new policy requiring reclamation performance bonding for all coal mining operations, publication of guidelines for sediment and erosion control, establishment of new water quality standards applicable to coal mines and a new policy for coal seam fire management are described. Policies developed through technical assistance provided during a series of science and technology exchanges between the U.S. Office of Surface Mining and Reclamation Enforcement and the Indonesian Ministry of Energy and Minerals provide evidence of technology diffusion and policy transfer in the absence of sustained diplomatic efforts or explicit threats.

The focus of the seventh chapter is improvement of regulatory program structure and performance for implementation of mining environmental policy in Indonesia. A proposed reorganization of the Ministry of Mines and Energy to provide a stronger organizational unit for mining environmental inspections, and recommended decentralization of inspection responsibilities to regional offices of the national Ministry are examined. The impact on these proposals of laws decentralizing national government mining regulatory and environmental functions to regional governments is analyzed.

In the eighth chapter, recent efforts to develop the human and technological resources of the Ministry of Mines and Energy to support more effective implementation of mining environmental policy are examined. These include enhancing the permanent capability of the Ministry to provide technical training to its employees, developing expertise of government employees in relevant technical subjects, transfer of state-of-the-art technologies for analysis of mining environmental problems, and training in the use of these technologies. All of these activities were instrumental in technology diffusion concerning mining environmental policy.

This book is informed by lessons learned from five years of effort pursuant to an innovative arrangement in which the Indonesian Ministry of Mines and Energy participated in science and technology cooperation with the U.S. Department of the Interior, funded by The World Bank, to develop institutional capacity for reformulation and implementation of mining environmental policy. From April 1995 to August 1998, a technical assistance program was conducted in Indonesia through a series of training activities and problem-solving seminars by joint teams comprised of both U.S. and Indonesian personnel. This lead to a second project conducted during 1998–2000 which helped establish a policy and capability to identify, inventory and extinguish coal seam fires, funded by the Southeast Asia Environmental Initiative of the U.S. Department of State (Whitehouse and Mulyana, 2004). A third project during 2000–2005 focused on issues of governance under a decentralized national structure in Indonesia (U.S.DOI/OSM, 2000). Chapter 9 describes the origins of the first project, its policy bases, and motivations of various institutions involved in delivering these international exchanges.

Provision of technical assistance to developing countries through bilateral partnerships with similar but more mature organizations in other parts of the world has proved to be an effective way to transfer expertise, train staff, build up management capabilities, and enhance institutional capacities (Cooper, 1984). Organizations which have participated in such relationships include railways, power companies, water authorities, port authorities, irrigation agencies, universities, and research centers. Management institutes, forestry and agricultural institutes, dairy development boards, municipalities, national development banks, and mortgage banks have also transferred expertise and technology through international partnerships (World Bank, 1985, pp. 253–254). Such partnerships constitute a mechanism for the planned diffusion of innovations, through which new ideas and knowledge are communicated by officials of one country to others seeking improvement in their capabilities. Accomplishments and shortcomings in implementation of the OSM project are described throughout this book. Consistency of program outputs with program intent is evaluated, and areas where additional effort is needed are identified. Conclusions concerning the significance of such efforts in advancing the concept of sustainable development are presented in Chapter 10.

This study is based on a wide assortment of documents and press reports, numerous interviews with high-ranking officials of government and mining operations, and direct observation conducted by the author during seven field visits to Indonesia spanning more than a decade. Much of the illustrative material concerns events and issues with which I was personally involved. Research on Indonesian mining environmental policy commenced in 1991 and visits to Indonesian mining operations and government offices for interviews on Borneo, Sumatera and Java began in 1992. Press reports are taken mostly from the daily *Jakarta Post*, and the highly-regarded weekly news magazine *Tempo*.

Acknowledgments

The author wishes to express special appreciation to Alfred E. Whitehouse, Richard O. Miller, Rosemary O'Leary, Kadar Wiryanto, Dibyo Kuntjoro and the many dedicated professionals employed by the Indonesian Ministry of Energy and Mineral Resources, the U.S. Office of Surface Mining Reclamation and Enforcement, and state surface mining regulatory authorities who shaped my thinking about mining environmental policy, reviewed portions of this manuscript, or participated in the activities described in this book. Without them, the world would be a much degraded place to live. And especially thanks to Al, for everything.

The professional activities that lead to this book were inspired by, and the book is dedicated to several special individuals. From Norman Wengert, Henry P. Caulfield, Jr., and Phillip O. Foss I acquired a stronger sense of public service, professional values and respect for human dignity; sound conceptual skills of policy analysis; and the notion academic research is of little worth unless related to the world around us and the people who live in it. From Ellen L. Moore Hamilton I acquired an abiding tolerance for others and unrepentant liberal values. From Harry S. Hamilton, I acquired a strong sense of duty, and the value of perseverance.

But most of all this book is for Carol Jean Boggis, the love of my life, who has put up with so much for so long with so little lament.

Nonetheless responsibility for the content and views expressed in this book is solely that of the author, and does not necessarily represent those of any organization with which the author previously or currently may be affiliated.

Although the current text is much revised and includes new data and in some cases new methods of analysis, earlier versions of portions of Chapters 5 and 8 appeared previously in "Natural Resource Policy and Underinvestment in Indonesian Mining Operations," *Natural Resource Management* 7 (September): 25–34 (2004) published by the Australian Association of Natural Resource Management; "Prospects for Increasing Profits by Improving Water Quality at Indonesian Coal Mines," *Minerals and Energy* 16 (February): 3–13 (2001) published by Taylor and Francis; "Policy Drivers: Estimating Capital Recovery Periods for Investments in Fine Coal Circuits at Indonesian Coal Preparation Plants," *Indonesian Mining Journal* 6: 69–77 (2000b); and "Lost Profits, Lost Royalties: Formulating Policy to Recover the Value of Lost Coal Fines from Indonesian Mining Operations," *Indonesian Mining Journal*, 4: 71–78 (1998a) published by the Indonesian Mining Association and used by permission.

Chapter 1

Introduction: Mining Environmental Policy Implementation in Two Countries

Environmental Impacts of Surface Mining

Wherever it occurs in the world, surface mining of coal completely eliminates existing vegetation, destroys the genetic soil profile, displaces or destroys wildlife and habitat, degrades air quality, alters current land uses, and to some extent permanently changes the general topography of the area mined (U.S.DOI/OSM, 1979). The community of micro-organisms and nutrient cycling processes are upset by movement, storage, and redistribution of soil. Generally, soil disturbance and associated compaction result in conditions conducive to erosion. Soil removal from the area to be surface mined alters or destroys many natural soil characteristics, and may reduce its productivity for agriculture or biodiversity. Soil structure may be disturbed by pulverization or aggregate breakdown. Removal may reduce soil permeability and increase bulk density. Temporary topsoil storage can result in loss of organic matter. Removal of vegetative cover and activities associated with construction of haul roads, stockpiling of topsoil, displacement of overburden and hauling of spoil and coal increase the quantity of dust around mining operations. This dust degrades air quality in the immediate area, can have adverse impacts on vegetative life, and may constitute a health and safety hazard for mine workers and nearby residents. The land surface, often hundreds of hectares, is dedicated to mining activities until it can be reshaped and reclaimed. If mining is allowed, resident human populations must be resettled off the mine site, and economic activities such as agriculture or hunting and gathering food or medicinal plants are displaced, at least temporarily. What becomes of the land surface after mining is determined by the manner in which mining is conducted.

Surface mining can adversely impact the hydrology of any region. Deterioration of stream quality can result from acid mine drainage, toxic trace elements, high content of dissolved solids in mine drainage water, and increased sediment loads discharged to streams. Waste piles and coal storage piles can yield sediment to streams, and leached water from these piles can be acid and contain toxic trace elements. Surface waters may be rendered unfit for agriculture, human consumption, bathing, or other household uses. Controlling these impacts requires careful management of surface water flows into and out of mining operations. Moreover flood events can cause severe damage to improperly constructed or located coal haul roads, housing, coal crushing and

1

Figure 1.1   Multiple coal seam operation, East Kalimantan
Source: Michael S. Hamilton

washing plant facilities, waste and coal storage piles, settling basin dams, surface-water diversion structures, and the mine itself. Besides the danger to life and property, large amounts of sediment and poor quality water may have detrimental effects many miles downstream from a mine site after a flood.

Ground-water supplies may be adversely affected by surface mining. These impacts include drainage of usable water from shallow aquifers; lowering of water levels in adjacent areas and changes in flow directions within aquifers; contamination of usable aquifers below mining operations due to infiltration or percolation of poor quality mine water; and increased infiltration of precipitation on spoil piles. Where coal or carbonaceous shales are present in the spoil, increased infiltration may result in increased runoff of poor quality water and erosion from spoil piles; recharge of poor quality water to shallow groundwater aquifers; or poor quality water baseflow to nearby streams. This may contaminate both ground water and nearby streams for long periods. Lakes formed in abandoned surface mining operations are more likely to be acid if there is coal or carbonaceous shale present in spoil piles, especially if these materials are near the surface and contain pyrites.

Surface mining of coal causes direct and indirect damage to wildlife. The impact on wildlife stems primarily from disturbing, removing, and redistributing the land surface. Some impacts are short-term and confined to the mine site; others may have far-reaching, long-term effects. The most direct effect on wildlife is destruction or displacement of species in areas of excavation and spoil piling. Mobile wildlife species like game animals, birds, and predators leave these areas. More sedentary animals like invertebrates, many reptiles, burrowing rodents and small mammals may be directly destroyed. If streams, lakes, ponds or marshes are filled or drained, fish, aquatic invertebrates, and amphibians are destroyed. Food supplies for predators are reduced by destruction of these land and water species. Animal populations displaced or destroyed can eventually be replaced from populations in surrounding ranges, provided the habitat is eventually restored. An exception could be extinction of a resident endangered species.

Impacts from surface mining on biological resources are of particular concern in Indonesia. Its species richness is of global significance, meriting recognition by the World Wildlife Fund as one of only six megadiversity countries that require special attention to address rapid environmental change, severe economic problems and lack of conservation resources that threaten its rich biodiversity (Mittermeier, 1988, p. 152–3). Indonesia has the greatest diversity of mammals in the world at 515 species, of which a third are endemic. It also has the greatest diversity of swallowtail butterflies, of which 44 percent are found only in Indonesia. Birds, reptiles, and amphibians are likewise highly diverse, and the country's 25 000 species of flowering plants ranks seventh globally. Indonesia's forests harbor the greatest diversity of palm species in the world at 477 species, half of which are found only there. Indonesia is home to more than 400 species of dipterocarp, the most commercially valuable trees in Southeast Asia. About 90 percent of Indonesians live in rural areas, and 40 million of them directly depend on biological resources for subsistence, using over 6000 species of wild and cultivated plants and animals (Stone, 1994, p. 131).

Many wildlife species are highly dependent on vegetation growing in natural drainages. This vegetation provides essential food, nesting sites and cover for escape from predators. Any activity that destroys this vegetation near ponds, reservoirs, marshes, and wetlands reduces the quality and quantity of habitat essential for waterfowl, shore birds, and many terrestrial species. The commonly used head-of-hollow fill method for disposing of excess overburden is of particular significance to wildlife habitat in some locations. Narrow, steep-sided, V-shaped hollows near ridge tops are frequently inhabited by rare or endangered animal and plant species. Extensive placement of spoil in these narrow valleys eliminates important habitat for a wide variety of species, some of which may be rendered extinct.

Broad and long-lasting impacts on wildlife are caused by habitat impairment. The habitat requirements of many animal species do not permit them to adjust to changes created by land disturbance. These changes reduce living space. The degree to which a species or an individual animal tolerates human competition for space varies. Some species tolerate very little disturbance. In instances where a particularly critical habitat is restricted, such as a lake, pond, or primary breeding area, a species could be eliminated. Large mammals and other animals displaced from their home ranges may be forced to use adjacent areas already stocked to carrying capacity. This overcrowding usually results in degradation of remaining habitat, lowered

Figure 1.2   Crushing plant and finished coal stockpiles, large open pit, wet floor in background, South Sumatera
Source: Alfred E. Whitehouse

carrying capacity, reduced reproductive success, increased interspecies and intraspecies competition, and potentially greater losses to wildlife populations than the number of originally displaced animals.

Removal of soil and rock overburden covering the coal resource, if improperly done, causes burial and loss of top-soil, exposes parent material, and creates vast infertile wastelands. Pit and spoil areas are not capable of providing food and cover for most species of wildlife. Without rehabilitation, these areas must go through a weathering period, which may take a few years or many decades, before vegetation is established and they become suitable habitat. With rehabilitation, impacts on some species are less severe. Humans cannot immediately restore natural biotic communities. We can, however, give nature a boost, through reclamation of land and rehabilitation efforts geared to wildlife needs. Rehabilitation not geared to the needs of wildlife species, or improper management of other land uses after reclamation, can preclude reestablishment of many members of the original fauna.

Degradation of aquatic habitats has often been a major impact from surface mining and may be apparent to some degree many miles from a mining site. Sediment contamination of surface water is common with surface mining. Sediment yields may increase 1000 times over their former level as a direct result of strip mining. In some circumstances, especially those involving disturbance of unconsolidated soils, approximately one acre-foot of sediment may be produced annually for every 80 acres of disturbed land (U.S.DOI/ OSM, 1979).

The effects of sediment on aquatic wildlife vary with the species and amount of contamination. High sediment loads can kill fish directly, bury spawning beds, reduce light transmission, alter temperature gradients, fill in pools, spread stream flows over wider, shallower areas, and reduce production of aquatic organisms used as food by other species. These changes destroy the habitat of some valued species and may enhance habitat for less desirable species. Existing conditions are already marginal for some freshwater fish in the United States. Sedimentation of these waters can result in their elimination. The heaviest sediment pollution of a drainage normally comes within five to 25 years after mining. In some areas, unrevegetated spoil piles continue to erode even 50 to 65 years after mining (U.S.DOI/OSM, 1979).

The presence of acid-forming materials exposed as a result of surface mining can affect wildlife by eliminating habitat and by causing direct destruction of some species. Lesser concentrations can suppress productivity, growth rate, and reproduction of many aquatic species. Acids, dilute concentrations of heavy metals, and high alkalinity can cause severe wildlife damage in some areas. The duration of acidic waste pollution can be long term. Estimates of the time required to leach exposed acidic materials in the Eastern United States range from 800 to 3000 years (U.S.DOI/OSM, 1979).

In certain situations, surface mining may have beneficial impacts on some wildlife. Where large, continuous tracts of forest, bush land, sagebrush, or grasslands are broken up during mining, increased edges and openings are created. Preferred food and cover plants can be established in these openings to benefit a wide variety of wildlife. Under certain conditions, creation of small lakes in the mined area may also be beneficial. These lakes and ponds may become important water sources for a variety of wildlife inhabiting adjacent areas. Many lakes formed in mine pits are initially of poor quality as aquatic habitat after mining, due to lack of structure, aquatic vegetation, and food species. They may require habitat enhancement and management to be of significant wildlife value.

Surface mining operations and coal transportation facilities are fully dedicated to coal production for the life of a mine. Mining activities incorporating little or no planning to establish postmining land use objectives usually result in reclamation of disturbed lands to a land use condition not equal to the original use. Existing land uses such as livestock grazing, crop and timber production are temporarily eliminated from the mining area. High value, intensive land use areas like urban and transportation systems are not usually affected by mining operations. If mineral values are sufficient, these improvements may be removed to an adjacent area.

Surface mining operations have produced cliff-like highwalls as high as 200 feet in the United States. Such highwalls may be created at the end of a surface mining operation where stripping becomes uneconomic, or where a mine reaches the boundary of a current lease or mineral ownership. These highwalls are hazards to people, wildlife, and domestic livestock. They may impede normal wildlife migration routes. Steep slopes also merit special attention because of the significance of impacts associated with them when mined. While impacts from contour mining on steep slopes are of the same type as all mining, the severity of these impacts increase as the degree of slope increases. This is due to increased difficulties in dealing with problems of erosion and land stability on steeper slopes.

Fires sometimes occur in coal beds underground. When coal beds are exposed, the fire risk is increased. Weathered coal can also increase ground temperatures if it is left on the surface. Almost all fires in solid coal are ignited by surface fires caused by people or lightning. Spontaneous combustion is caused when coal oxidizes and air flow is insufficient to dissipate heat, but this occurs only in stockpiles and waste piles, not in bedded coals underground. Where coal fires occur, there is attendant air pollution from emission of smoke and noxious fumes into the atmosphere. Coal seam fires may burn underground for decades, threatening destruction of forests, homes, schools, churches, roadways and other valuable infrastructure. Spontaneous combustion is common in coal stockpiles and refuse piles at mine sites.

Adverse impacts on geological features of human interest may occur in a surface mine area. Geomorphic and geophysical features and outstanding scenic resources may be sacrificed by indiscriminate mining. Paleontological, cultural, and other historic values may be endangered due to disruptive activities of blasting, ripping, and excavating coal. Stripping of overburden eliminates and destroys all archeological and historic features unless they are removed beforehand. Extraction of coal by surface mining disrupts virtually all esthetic elements of the landscape, in some cases only temporarily. Alteration of land forms often imposes unfamiliar and discontinuous configurations. New linear patterns appear as material is extracted and waste piles are developed. Different colors and textures are exposed as vegetative cover is removed and overburden dumped to the side. Dust, vibration, and diesel exhaust odors are created, affecting sight, sound, and smell. Some members of local communities may find such impacts disturbing or unpleasant.

Due to intensive mechanization, surface mines may require fewer workers than underground mines with equivalent production, although this does not appear to be true in Indonesia. The influence on human populations from surface mining is therefore not generally as significant as with underground mines. In low population areas, however, local populations cannot provide needed labor so there is migration to the area because new jobs are available at a mine. Unless adequate advance planning is done by local government and mine operators, new populations may cause overcrowded schools, hospitals and demands on public services that cannot easily be met. Some social instability may be created in nearby communities by surface coal mining.

Many impacts can be minimized but may not be eliminated entirely by use of best mining practices either voluntarily or to comply with government regulatory programs. Financial incentives to minimize costs of production may minimize use of best mining practices in the absence of effective regulation. Some temporary destruction of the land surface is an environmental price we pay for utilization of coal resources. The scale of disturbance, its duration, and the quality of reclamation are largely determined by management of the operation during mining.

Is Sustainable Development Possible?

For some newly industrializing states, mineral resources are development capital. Coal and other mineral resources can provide significant sources of capital for economic and social development, which in turn provide the means for improving the quality of life for people. Improving quality of life includes improving both the standard of living and environmental quality where people live and work.

A crucial policy issue in development concerns the need to balance actions which support continued economic growth with those that maintain or improve environmental quality, to achieve sustainable development. "Meeting the needs of the present without compromising the ability of future generations to meet their own needs" (World Commission on Environment and Development, 1987, p. 43), whatever that may mean, seems clearly to imply some net gain, or at least no net loss in productivity should be evident in the resulting intergenerational calculus of resource use. Although there seems to be little agreement among economists concerning an appropriate accounting for national wealth or intergenerational transfers of it (Mazmanian and Kraft, 2001, pp. 16–18), it seems almost axiomatic that for economic development to be sustainable over time, total benefits must accumulate more rapidly than total costs. If unpaid environmental costs of development – such as disturbed and unproductive land, toxic waste dumps, or polluted air and rivers – are allowed to exceed benefits, demands for payment of these costs will eventually divert capital and labor from new development to cleanup: Development will not be sustainable in the long term. If expenditures necessary to protect the health, safety and welfare of people exceed income or significantly deplete accumulated investment capital, development will stall.

The costs of development must always be paid, if not today, then tomorrow. There really is no such thing as a free lunch. It is the nature of development costs that they will eventually demand payment, if not in currency, then in reduced health, safety and welfare of people who live near wasted land, breathe polluted air, or consume contaminated fish or other foods. In the United States, this is the lesson of Love Canal, Superfund cleanups, radioactive waste disposal, and abandoned mined lands. Like some macabre wraith, unpaid environmental costs of development eventually come back to haunt us. In the United States, we continue to find and clean up sites filled with materials toxic to human life, in dumps created during periods of our most rapid economic expansion. We have already spent several billion dollars, and will spend hundreds of billions more to clean up rivers and hazardous waste dumps contaminated by minerals development, manufacture and use of industrial chemicals, even the curing of animal hides for leather in the 1800s. These are expenditures that would have been smaller if our predecessors had been wise enough to understand such costs can only be postponed, but must eventually be paid. Their grandchildren and our grandchildren will likely continue paying these costs for a long time to come.

This has been a hard lesson for many industrialized countries to learn. Funds that might have contributed to even greater economic growth have gone instead to clean up hazards to human health and welfare produced by previous development activities, spending amounts greater than would have been necessary if expenditures had been made earlier. Because the costs of labor generally increase in a growing economy, previously unpaid environmental costs are usually more expensive to pay today than they were yesterday, and will almost certainly be more expensive to pay tomorrow than they are today. Governments throughout the world are only beginning to realize that minimization or prevention of pollution and environmental degradation before they occur is often less expensive than cleaning up the mess afterwards. Increasingly this is being found to be the case for many forms of land, water, air and toxic materials pollution. It is generally true for mineral development, where mining practices used today determine both the expense for reclaiming land for other postmining uses and the long-term efficiency of mining operations, as illustrated in Chapter 5.

In 1977, it was reported by the Senate Committee on Energy and Natural Resources that there were over four million acres already disturbed in the U.S. by surface mining of all minerals, about 43 percent of which was from coal extraction, which continued to disturb 1000 acres of land per week. At that time, only about half of 1.3 million acres of surface mined land in eastern coal fields had been reclaimed (Congressional Quarterly, Inc., 1977, p. 618). Coal production increased by over 57 percent from 1977–2002, and surface mining today accounts for about 67 percent of production, compared to less than 62 percent in 1977 (U.S. DOE, 2002, p. 203). Thus, it is reasonable to surmise that more than 1000 acres per week are probably disturbed by surface coal mining today. Fortunately, most of it must now be reclaimed shortly after it is disturbed, due to regulatory requirements for contemporaneous reclamation which have been in place since 1978.

No one knows how much it may eventually cost to return previously mined lands in the United States to other productive uses, but some have estimated it may run as high as US$35.2 billion in 1978 dollars just to reclaim lands damaged by coal extraction (Johnson and Miller, 1979, p. 28; see also Baxter, et al., 1983). Adjusted for inflation,[1] this would be about US$54.2 billion in 2004. Since enactment of the Surface Mining Control and Reclamation Act of 1977, the United States government spent a little under $5.5 billion for reclamation of abandoned coal mined lands through 30 September 2004, leaving nearly $1.7 billion unspent in the Abandoned Mine Land Fund (U.S.DOI/OSM, 2004). At this rate, it will take over 230 years to reclaim the remaining abandoned coal mine lands in the United States, and costs will continue to escalate at approximately the rate of inflation. Moreover, this estimate does not include the cost of reclaiming over two million acres of land disturbed by surface mining minerals other than coal.

Similarly, no one knows how much it may eventually cost to clean up all the hazardous waste dumps in the United States. Industry estimates that it might require $120 billion in 1995 dollars to complete necessary cleanups (Sablatura, 1995) have already increased in value by the passage of time, and do not include sites discovered in the interim, or yet to be discovered. From 1981–2003, nearly $25.9 billion was appropriated by the U.S. government for Superfund cleanups, including about $20.6 billion recovered from settlements with responsible parties (U.S.EPA, 2003a, 2003b). These figures do not include the costs of cleanups done without Superfund participation, which may run as much as another $15 billion from 1981–2003.

By the end of 2002, removals of hazardous waste had been conducted at 5253 sites, construction of long-term remedial facilities was completed at 846 sites. A bit fewer than 44 500 sites had been assessed, with 11 312 sites remaining in the site assessment program or on the National Priorities List (NPL) for cleanup (U.S.EPA, 2003a). According to Timothy Fields, former EPA Assistant Administrator of Solid Waste and Emergency Response, new sites are added to the NPL at the rate of 30–40 per year (Fields, 2002). New sites continue to be added to the list yet to be assessed as we discover ones previously unknown. A recent assessment by the U.S. Environmental Protection Agency estimates as many as 355 000 sites in the United States contaminated with hazardous waste and petroleum products will need to be cleaned up, at a potential cost of US$250 billion in 2004 dollars (U.S.EPA, 2004). Recent levels of cleanup activity in the U.S. have been in the range of US$6–8 billion per year, so it may take 35 years to complete the work, and costs will continue to escalate at the rate of inflation.

Belated expenditures for cleanup of land disturbed and contaminated by previous economic activities do provide current employment opportunities but represent a drain on capital available for investments in sustained economic development. Thus, effective management of the environmental impacts of mining and other industrial activities is essential for sustainable development. This is probably true in all countries, at every stage of development, in all parts of the world.

Today, knowledge of mining and the environment have advanced to the point that, with few exceptions, land disturbed by mining activities can be reclaimed and returned to other productive uses. Efforts to reclaim and restore mined land in the U.S. today are achieving greater than 95 percent success rates, although success rates of 10 percent or less are common in developing countries. Erosion and acid mine drainage at new mines can be prevented or controlled, largely through careful mine planning and sound earthmoving practices. Stability of fills constructed with excess overburden materials can be assured through proper design and construction practices, even in areas of heavy rainfall. Vegetation can be re-established in an appropriate growing medium even in areas having thin topsoils.

Common uses of previously mined land in the U.S. today include: cropland agriculture, commercial forestry, recreation and tourism; public works such as airfields, roads, housing developments, and industrial sites; fish and wildlife conservation on shrub lands, woodlands and wetlands, for hunting and fishing, and to conserve biological diversity. With only a few limitations, similar results can be produced in most other areas of the world where mining occurs. In some cases, land may be reclaimed to permit a more valuable use than was previously possible. Removal of tree stumps and large rocks during mining improves land for tillage agriculture, building sites, and some recreational uses such as golf courses. Graded or terraced slopes provide improved tillage and erosion control. Leveled areas provide more stable areas for housing, industry, airports and agriculture. Reforested land establishes a foundation for the return, over time, of native species and biological diversity.

What are the limitations on effective reclamation of mined lands? First, reclamation, if it is to be done successfully, must be based on a sound mining plan and environmental assessment, prepared in advance of mining operations. Why? Because the nature of a postmining land use, whether for agriculture, forestry, conservation of biological diversity, or community development, determines how and where topsoil and overburden are placed during mining, how expensive it is to move them, and the final contours of reclaimed land. Moving heavy materials once is less expensive than moving them twice. A plan for removal, storage, restoration, and stabilization of the surface is essential if reclamation is to succeed. Lack of such a plan, and failure to segregate and preserve topsoil or some other growing medium, are what makes reclamation of previously mined but abandoned lands so difficult and expensive.

Second, revegetation of land in extremely harsh climates is unlikely to achieve much success. Desert and alpine climate conditions are not conducive to rapid plant growth, which is necessary to stabilize reclaimed soil. The suitability of mining in such areas must be carefully examined in terms of the expected postmining land use. In some areas, non-mineral features of land may be more valuable to sustained development than the mineral resource. Yet, in the U.S., the areal extent of mineral resource lands designated unsuitable for mining under legislative criteria is infinitesimal compared to the area available for mining.

Third, adequate water and labor must be dedicated to reclamation to ensure effective revegetation. In the Eastern U.S., this takes about five years; in the more arid West, about 10 years. Climatic conditions in each region of the world – especially rainfall – determine how long it takes to achieve reclamation success. Shortages of water or labor reduce effectiveness of reclamation efforts. Even rain forest environs, despite their relatively thin soils, may be reclaimed if an appropriate growing medium can be developed from mineral rich subsoils found below the leached zone. In rain forest, segregation of unweathered mineral-rich subsoils from leached overburden, and effective use of composted forest materials may be as important to successful reclamation as preservation of shallow rain forest topsoils. Introduction of fast-growing, sun-tolerant tree species may allow establishment of a successional forest which will return to rain forest. Some sun-tolerant species provide sufficient cover to encourage invasion by shade-loving native plant and animal species. This may allow re-establishment of rain forest vegetation while shading out seedlings of sun-tolerant exotics. In the forty years it takes some sun-tolerant species to reach maturity, die and contribute their organic material to the forest floor, a young rain forest, complete with understories, groundcover and native animal species can grow up in the shade, retaking the land. Invasion by rain forest species was well established about 18 months after planting fast-growing sun-tolerant trees on a test plot of unconsolidated quaternary soils from a coal mine in East Kalimantan, Indonesia visited by the author in 1992. Mine operators there were unaware they had successfully established a successional rain forest, and were attempting to determine if the trees would be useful as a crop for pulp and paper production.

Fourth, effective reclamation requires some investment of political and financial capital. The political will must be found to ensure that mining operations prepare, finance, and follow mining plans necessary for effective reclamation. Use of modern mining practices and technology can increase production and efficiency of coal mining operations while reducing offsite environmental degradation of land, water and air quality. But the greatest potential payback for this investment lies in long-term sustained productivity of the land after mining has ended.

What does this mean for sustainable development? First, countries which must rely on extraction of mineral resources to provide capital and raw material inputs for economic development may do so without permanent loss of productivity of the land. Mining is only a temporary use of the land if managed properly. Second, significant sources of expert technical assistance are available to developing countries to mitigate the undesirable effects of surface mining, as well as the surface effects of underground mining. Expertise and technology are available to prepare land during mining for other uses after mining has ceased. Finally, development based on extraction of mineral resources may be sustainable, provided the economic value of these resources is converted to usable capital and invested in infrastructure for clean drinking water supplies, sanitation facilities, medical care, education, and less extractive economic enterprises that provide employment, such as manufacturing and services.

Thus, sustainable development and coal mining can be compatible, if mineral development is conducted in a manner that: (1) effectively controls environmental effects and costs during mining, (2) reclaims mined land for other productive uses, and (3) provides revenues from mineral extraction for investments in infrastructure and enterprises which sustain economic and social development and promote diversification as mineral reserves are depleted.

Laying a sound foundation for sustainable development of mineral resources requires successful implementation of effective policies to minimize both the short-term and long-term environmental costs of mining, while ensuring the future productivity of mined land for non-mining uses. Whether Indonesian mining environmental policy has been successful in meeting the first two conditions described above is one focus of this book. Although there is evidence Indonesian investment policies have been at least partially effective in promoting diversification of the national economy and expansion of infrastructure and other non-mining enterprises which might sustain development, that complicated and weighty analysis is beyond the scope of this book. In any event, fulfilling the third condition would not be sufficient if the first two more fundamental conditions were not attained.

Potential consequences during policy implementation include various degrees of success and failure. The circumstances of policy failure during implementation have been explored in the United States (Ingram and Mann, 1992). Several ways in which policies may be compared between countries have also been examined (see generally Antal, Dierkes and Weiler, 1987; Heidenheimer, Heclo and Adams, 1990; Heady, 1996, pp. 48–51; Honadle, 1999).

Four sets of causal factors based on contextual, political, legal and organizational circumstances have been distinguished which help explain variation in decisions made by regulatory officials in the United States. Contextual factors include political culture, public opinion, technological capability to regulate the activity of concern, and economic conditions, especially as they affect the perceived prosperity of regulated industries and fiscal resources available to regulatory agencies. Political variables include constituency balance, intergovernmental relations and interparty competition. Legal resources and constraints include the nature and clarity of policy directives, and adequacy of sanctions. Organizational variables are shaped by contextual factors, legal resources and political variables, and include administrative structure, financial and personnel allocations, access of outsiders to decision makers, patterns of communication within an agency, decision making procedures, dispositions of decision makers, leadership ability, and interpretations of policy used during policy implementation (Hamilton, 1990, pp. 9–21).

Do these factors function in similar ways to influence regulatory decision making in Indonesia, a very different culture and form of government than that in the United States? Using a comparative approach, this book explores to what extent and how contextual factors, legal resources, and organizational variables function to determine regulatory decision making outputs in Indonesia. It examines cultural, structural and institutional reasons why Indonesian mining environmental policy implementation suffers from frequent failures to attain policy goals.

The Prismatic Society of Contemporary Indonesia

Structural and cultural aspects of public policy and administrative decision making are functionally significant factors during policy implementation in Indonesia, as they are in the United States. How these factors function in Indonesia is often less than obvious. Although Indonesia has recently obtained much of the technology and trappings of industrialized countries, government administration there still displays many attributes of more traditional societies. Things are not always as they seem in such "prismatic" transitional societies (Riggs, 1964). Their governments tend to enact policy goals without clear operational directives, and administrators often make decisions on a case-by-case basis which do not appear to be consistent with policy goals or guidance. Moreover, Westerners often mistake for corruption behaviors that are culturally legitimate practices for initiating accepted social relationships in Indonesia (Jackson, 1980, pp. 229–230).

Overlapping Functions

Contemporary prismatic societies contain a broad mixture of attitudes, practices and social relationships, including some residual norms held over from earlier periods which continue to affect the operation of government institutions in fundamental ways. Western models of public administration lead us to expect government agencies in industrialized countries perform a single function of implementing public policy. Yet government agencies in prismatic transitional societies often serve multiple, overlapping functions simultaneously during policy implementation, while displaying an illusion of autonomy from traditional social, economic, and religious systems in which they remain deeply enmeshed. Because government administration occurs in the context of these behavior systems, policy implementation is influenced by them in complex, often subtle ways (Riggs, 1964, pp. 12–15; King, 1995). In Indonesia, overlapping functions may influence patterns of communication within an agency, dispositions of decision makers, and interpretations of mining environmental policy during implementation in unpredictable ways.

Form Lacking Substance

In prismatic societies where overlapping functions are performed by administrative agencies, we may expect discontinuities to occur between formal policy goals and results of administrative implementation. Policy goals may be enacted which are not enforced by administrators because some policies set in motion secondary effects that conflict with other policies, social practices or cultural norms (Riggs, 1964, pp. 15–19). In effect, policy makers enact the formal appearance of a policy, but administrators are either unable or unwilling to implement the substance of it. A general statement of policy intent is expressed, but effective regulatory enforcement does not occur: actual behavior does not conform to the prescribed norm. The resulting policy statement is mostly form and little substance; the form provides an appearance of policy intent that is contradicted by observations of reality during policy implementation.

When this happens, the formal policy goal is in effect merely an expression of objectives or wishes it is hoped will be attained. Often stated in general terms, without elaboration of specific requirements or behaviors which might be monitored during implementation, such goal statements confer considerable discretion on enforcement personnel to decide when a regulated entity is in compliance with a policy goal, without fear of being held accountable for their decisions. Enforcement personnel may then rely on their own personal value preferences – possibly including convenience or profit – rather than carrying out a policy as directed by its terms. In effect, they are free to make policy decisions individually on a case-by-case basis, without fear of being held accountable for such decisions, and their actions may be influenced by persons they are supposed to regulate (Riggs, 1964, pp. 334–336). Thus, the nature and lack of clarity of policy directives, decision making procedures, and dispositions of decision makers may be expected to greatly influence interpretations of mining environmental policy during implementation in prismatic societies such as Indonesia.

Criteria for Decision

To be effective in limiting or channeling the discretion of personnel implementing a policy, it must contain criteria a decision maker must use to decide what to do or what not to do in a given factual situation (Caulfield, 1984). These criteria provide guidance to decision makers concerning how a policy is to be interpreted and applied, thereby limiting the discretion of administrators to substitute their own value preferences. The more specific the criteria, the less discretion is left to enforcement personnel, and the easier it is for superiors and external institutions like courts to hold them accountable for their decisions.

In regulatory policy, decision criteria may be stated as requirements for action that must be taken by government officials or persons engaging in permitted actions; in specific decision rules for administrative decision making; or in the form of performance standards or design standards that regulated entities must meet for permitted activities. In Indonesia, the nature and lack of clarity of policy directives combine with inadequacy of sanctions to greatly influence dispositions of decision makers, administrative structures, patterns of communication within and between agencies, and decision making procedures during implementation of mining environmental policies, as discussed throughout this book.

Status and Contract

In traditional societies, interpersonal relations are governed predominantly by the relative status of individuals, and existing rules of conduct tend to be local customs or commands of rulers. The desire for increased trade and fear of conquest in transitional societies cause a slow and sometimes painful shift towards exchange relationships based on contracts and constitutional ideas found in industrialized societies. Business concessions are granted and new legal and judicial regimes are created to enforce trade agreements and laws. In prismatic societies, interpersonal relations are based on a mix of behavioral norms concerning status and contracts which may clash during interpretation and implementation of policy (Riggs, 1964, pp. 46–47).

Attitudes toward power and authority in Indonesian political culture reveal great deference to social status, often manifest in patron-client relationships based on traditions of *bapakism*.

> *Bapakism* (literally, fatherism) is a term frequently used by Indonesians to describe this complex system of social relations. The *bapak* (father) is the leader of a circle of clients and is expected to care for the material, spiritual, and emotional needs of each of his followers who are called *anak buah* (children). In return the anak buah are expected to be at the bapak's beck and call to pay deference, make contributions, participate in family ceremonies, join or leave political parties, and in some cases

even defend the patron's life. These diffuse, personal, face-to-face, enduring, noncontractual relationships are the primary social cement integrating Indonesian organizations to the limited degree that they are integrated at all. Only when a bureaucrat is a bapak can the official be sure that his orders will be carried out.

(Jackson, 1978b, p. 35)

Thus, Indonesian social life is composed of a multitude of highly personal groupings, each with its own elite and mass elements bound together by an elaborate set of personal relations based on a diffuse sense of personal reciprocity between patron and client, but only tenuously connected to surrounding social groupings.

Bonds between peers that transcend time and immediate interests are even rarer in Java than in supposedly cold and impersonal industrial societies (Jay, 1969, pp. 201–206). The concern for embedding oneself in secure bapak-anak buah dependency relations flows from the absence of other enduring relationships in the culture (Jackson, 1978b, p. 36). Moreover, anak buah remain open to formation of dependency relations with new bapaks whenever the opportunity arises. Patrons cultivate new clients, and clients having multiple patrons frequently switch their major allegiances as the political winds change (Jackson, 1978a, p. 14; 1978c).

Status rather than wealth is the major social objective in Javanese culture, which has long dominated the national government of Indonesia. Status is defined not by wealth but by having a large circle of followers. Material wealth is converted into status and respect through a process of exchange of gifts in return for loyalty and obligation.

Gifts are bestowed on many occasions and, along with the use of respect terms in Javanese and Sundanese, they provide a means for stratifying new relationships and for extinguishing status ambiguity. What Westerners often mistake for corruption is a culturally legitimate way of initiating a dependency relationship. For instance, on meeting a village headman for the first time it is not improper to bring a gift, and the presentation is not perceived as bribery. But in accepting the gift the official is obliged to reciprocate in some manner. To give a gift is to initiate a potential dependency relationship; if the recipient does not give some token, and preferably more valuable, gift in return, it is an implicit admission of the recipient's subordinate position. What is important is not the nature and magnitude of the gift, but the act of giving.

(Jackson, 1980, p. 229)

A strong tendency of local villagers in Teluk Lingga, East Kalimantan, to seek the benefits of a patronage relationship with managers of the nearby mining operation of PT Kaltim Prima Coal has been well documented (Kunanayagam and Young, 1998, pp. 139–158). These cultural norms, and the resulting attitudes and social relationships, are pervasive influences in Indonesian social life, government administration and regulatory policy implementation.

The cultural norms surrounding bapakism lend themselves easily to the use of patronage in political life, and function to inhibit and constrain enforcement actions during implementation of mining environmental policy in Indonesia. Mine managers are perceived as making significant contributions to national economic development, are well compensated for their management skills, and are accorded high social status and authority, especially at remote mining operations. Mine inspectors and technical personnel in regulatory agencies are less well compensated for their efforts, are generally accorded lesser status, and may therefore be reticent to make demands on mine managers or their representatives. The standard practice of requiring mine operators to pay travel expenses of government inspectors to the mine site, or informal offers of small amenities from mine managers, cannot easily be reciprocated by mine inspectors or technical staff, thereby accentuating their subordinate positions to those they are obliged to regulate, and inhibiting stringent regulation of mining operations.

Salaries and Honoraria

The dynamics of social change in transitional societies induce a process of centralization and bureaucratization of administration which may eventually provide administrators sufficient income from salary to enable them to live on their earnings without seeking external income from fees, gifts, rents, tributes or other payments not directly distributed from a central treasury. Differentiation of structurally specialized agencies staffed with specialists and limited in scope of activity accompany the shift to salaried professional personnel. However opportunities and temptations for officials to augment their incomes remain pervasive, unless sharply curtailed by political and judicial control systems (Riggs, 1964, pp. 44–45).

Salaries for Indonesian government employees are generally paid in cash and do not provide a living wage. In the countryside, traditional practices of gifting the village headman before requesting a decision within his jurisdiction continue as they have for hundreds of years. In congested commercial sectors of urban areas, every street and parking lot has its *orang parkir* (e.g., parking attendant) who receives a small wage from the municipal government to stop traffic while drivers park their cars, and who expects a "tip" of "small-money" for providing this service. Automobile ash trays often contain coins and 100 rupiah (about US$0.88 at 2003 exchange rates of 8800 rupiah per dollar) notes for this purpose. To avoid traffic jams on the main roadways of Jakarta, one must often pay a small-money toll to *pak ogah* (Mr. Thief), often a local tough, for safe passage of an automobile on narrow back streets or through hotel parking lots from one street to another.

Following practices dating at least to the Dutch colonial period, a parallel tradition of cash side-payments evolved throughout the national government (and perhaps throughout the society), whereby small "honoraria" are paid by members of one subunit of government to members of other subunits who attend meetings of task forces, committees and workshops of all sorts (Gray, 1979). Payments are usually shared with members of the subunit represented by the receiving individual, ostensibly on the rationale that subordinate staff must work harder to compensate for absent superiors. Because attendance at meetings earns extra compensation, a superior who is able to get invited to many meetings is likely to have a devoted staff. Honoraria are paid for attendance, not commitment to decisions discussed at meetings, but disagreements may be expected to be subsumed somewhat by a desire to be invited to future meetings. Amounts are small by American standards; an office head in middle management may garner the equivalent of US$20.00 to $50.00 per meeting, and may divide shares among eight or more subordinates. Higher officials receive proportionately greater honoraria, and meetings where serious decisions are expected may earn considerably greater sums. In their more candid moments, some successful Indonesian officials estimate perhaps half of their annual income may be derived from distributed shares of honoraria.

Thus, some Indonesian government officials may spend considerable time attending meetings, and little time supervising work of their own offices. Networks of reciprocal invitations develop, and influence is extended to broader cohort groups, who tend to follow each other in promotions and career paths. Because most of these funds probably originate as budgeted revenues, accountability for expenditures from general revenue sources is limited by the practice of paying honoraria. Although acceptance of side-payments from individuals outside government is officially discouraged, top government officials often hold two or more positions simultaneously in government, private firms, or universities. Various gratuities continue to be offered to administrators for attending meetings and for services rendered, and news stories about the collection of "illegal taxes" and "user fees" are common in Jakarta and other large cities. Political and judicial controls in Indonesia do not effectively curtail these practices.

Enforcement personnel inspecting mines are instructed to decline any payments offered by mine operators, but it is not possible to determine whether they do so consistently. Some mine inspectors report suspicions that other inspectors may accept such payments frequently, while asserting their own refusal to ask for or accept such gratuities. Mine managers often complain about the lack of consistency in inspections conducted by different mine inspectors who visit their operations, indicating a diversity of attitudes among inspectors, and perhaps differential responses to offers – or unexpected lack of offers – of side payments.

Subtlety and Indirectness

Indonesian political culture displays attributes of a well-known Asian aversion to conflict, accentuated by a dominant Javanese subculture that strongly values subtlety and indirectness in interpersonal relations. Consequently, when questions are asked in Indonesia they rarely receive a direct response of "no" even when it might seem appropriate. A sense of social harmony being the ultimate cultural goal of interpersonal communication there, a more positive "yes" is the response an Indonesian prefers to give to inquiries and requests (Jackson, 1978b, p. 38). A direct reply of "no" would seem too confrontational for Javanese norms of interpersonal behavior. Opportunities to make a negative response encourage evasive action. Often one cannot tell whether "yes" really signifies agreement or merely a desire not to seem disagreeable.

Of course, persons with superior social status are less reluctant to say "no" than persons of subordinate status, especially when the former is addressing the latter. Yet, when a negative response seems unavoidable, it is often expressed in more tentative terms, using the term "*belum,*" which roughly translates as "not yet." Often this softly spoken response is accompanied by the kind of smile that might be considered modest or apologetic in the United States. It suggests that, although a negative response might appear to be appropriate at present, it may be premature to reach a firm conclusion, implying unstated possibilities may yet occur in the future which would alter circumstances and justify a more positive eventual conclusion. Yet every Indonesian understands "belum" effectively means "no."

This cultural aversion to conflict undermines attempts at command and control regulation of human behavior. Often based on Western models, rules and regulations designed to control or channel behavior away from undesirable actions are often poorly designed for Indonesian political culture, stated in very general or ambiguous terms. Efforts to enforce them suffer both during interpretation and application in the field. Consequently, decision makers are seldom predisposed to stringent enforcement of regulations in any area. Empirical observations of lax enforcement in the realm of mining environmental policy implementation are illustrated throughout the text below, especially in Chapters 2, 4 and 5.

Compensation and Prohibition

Notions of economic development and environmental regulation were imported, largely from European sources, and imposed on indigenous practices with which they sometimes do not fit well. A general cultural aversion to conflict is reinforced by the dominant role accorded compensation for harm done as a preferred remedy, instead of binding prohibition of certain types of behavior. This is evident in traditional *adat* law concerning the uses of land and associated natural resources. The constitutional framework of Indonesian environmental policies has been interpreted by jurists in the context of customary adat law, which emphasizes resolving potential conflicts by reaching consensus (*mufukat*) through mutual consultation (*musyawarah*) and negotiations (Petrich, 1994, p. 7), rather than the European approach of prohibiting certain actions unless conducted in accordance with prescribed rules of behavior, enforced by government sanctions.

A system of recorded private ownership of land, created on a European model, exists side-by-side with indigenous patterns of communal rights to land near villages and communities. A strong pro-development bias is evident in government policies which provide that traditional land rights of local peoples should not impede national development. Prohibitions against incursions on the land rights of others are uncommon and such rights are often ignored by developers. Government interpretations of adat law have divided it into two varieties, one of which requires compensation.

Hak adat concerns private land rights of ownership acquired by individual members of a community based on historical usage that was not officially recorded. Visible evidence of hak adat may be found in the form of some sort of investment of effort in *tanah adat* (traditional land), such as forest clearing, terracing and construction of a hut, which need not be a permanent residence. The marking or planting of a tree is sufficient to qualify the land on which the tree stands as tanah adat (Peluso, 2000, pp. 147–166). When tanah adat is used for development of a mining operation, compensation must be paid for the value of the land.

Hak ulayat concerns government-owned land located within the territory occupied by some tribal community without their having a right of ownership, based on historical use by the community. Claims of hak ulayat are based on traditional uses of *tanah ulayat* (used land) for hunting, gathering plant products for food and traditional medicines, and shifting agriculture. Unlike some South American patterns of shifting agriculture, it is generally not nomadic in Indonesia, but centered on relatively stable village communities, which continue to use land after it is no longer useful for growing rice. Such uses include gathering small game, non-rice foods, wood and medicinal plants through various stages of forest regrowth and succession (Colfer, Peluso and Shung, 1997). When tanah ulayat is taken for mining development, direct compensation to individuals is not required, but *recognisi* (recognition) of historical use is generally provided to the community in the form of houses, public utilities and infrastructure such as schools, health clinics and roads.

Because local people often believe payments for recognisi of tanah ulayat are significantly less than the value of land taken, mining operations may find themselves at odds with local communities for failing to pay compensation for what local leaders consider tanah adat. Thus, there are substantial differences in interpretation between the government and local people concerning what is hak adat and what is hak uluyat. Some observers have suggested these disagreements may impede development of local communities, and it might be less expensive in the long run to pay compensation rather than recognisi, thereby avoiding the costs of conflict and poverty alleviation after development proceeds (Soemarwoto, 1996, p. 79).

Complicating matters further, a third category of land outside adat law concerns unencumbered government-owned land where there is no pre-existing communal right to use. Usually located outside any tribal territory, in undeveloped areas with very low human populations, much of this type of land is situated in mineral-rich areas of Kalimantan and Papua (formerly Irian Jaya). National government policy is to pay neither compensation nor recognisi when this land is used for development, and to reveal the location of proposed developments only after the land has been surveyed by mine inspectors to determine if anyone lives there. Mine inspectors tell stories of visiting uninhabited locations identified as potential mine sites before mining proposals are announced, and returning shortly after a proposal is made public to find the land has been cleared and plantations established virtually overnight, sometimes with crops such as coffee planted upside down, roots in the air. Consequently, the government is wary of the potential for fraud in claims for compensation based on hak adat, and somewhat predisposed against them.

Thus, the culturally preferred remedy, to the extent there is a remedy at all, tends to be some form of compensation after harm is done, rather than any binding prohibition of harm before it occurs, or regulation to minimize harm. Even when issuing regulations, Indonesian political culture favors the use of compensation over regulation. Environmental laws that express a preference for payment of compensation for environmental harms, and settlement of disputes through negotiations rather than litigation, reflect cultural grounds for government reluctance to take enforcement measures or impose sanctions on violators. This preference is evident in public statements by prominent regulators, such as the Minister of Environment and Population favoring use of economic incentives and self-regulation by industrial firms (Petrich, 1994, p. 8).

Pancasila and Legitimacy

A national creed or ideology of tolerance for social and religious diversity is apparent in Pancasila, a set of basic principles or goals articulated at its founding by President Sukarno as a political compromise to unify an ethnically and culturally diverse new Republic of Indonesia, while deflecting demands for creation of an Islamic state. Included in the Preamble to the Indonesian Constitution of 1945, the five principles of Pancasila are: belief in God, a just and civilized humanitarianism, national unity in diversity, democracy through consultation and consensus, and social justice (Ramage, 1995). Because national unity would be threatened if one religious ideology was enshrined as the basis of a culturally diverse nation-state, Pancasila provided a secular *desar negara*, or philosophical basis for the new state as an alternative to Islam. Thus although the state is philosophically based on religious belief, Pancasila does not endorse any particular faith, implying that a legitimate government respects the religious diversity of its citizens. In the state ideology of Pancasila, national unity is more important than adherence to a single religion.

The Pancasila principle of a just and civilized humanitarianism originally recognized the place of Indonesia as an equal member in the international community of nations, indirectly asserting sovereignty over its territory and a fundamental right to exist and be treated with respect and human dignity according to the ideal of humanitarian behavior between all peoples. During the New Order of President Soeharto, this meaning lost many of its international connotations and shifted towards an emphasis on domestic economic development and political stability while fostering tolerance and mutual respect between all Indonesians (Ramage, 1995, pp. 12–13). This brought the Pancasila principle of humanitarianism closer to the principle of social justice, which posits a goal of economic and social egalitarianism and prosperity, a sort of economic democracy for all Indonesians through sustained economic development and rising standards of living as a result of an active government role in the national economy (Ramage, 1995, p. 13).

But the greatest genius of Sukarno in articulating Pancasila may lay in its incorporation of idealized concepts of traditional village governance to describe an Indonesian form of democracy based on the Islamic concepts of *musyawarah* and *mufakat*, or consultation and consensus: "The ideal is that decisions are reached only after all members of a community have had an opportunity to present their opinion (consultation) and then, only after all participants unanimously agree, is a consensual, harmonious decision reached" (Ramage, 1995, p. 13). Expressions of the ideal often fail to recognize the realities of consensus decision making, where more respected, charismatic, or assertive individuals inevitably emerge from a group and exercise more influence over decisions than do most members. In Indonesian village governance, where respected *bapak* leaders are known long before issues are discussed, it is generally those same leaders who formulate decisions and may not find it difficult to secure unanimous agreement with their formulations from *anak buah* before announcing a "consensual, harmonious decision" by decree. Incorporation of traditional village decision making procedures into Pancasila as a basis for government decision making gave cultural legitimacy to the new state, while allowing it to be dominated by social elites. Moreover couching Indonesian democracy in Islamic terms assured those seeking an Islamic state that their concerns and beliefs would be accommodated in a Pancasila state (Ramage, 1995, p. 13).

Embraced and promoted vigorously since 1945, Pancasila was over time embedded in Indonesian society as the sole source of ideological and political legitimacy not only for the government, but for regime critics as well, be they Muslim religious leaders, secular nationalists, or members of the armed forces (Ramage, 1995, pp. 1–11). Pancasila provided a sense of legitimacy, of tradition for a young state comprised of island cultures with a long history of social and religious diversity. Pancasila's success and longevity are attributable to its articulation in the most general and ambiguous terms of seemingly universal indigenous social values and themes. Although its principles are subject to differing interpretations, political, military, religious and intellectual leaders used Pancasila to discuss sensitive and controversial issues throughout the 1990s because it made open debate on such issues legitimate even under authoritarian rule. Evidence of its continued political and social importance after the downfall of President Soeharto is apparent in repeated references to it in the sweeping decentralization laws of 1999, discussed in Chapter 2, and the fact heads of regional governments must take an oath of office swearing obedience in applying and defending Pancasila as well as pledging to uphold the laws of the Republic of Indonesia (Republic of Indonesia, 1999b).

A Presidential Bureaucratic Polity

Throughout the 1990s, the Indonesian government functioned as a bureaucratic polity in which national policy making is largely insulated from social and political forces outside the highest elite echelons of the capital city. A presidential bureaucratic polity is a political system headed by a president in which power and participation in national decisions are limited almost entirely to employees of the government, particularly the military officer corps and the highest levels of the bureaucracy. Power is obtained through interpersonal competition in the elite circle in closest proximity to the president. In bureaucratic polities, there is no regular participation or mobilization of the people (Jackson, 1978a, pp. 3–22; Huntington, 1968; Riggs, 1966).

In the Indonesian bureaucratic polity, the military and bureaucracy are not accountable to other political forces such as parties, interest groups, or organized communal interests. Power does not result from articulation of interests from the social and geographic periphery of society. Parties, to the extent they exist at all, neither control the central bureaucracy nor effectively organize the masses at the local level. They do not represent long-term movements with roots in organized, enduring social, economic, or religious cleavages in society. Electoral campaigns are mobilized in the short-term by elites from above, and elections are held to legitimize, through democratic symbolism, the power arrangements already determined by competing elite circles in Jakarta (Jackson, 1978a, pp. 4–9; see also Emmerson, 1978). This may have begun to change with direct election of a President for the first time in 2004, but there are strong indications campaigns were organized, directed and funded by factions of the national elite centered in Jakarta, and it is as yet too early to conclude Indonesia has departed substantially from the familiar model of a bureaucratic polity.

Despite recent efforts at policy reform and decentralization, Indonesia has functioned as a Presidential bureaucratic polity at least since 1957. Although not a military man, B.J. Habibie was a protégé of President Soeharto who served in several prominent government offices, was Minister of Technology and Research for twenty years before rising to Vice President, and then President. Before his surprising ascendency to the Presidency, Abdurrahman Wahid was for fifteen years General Chairman of the moderate Nahdlatul Ulama, the largest Muslim organization in the country, an organization founded by his grandfather in 1926 and traditionally led by members of his family.[2] And being the eldest daughter of the founding national hero, President Sukarno, certainly conferred special status on President Megawati Sukarno-putri, comparable to family of President John Quincy Adams in the United States (Javanese do not have surnames: "Sukarnoputri" is a descriptive term meaning "daughter of Sukarno"). She grew up in the immediate presence of governing power, and moved easily in elite social circles. The first directly elected President Susilo Bambang Yudhoyono (popularly known as SBY) is a retired Army General known in the news media as "the thinking general" who served briefly as Minister of Mines before moving to Minister for State Security and Political Affairs in the Wahid administration. In 2001, when President Wahid was facing impeachment, he called on Yudhoyono to declare a state of emergency to shore up his position against the MPR. When Yudhoyono refused this invitation to a military coup, he lost his job and gained a reputation for integrity that served him well during the 2004 Presidential elections. In the interim, Yudhoyono was quickly reappointed Minister for State Security and Political Affairs by the Megawati administration. Thus, all of Indonesia's Presidents have been members of various factions of a national bureaucratic elite. Changes in electoral procedures towards greater democratization have not yet produced election of a popular "outsider" to the Indonesian presidency.

Bureaucratic polities are largely impervious to social and political currents in their own societies and may be more responsive to external influences from foreign sources. To the extent that outside aid and private capital are perceived as necessary to meet regime needs, domestic policy making will be influenced by the preferences and technologies of potential donors and investors. The necessity of outside aid has affected substantial policy decisions in Indonesia. The most obvious examples of foreign influence are supplied by conditions imposed on Indonesia by the IMF after the fall of Sukarno and again after the economic collapse of 1998 for resumption of large-scale financial assistance, and by The World Bank and Western donor countries in return for grants, loans and technical assistance such as that described in this study. An axiom of bureaucratic polity is that the greater the reluctance or inability to mobilize non-elite groups into domestic political and economic decision making, the greater the dependence on foreign resources for financial and technical inputs necessary to stimulate economic development (Jackson, 1978a, pp. 9–10). This generates opportunities for diffusion of technological and policy innovations into transitional societies, and internal circumstances that favor the politics of getting.

The Politics of Getting

The necessity of acquiring foreign funding and technology encourages bureaucratic polities in developing countries to seek international development assistance and makes them particularly receptive to influence from other countries. Harold Lasswell's definition of politics as getting (1958), and Norman Wengert's suggestion that government officials participate actively in the pursuit of non-local sources of funding (Wengert, 1979, p. 17) may find no better validation than in the quest for foreign financial assistance by governments of developing countries, and especially bureaucratic polities. Entrepreneurial administrators in mission-oriented bureaucracies of developing countries will get as much foreign assistance as they can for their governments, to the extent they are not likely to be held directly accountable for the costs-that is, when they are not directly identified with raising domestic taxes or other adverse local effects. What better way for a government official to do this than securing grants, loans or technical assistance from a foreign government or international development bank, to be repaid by the general treasury or user fees for services at some future date?

Of course, an outright grant will be the highest priority, but a loan will often be acceptable, especially if those allowed to spend substantial portions of the money are located in government organizations separate from those who must raise the funds necessary to repay the loan. Thus when an agency with a specialized mission such as regulation and development of mineral resources is able to secure grants or loans for technical assistance projects it will implement, and any loans will be repaid from general revenues or dedicated fees or charges collected by a different agency, there is a strong immediate incentive for behavior directed towards securing such funding.

Such grants or loans generally are provided only for specific purposes or projects that are consistent with the goals and purposes of funding organizations. Entrepreneurial bureaucrats in search of such funding must therefore generate proposals for projects that appeal to the goals and purposes of organizations providing such funding. Construction of major industrial development projects such as hydroelectric or coal-powered generating stations, and major land reform or agricultural development projects designed to stimulate economic growth often provide opportunities for obtaining foreign financial assistance.

Diffusion of Innovations

Major construction projects funded by donor countries and international development banks often include substantial programs designed to transfer technology to the recipient country and develop human resources through technical training activities, advice on program management or regulatory program improvement, and assistance in solving environmental problems attending development. Because large projects often affect specialized jurisdictions of several agencies or executive departments, they provide opportunities for entrepreneurial officials in one agency to propose relatively small technical assistance programs be "added on" to much larger development projects secured by a different agency. Provided required approvals can be obtained from superiors in governments receiving the funding, donor countries and international development banks are often receptive to such proposals from their clients.

Consequently, the introduction of new technologies, ideas and policies for consideration concerning a broad range of subjects is often willingly embraced in proposals designed to attract funding for development projects from foreign sources. Recently, successful proposals for international development assistance submitted to the United Nations Environment Program, The World Bank, the Global Environmental Facility and others involved importing new ideas to developing countries concerning the structure and management of financial institutions, markets, taxation, social welfare, commodity production, and government organizations. Vehicles for diffusion of innovations in these and other areas have included proposals for science and technology cooperation and provision of bilateral and multilateral technical assistance concerning development of new policies for management of natural resources and the environment. A case study of entrepreneurial administrative behavior in the Indonesian bureaucratic polity which illustrates the politics of getting foreign aid in international technical assistance projects is provided in Chapter 9 below.

A Note on Codification of Laws

Two significant difficulties attend comparative studies of policies in countries so different as the United States and Indonesia: language and understanding systems of official record keeping. Fortunately, English language translations are available for all major laws, regulations and decrees of the Indonesian government, published by the government and by various commercial entities for use by businesses. Understanding the Indonesian system of official record keeping is a bit more challenging. Compilations of Indonesian policies are organized quite differently than in the United States, and documents accumulate and relate to each other in a manner rather more confusing than the familiar method. Consequently, this deserves some brief explanation.

In the United States, statutes and regulations of the national government are published as they are created and regularly codified. Statutes first appear as paper slip laws in pamphlet form when enacted, then are published in chronological order at the end of each year in new hard-bound volumes of *Statutes at Large*. Thereafter, the text of each new statute is integrated and codified section by section with previously enacted statutes concerning similar subject-matter in the *United States Code*, first by insertion of "paper part" supplements to existing hard-bound volumes, and subsequently about every five years by publication of new hard-bound volumes replacing earlier versions. Provisions of law repealed by later enactments are deleted, language of amended provisions is changed, and new provisions are located with previous enactments on the same or similar subjects in the *U.S. Code*, providing a single, authoritative, relatively up-to-date text of current law. These volumes are extensively indexed, with tables of cross-references to related statutes, for ease of use.

A similar process is applied to regulations of the national government, which are first published in the *Federal Register*, the official daily publication of the Executive branch of the U.S. Government. Each year these regulations are integrated and codified with the text of previous regulations of the same agency concerning similar subject matter in the *U.S. Code of Federal Regulations*, which is bound in a set of paper back volumes. After codification, the result is a single set of current regulations issued by each agency on particular subjects, also extensively indexed, with tables of cross-references to related statutes and other regulations.

Codification of statutes and regulations is not practiced in Indonesia, which publishes both in chronological order. When a new statute substantially replaces and amends a previous law, the replacement text substantially repeats those portions of the previous enactment which are to be kept in force, omits what is repealed and adds whatever new policy is created. Indonesian statutes generally require elaboration in the form of Presidential Regulations, which in turn require sector-specific elaboration in Ministerial Decrees, which may in turn require elaboration in more specific decisions or decrees of subunits of a Ministry, such as the Director General of General Mining in the Ministry of Energy and Mineral Resources. Consequently, each new policy generates a ripple-effect of subsidiary pronouncements, usually a year or two later.

Generally there appears near the end of new Indonesian laws a statement to the effect that all existing laws and regulations on the same subject continue to apply to the extent they do not conflict with and are not replaced by the new enactment. This may or may not be followed by a more specific statement saying an earlier version of the law is declared no longer to be in force, or is void. Sometimes there is a statement saying a previous law or regulation is supplemented by the new one. Thus, it is largely left to the reader to figure out the extent to which existing laws and regulations do not conflict with and are not replaced by the new enactment. Explanations of new statutes which are published with them focus on interpreting the accompanying language, but seldom say whether new provisions conflict with or modify a previous law or regulation.

The principle result of this lack of codification of statutes and regulations is that there is no single, coherent, authoritative text one may refer to in order to find out what the current rules are in Indonesia. One must examine the most recent statute, applicable regulations and decrees and compare them to earlier versions before one may discern which previous rules are no longer in force, and what new requirements have recently been imposed. Sometimes language in previous decrees "moves up" to reappear in later regulations, and provisions in previous regulations reappear in new legislation that replaces previous requirements. Consequently, opportunities for human error in promulgating new regulations (e.g., omission by mistake), and interpreting the results are legion.

Often there remains some uncertainty, because it may take one or a few years after a law is enacted before new regulations and decrees are issued. Do the old regulations and decrees still apply if provisions in a previous statute were omitted from the newest version? Sometimes old regulations based on omitted provisions of previous statutes find new life in more recent enactments. Sometimes deleted provisions of previous statutes resurface later in entirely new statutes. Moreover, translations of the same provisions by different translators often provide different words with slightly different shades of meaning. One may understand the general sense of a requirement, but it is risky to quote directly from a single translation.

Consequently, there is often some ambiguity on the part of government officials concerning what current requirements are, and some uncertainty on the part of persons and firms – especially foreigners – who may have to meet regulatory requirements. In the case of Indonesian mining environmental policy, where the Basic Mining Law has not been formally amended since 1967, except indirectly by the decentralization laws of 1999, and several environmental laws applicable to mining operations have been enacted since 1982, there is much confusion over the appearance of many inconsistent or outdated requirements. This is not a formula for stringent application of the rule of law, and implementation of any policy may suffer as a result. This will become evident in the pages that follow.

Notes

1 Using U.S. inflation rates 1978–2004 totaling 153.86 percent available at InflationData.com/Inflation/Inflation_Rate/InflationCalculator.asp (accessed December, 2004).

2   Family leadership of Nahdlatul Ulama was weakened and the stage set for schism in December 2004 when Gus Dur's candidates were defeated at the 31st NU Congress (Wijayanta, et al., 2004).

Chapter 2

Coal Mining Regulatory Policy in Indonesia

Mining and Economic Development

Mining has a long history in the economic development of Indonesia, an archipelago of over 17 000 islands covering over 735 000 square miles, located south of the Philippines and north of Australia along the Equator. Tin mining on Bangka Island started in 1709 and coal mining in West Sumatera in 1849. Other significant mining activities include gold, silver, copper, nickel, aluminum, manganese, iron sand and industrial minerals such as limestone, clay, marble, and quartz sand (Kartasasmita, 1991; Kuntjoro, 1993). Indonesia has the ninth largest reserves of coal in the world, large quantities of which have higher heat content and less sulfur than low-sulfur coal from the Western United States. Recoverable reserves of coal are estimated at about 5.2 billion metric tonnes from a potential but unproven resource base of perhaps 38 billion metric tonnes (Ministry of Mines and Energy, 1999a, p. 9). However only 80 percent of the land area of Indonesia has been geologically mapped (Kuntjoro, 1993), and there are large areas still unexplored where conditions are favorable for the occurrence of mineral deposits (Sigit, 1988).

Coal production jumped from 10.6 million tonnes in 1990 to nearly 114.3 million tonnes in 2003 (Ministry of Energy and Mineral Resources, 2004, p. 13; Ministry of Mines and Energy 1999b, pp. 66, 241; U.S. Energy Information Administration, 2001, p. 35). This is phenomenal growth in a period of twelve years, making Indonesia the eighth largest producer of coal in the world. Low labor costs and access to deep water ports make the delivered cost of this coal competitive with that from other world suppliers, especially in Pacific Rim markets. The Government of Indonesia views expanding coal production as necessary to support development of electricity supplies for use in stimulating growth of industry and to replace declining oil export revenues. Indonesia is attempting to displace some domestic oil consumption with coal, so more of its dwindling supply of oil can be exported. The World Bank has financed development of the Indonesian electric power sector, including several hydroelectric and coal-fired power projects.

The oil, gas and mining sectors account for over 25 percent of all government revenues, 20 percent of exports, and almost 40 percent of total direct foreign investment in Indonesia (Yusgiantoro, 2003). As oil production declines, coal is expected to eventually displace oil as the principal energy source for domestic use (Kuntjoro, 1993; Mahmud, 1996). Indonesia exports about 60 percent of its annual coal production. Coal exports in 2002 were almost 74.2 million tonnes, up from 4.5 million tonnes in 1990, producing an estimated $2.2 billion in export receipts (Ministry of Energy and Mineral Resources, 2003, pp. 19, 31). Indonesia emerged as the fifth largest steam coal exporter in the world in 1993, shipping the majority of coal exported to Asian countries, principally Japan and Taiwan (U.S. Department of Commerce, 1994). Coal exports to countries in the region provide an increasingly important source of foreign exchange.

Most coal production comes from 25 large mines having annual production between 200 000 to 22 million tonnes. Both surface (open pit) and underground mining are used to extract coal, but open pit mines are most numerous, and more are expected in future years. The largest operation, PT Adaro Indonesia in South Kalimantan, produced over 22.5 million tonnes in 2003, almost 61 percent for export; the second largest producer, Kaltim Prima Coal in East Kalimantan, produced nearly 17.6 million tonnes in 2002, nearly 95 percent for export (Ministry of Energy and Mineral Resources, 2004, pp. 11, 18).

Technical Challenges

The technical challenges of Indonesian surface mining operations are substantial. Vegetation and soil are removed to expose coal or other minerals, disturbing areas covering hundreds of hectares at each mine. Average annual precipitation is about 71 inches per year, with over 200 inches per year in some mining areas of Kalimantan, and sporadic heavy rains in all areas of the country. Rain forest soils are thin (average 4cm), highly erodible, and often infertile in Kalimantan, producing high volumes of storm water runoff and heavy sediment loads in rivers and streams. Volcanic soils are often abundant on Java and Sumatera, producing less erosion. Toxic overburden strata are common, producing acid mine drainage, the effects of which on reclamation and water supplies are poorly understood in Indonesia. Consequently, rivers used for human drinking water and food supplies are often contaminated by mining operations, carrying their load of soil and fine coal dust downstream, degrading wetlands, mangrove swamps, shallow seas and valuable near-shore fisheries.

Indonesia's natural wealth is some of the greatest in the world, and its biological resources rival its mineral resources. Indonesia ranks second behind only Brazil for largest area of tropical forest, greatest primate diversity (about 35 species), and greatest number of primates found nowhere else (about half) (Mittermeier, 1988, pp. 148–150). Though Indonesia covers only 1.3 percent of the world's land area, it contains 10 percent of the world's flowering plant species, 12 percent of mammals, 16 percent of reptiles and amphibians, 17 percent of birds and 25 percent of fish species (Stone, 1994, p.136). Surface mining in rainforest offers special challenges in preserving biological diversity for future medical, scientific or ecotourism uses.

Although a hot, humid climate allows reestablishment of many vegetation types with relative ease, there is in Indonesia little comprehension of the importance of topsoil or the effects of soil compaction on efforts to reclaim mined land. An apparent cultural bias in favor of gardening and forestry (e.g., fruit trees and tree crops) appears to limit the effectiveness and monitoring of reclamation efforts, which display little use of grasses to secure soil against erosion and scant understanding of ecological principles in the absence of a designated postmining land use (Farrel, 2000, p. 17). Thus limited technical capabilities of government and mine personnel challenge effective environmental controls and reclamation of mined land in Indonesia. Visitors to five mines in 1992 found substantial difficulties with control of soil erosion and sediment, contamination of off-site water bodies, and minimal reclamation of land for postmining uses (Hamilton, Whitehouse, and Tipton, 1992).

The United States and the Republic of Indonesia both regulate these environmental effects of mining, although they are fundamentally different forms of government. Both governments are based on written Constitutions, but the United States is a federal system comprised of a national government and 50 states which have some independent sources of power, while Indonesia is a unitary system in which all power flows from a single source, and subunits are dependent on the national government for their authority.

A Unitary Government

The 1945 Constitution of the Republic of Indonesia describes it as a unitary form of government, in which all power flows from a single source, the national government. Unlike the federal system of the United States, the Indonesian Constitution provides no independent sources of power for its subdivisions. Indonesia has a central government based in Jakarta with the nation divided into 31 provinces (*propinsi*), which are in turn subdivided into 349 regions or regencies (*kabupaten*), 91 municipalities (*kotamadya*), and numerous village (*desa*) governments. The largest cities, Jakarta and Yogjakarta are special units counted as provinces. Although both provinces and regions had their own partly elected, partly appointed legislatures, historically these were largely consultative assemblies, dependent on the national government for funding and subordinate to central policy and regulation. They had no independent source of power, and were dependent upon national direction to determine their authority.

From its founding in 1945 by President Sukarno until its second President Soeharto was forced to resign after 32 years in office in 1998, the powers of policy initiative and decision rested almost exclusively in the Executive. Indonesia's national legislature, the House of Representatives (Dewan Perwakilan Rakyat – DPR) served in a consultative role, ratifying decisions of the President, but displaying little policy initiative of its own. Most of its laws were prepared in the Executive and ratified by the DPR with few modifications. The DPR must not be confused with the People's Consultative Assembly (Majelis Permusyawaratan Rakyat – MPR), a super-legislative assembly at that time comprised of approximately 600 members, including the 500 elected members of the DPR and additional "functional" representatives of provincial governments, political parties, the military and police, and selected civic groups.

Traditionally, the MPR met annually in August to hear the President's "accountability report," and briefly each five years to elect the President and Vice President, a largely ceremonial function similar to that performed by the U.S. Electoral College. Indonesia's 1945 Constitution requires the President to present an annual report, somewhat analogous to the U.S. President's State of the Union message, except it focusses more on the previous year, and less on priorities for the future. Historically the MPR routinely approved the President's accountability report and adjourned to participate in festivities surrounding Indonesian Independence Day, 17 August. Although the MPR is the only body empowered to amend the Indonesian Constitution, like the DPR it functioned mostly to legitimize previous actions and proposals of the President.

These traditions had begun to change by late 1997, when the Indonesian economy abruptly crashed along with others in the Asia region. Public sentiment turned strongly against the wealthy, and the Soeharto family stood out at the top of the list. As the economy of Indonesia had grown over the previous thirty years, so had the wealth and business interests of Soeharto's family, often due to favorable treatment by the government. In 1998, each of Soeharto's six children reportedly controlled assets in excess of $100 million (one over $2 billion), including a government-legislated monopoly on cloves, the only road from Jakarta to its international airport (charging tolls), a "national car" manufacturer subsidized by the government, and a failed national bank that had previously been bailed out by the government.

Soeharto's Golkar Party won substantial majorities in Parliamentary elections held in May 1997, but his formal reelection to the Presidency in March 1998 by the MPR was marred by criticism of economic performance, his age (77), and his family's wealth. During violent demonstrations in May 1998, rioters trashed some business offices of the Soeharto family, students seized the Parliament building, demonstrated against nepotism and government corruption, and called for Soeharto to resign. Rumors surfaced in the news media that the MPR might not accept his accountability speech in August.

After four students were shot to death by snipers during demonstrations at Trisakti University, President Soeharto found his support had evaporated, but insisted on a Constitutional transition of power to his Vice President, B.J. Habibie. Soeharto resigned the Presidency on May 20, 1998, after serving 32 years, the longest of any Asian leader in modern history. As the economy foundered and unrest continued, President Habibie attempted to defuse ethnic violence and demands for independence in four different parts of the country – Aceh, North Sumatera in the west; the Moluccas (Spice Islands) in the north; East Timor and Papua in the east – by granting a measure of autonomy to provincial governments. Independence for East Timor played a central role in the struggle for succession to the Presidency. A price was paid not only by the East Timorese who were displaced or killed seeking independence, but also by President Habibie, who gambled allowing East Timor to vote on independence would resolve an old and embarrassing political problem. However this decision was not widely popular in Indonesia outside East Timor. Indonesians are fiercely nationalistic and embrace the principal of unity as part of their national creed. Many Indonesians saw allowing an independence vote in East Timor as unpatriotic, an insult to the national honor, and a direct threat to national unity.

Habibie's Golkar Party placed a surprising second in national elections to the People's Consultative Assembly in May 1999, receiving about 22 percent of the vote. Megawati Sukarnoputri, the daughter of former President Sukarno, founding President of Indonesia, received about 34 percent of the vote. Abdurrahman Wahid, an aging, moderate Muslim cleric better known as Gus Dur (a play on the second syllable of his first name), placed a distant third with only 12 percent of the vote. Habibie's poor showing in the election eroded his standing with the military and in other government circles where he previously enjoyed support. After the People's Consultative Assembly refused in October 1999 to endorse Habibie's performance as chief executive during the previous 17 months, he had little choice but to withdraw from the Presidential race. It was unlikely the same group would elect him President, although they did grudgingly accept the results of a U.N.-supervised independence vote in East Timor.

When representatives of conservative Muslim parties balked at electing a woman President, Gus Dur was elected by the MPR, and Megawati was named Vice President. Gus Dur was a weak President whose support was based on a fragile coalition of several smaller, socially conservative parties and the establishment Golkar Party of deposed President Soeharto. His more popular Vice President represented the Indonesian Democratic Party of Struggle and advocated democratic reforms. The domestic situation did not improve greatly during 2000, and by early 2001 there were moves to oust Gus Dur. Suffering from an economy in shambles and civil unrest continuing in three separate regions of the country, Indonesian legislators resisted pressures to create an Islamic state. Instead, they deposed President Wahid using the constitutional mechanism of impeachment, and elevated Vice President Megawati Sukarnoputri to the Presidency.

Thus from March 1998 during the closing year of President Soeharto's reign, through the truncated terms of President Habibie and President Wahid to July 2001, the DPR and MPR became more politically assertive. In a period of less than three and one-half years, what began with risky criticism of an aging, long-term authoritarian President shifted to refusal to accept the performance of his successor, President Habibie, and culminated in impeachment of President Wahid. Such a rapid – and relatively but not entirely peaceful – transition from authoritarian rule towards a more democratic form of government is unprecedented in the history of the Asia region.

During the period 1999–2003, significant reforms which moved Indonesia towards democratization were achieved through a series of amendments to the 1945 Constitution by the MPR. The national police were separated from the military, and the "functional" seats in the DPR which were reserved for the police, the military and other special groups and appointed by the President were abolished, making it a wholly representative body of 550 members. A new legislative body was created, the Regional Representatives Council (Dewan

Perwakilan Daerah – DPD), consisting of four members each, directly elected from the 32 provinces, totaling 128 members. The number of members in the DPD may change if the number of provinces changes, but cannot exceed one-third the number in the DPR. The DPR continues as the principal legislative body. It approves all laws and may submit draft bills for approval by the President. The authority of the DPD is limited to regional issues. It may propose to the DPR bills relating to the relationship between national and regional governments.

The MPR now consists of the DPR plus the DPD, and its powers have been limited to amending the Constitution, inaugurating (not electing) the President and Vice President, and impeaching them for specified violations. The amendments provided for direct popular election of the President and Vice President for a term of five years beginning in 2004, created an independent, national election commission, and set forth basic human rights protections. The judiciary was separated from the executive branch and given independent Constitutional status, which it has begun to exercise. These substantial structural changes, part of the reform movement called *Reformasi*, have been aptly described by one observer as brilliant (Rieffel, 2004).

Still, the Republic of Indonesia retains the unitary form of government described in its 1945 Constitution. Legislation supported by President Habibie in May 1999 during this turmoil began a limited process of decentralization of power from the national government to the provinces and kabupaten, a transformation as yet incomplete. This legislation dramatically unsettled both mining and environmental policy implementation in Indonesia in ways that are as yet poorly understood, as discussed below.

Regulatory Policy for Coal Mining

Before Indonesian independence was declared in 1945, mining activities were governed by Dutch colonial laws and regulations. Principal among these was a safety regulation, the Mine Police Regulation of 1930. This regulation focussed on safety matters, although it contained a few articles applicable to control of pollution from mines (Subagyo, 1994). Article 33 of the 1945 Constitution of the Republic of Indonesia states: "The land, the waters and the natural riches contained therein shall be controlled by the State and exploited to the greatest benefit of the people," thereby asserting government ownership over the natural resources of the country in behalf of the people.

After revisions of colonial laws and regulations, a Basic Mining Law was promulgated in 1967. The Basic Mining Law of 1967 requires authorization by the Minister "whose tasks cover mining affairs" (currently the Minister of Energy and Mineral Resources) before exploration or development of any mineral resource; empowered the Minister to issue regulations governing mining operations; required that holders of Mining Authorizations (Kuasa Pertambangan, or KP) pay compensation to surface owners and users before mining begins; prohibited mining in certain areas; required restoration of the land after mining; and encouraged foreign mining firms to assist in developing

Indonesia's mineral resources through agreements with the government (Republic of Indonesia, 1967). Foreign investors were initially required to enter into Coal Cooperation Contracts with the government-owned coal mining company PT Tambang Batubara Bukit Asam, but enactment of Law No. 20/1994 authorized joint ventures with privately-owned Indonesian firms and foreign ownership of mining operations under Contracts of Work (CoW) with the Ministry of Mines and Energy (Sunardi, 1997). Every application for a KP or CoW is accompanied by an application fee, portions of which contribute to an informal "mystery budget" which may provide honoraria for technical teams and decision makers involved in evaluating and approving the application. A lack of accountability in payment of honoraria to technical teams and decision makers may have facilitated approval of KP and CoW which contain information providing an inadequate basis for inspection or enforcement of their terms.

Almost all coal Contracts of Work and Mining Authorizations are for locations in tropical rain forest, which covers about two-thirds of Indonesia, comprising about 10 percent of all rain forest in the world, and nearly 60 percent of the tropical forest in Asia. The extent of rain forest in Indonesia makes it a resource of global, not just national significance. Although changes to the mining law were under consideration during 2000–2002, it has not been revised since 1967. Draft revisions considered would have transferred authority to issue Mining Authorizations to provincial governors unless the mine operation was controlled by foreign investors or the mine location crossed borders between provinces, but included no more about environmental protection or reclamation than the Act of 1967.

The Basic Mining Law echoed the 1945 Constitution in stating: "All minerals found within the Indonesian mining jurisdiction in the form of natural deposits as blessing of God Almighty are national wealth of the Indonesian people and shall, therefore, be controlled and utilized by the State for the maximum welfare of the people" (Republic of Indonesia, 1967, Article 1). It is noteworthy neither the 1945 Constitution nor the Basic Mining Law of 1967 limits the scope of what is meant by "greatest benefit of the people" or "maximum welfare of the people" to economic or financial benefits, allowing interpretation of these phrases to include public health and environmental benefits.

Mining Authorizations

The Basic Mining Law of 1967 delegated broad authority to the Minister to regulate mining operations and grant Mining Authorizations to Indonesian mining firms, government enterprises and cooperatives for general surveys, exploration, exploitation, processing and refining, transportation, and sales of Indonesian mineral resources. The Minister was authorized to issue regulations governing application procedures for Mining Authorizations, their contents and terms; to include conditions in Mining Authorizations in addition to those in applicable regulations; and to cancel Mining Authorizations if a mine operator failed to fulfill requirements or disobeyed instructions and guidelines.

Holders of Mining Authorizations are required to pay land rents, royalties and other payments to the government as determined by regulation (Republic of Indonesia, 1967, Articles 15, 17, 22, 28). Portions of the royalties and land rents received each year go into the budget from which salaries of mine inspectors and decision makers are paid, and into an informal "mystery budget" which provides honoraria for technical teams and decision makers involved in evaluating and approving applications for KP and CoW. Nothing in the Basic Mining Law prohibits the issuance of Mining Authorizations to persons or firms under the ownership or control of individuals who have outstanding violations of the terms of previous Mining Authorizations.

At the end of 1999, 796 coal Mining Authorizations had been issued, including 514 general survey and exploration authorizations and 282 exploitation authorizations. Of these, 10 private companies and 14 KP Cooperatives were producing coal, and 17 private companies were conducting exploration activities (Ministry of Energy and Mineral Resources, 2001, pp. 60, 76–77, 81). In 2003, only 12 private firms and 12 cooperatives with Mining Authorizations produced coal (Ministry of Energy and Mineral Resources, 2003, p. 12).

The Ministry of Energy and Mineral Resources is organized into five Directorates General for Geology and Mineral Resources, Oil and Gas, Electricity and Energy Development, the Energy and Mineral Resources Research and Development Agency, and the Energy and Mineral Resources Education and Training Agency. The Directorate General of Geology and Mineral Resources has national regulatory authority over mining operations, and is further subdivided into five Directorates of Technical Mining, Coal and Mineral Business, Mineral Resources Inventory, Volcanology and Disaster Mitigation, and Geology, Environmental Space and Mining Regions. The Directorate of Technical Mining (Direktorat Teknik Pertambangan Umum – DTPU) is responsible for mine inspections and provides technical team expertise for environmental assessments on mining proposals, as discussed in Chapter 4. The Directorate reviews applications for Mining Authorizations and Contracts of Work. Thus both regulatory and commodity development functions are located in the same Ministry.

Historically, the Directorate General for Mines (now Directorate General for Geology and Mineral Resources) had national responsibility for supervising the application process for Mining Authorizations, regulating mining operations for coal and other ferrous and nonferrous minerals, and negotiating CoW with foreign companies. It maintained a cadre of mine inspectors in Jakarta and in Kantor Wilayah (Kanwil) regional offices of the Ministry located in 14 provinces where mining was most active. Regency (kabupaten) mine inspectors appointed by Provincial Governors were responsible for regulation and inspection of industrial construction minerals such as sand and gravel (Subagyo, 1994, p. 87).

Regulations clarifying and in some cases constraining this authority were issued by President Soeharto in 1969. For example, the scale of maps required in applications for Mining Authorizations was specified at 1:200 000 for General Surveys, 1:50 000 for Exploration, and 1:10 000 for Exploitation

(Republic of Indonesia, 1969, Article 13(3)). Especially for Mining Authorizations for Exploitation, these maps are not of sufficient detail to allow effective assessment of environmental impacts or calculation of the costs of reclaiming disturbed lands.

Surface coal mining permit applications in the United States require a map scale of 1:500, with more detailed construction drawings or detail plans for specific facilities, which show topographic features, excavation sites, drainage, mining facilities, and stationary equipment with sufficient accuracy to allow effective assessment of the environmental effects of coal mining and reclamation plans, and evaluation of compliance with the terms of the permit during inspections. Specification of the scale of maps in a Presidential Regulation made it difficult for the Indonesian Minister of Mines to require maps of a more accurate scale be included in applications for Mining Authorizations.

Moreover, Presidential Regulation 32/1969 contained a provision which allows a map for an application for Mining Authorization for Exploitation to be presented within a period of six months at the latest, if it cannot be attached at the time of the application (Republic of Indonesia, 1969, Article 13 (5)). This means an application for Mining Authorization for Exploitation may be approved by the Minister without any detailed map showing locations of mining facilities, excavations or equipment. Such an application would not provide a sound basis for decision making about environmentally disruptive mining operations, or calculation of reclamation performance bonds.

The Minister was authorized by the Basic Mining Law of 1967 to supervise production and other activities in mining related with the public interest (Republic of Indonesia, 1969, Article 29). Presidential Regulation 32/1969 provided that administration, supervision and regulation of mining operations would be centralized in the Ministry in charge of mining. The procedure for supervision and regulation was to be specified in regulations designed for security, labor safety, and the efficiency of the implementation of mining operations (Republic of Indonesia, 1969, Article 64). These provisions were interpreted by the Ministry as authorizing mine inspections for labor safety, and later inspections of the environmental effects of mining operations. However the lack of explicit authorization for examination of environmental effects provided a tenuous basis for environmental inspections unless such effects could be considered inefficiencies of mineral production.

Contracts of Work

The Basic Mining Law of 1967 encouraged participation by foreign investment in development of Indonesia's mineral resources by authorizing the Minister to enter into Contracts of Work (CoW) with foreign mining firms. These are lengthy legal documents which must be approved by the Indonesian House of Representatives, and therefore may be considered to have the force of law. In effect, they are each *ad hoc* amendments to the Basic Mining Law of 1967 that substitute their own language for regulations and requirements otherwise applicable to Mining Authorizations.

To the end of 1999, the Minister had signed Contracts of Work with 235 companies, of which 103 contracts were still active (e.g., not terminated by the government for nonperformance or by the contractor). At the end of 2002, 23 CoWs were producing coal (19 had significant production), seven were at the construction stage, 15 were preparing feasibility studies, and 68 at the exploration stage (Ministry of Energy and Mineral Resources, 2003, pp. 33–35). A little over 84 percent of the 103.4 million tonnes of coal produced in Indonesia in 2002 was produced by mining firms with Contracts of Work (Ministry of Energy and Mineral Resources, 2003, p. 12).

In general, a CoW describes the contractor's right to explore for minerals in the contract area, to develop and mine any mineral deposit found in that area, to process, refine, transport, market or dispose of all production, either inside or outside Indonesia, for a period of 30 years from commencement of actual operations. As with the Mining Authorization, this term may be extended for an additional 20 years by negotiated agreement. CoW specify land rents, royalties and other payments to the government. The CoW also describes whatever conditions and environmental obligations are imposed on the contractor, which are generally fixed for the term of the agreement.

Some of the more recent CoW contain requirements for environmental management; for example: operations must be conducted "so as to avoid waste or loss of natural resources, to protect natural resources against unnecessary damage and to prevent pollution and contamination of the environment within reasonable limits" (Farrell, 2000, p. 13). However such general statements lack the specificity required to allow effective inspection and enforcement of their terms. What is "unnecessary" damage? Who decides? Use of imprecise qualifying language such as "within reasonable limits" makes such ambiguous statements exceedingly difficult to enforce. Failure to avoid black water discharges of fine coal particulates from washing plants at coal mining operations as described in Chapter 5 would certainly constitute a "waste or loss of natural resources" under these terms, but no enforcement action has yet been taken against a CoW mining operation in Indonesia on these grounds.

Broad discretion granted by statute to the Minister to negotiate CoW agreements, the *ad hoc* nature of these documents, and the lack of statutory guidance for their preparation introduce a possibility of many different requirements being imposed on foreign firms conducting mining operations simultaneously in Indonesia. In 2003, Indonesia was using the seventh version of a "standard" CoW in negotiations with foreign mining firms. An eighth version was drafted but not finalized because its terms were considered too burdensome by foreign investors. Also, regional governments would have been authorized by the eighth version to sign a Contract of Work, but foreign investors considered such agreements less secure and were afraid banks would be less willing to lend investment capital based on them, instead of a contract with the central government. To date, an eighth version CoW has not been adopted.

Thus a strong pro-development bias is evident in the Basic Mining Law of 1967. Negotiating a CoW with foreign investors requires the Ministry to promote development of Indonesian mineral and coal resources, at the same time it is charged with regulating that development. A pro-development bias has been evident in the published writings of top officials in the Ministry of Mines and Energy. They often touted the high ranking of Indonesia in surveys of the international mining industry as "offering the most attractive framework of mining laws, fiscal conditions and political risk along with good geological prospectivity" (Mangkusubroto, 1996a, p. 45), and promised "simplification and speeding up of administrative procedures for contract and mining right application," reducing the time period to no longer than six months for CoW and three months for KP permits, and proposing "setting maximum time limits for processing of other permits required during operations"(Mangkusubroto, 1996b, pp. 66–67). A similar bias was evident in statements by Indonesia's Vice President and other government and military leaders in the early 1990s, who repeatedly referred to environmentalists as "new traitors" (*penghianat baru*) threatening Indonesian national security (Warren and Elston, 1994, p. 16).

Enforcement

The Basic Mining Law provides penalties for mining without a Mining Authorization or CoW of imprisonment not exceeding six years and/or a fine not exceeding five hundred thousand rupiah (about US$57.00[1]). Mine operators acting in violation of the Basic Mining Law, regulations issued pursuant to it, the terms of their Mining Authorization, or orders and guidelines of "the competent authorities" are subject to detention not exceeding three months and/or a fine not exceeding ten thousand rupiah (about US$1.15).

Imprisonment for violation of the Basic Mining Law is unheard of, and the fines are considered by mine operators to be insignificant. Prior to 2001, inspections were often conducted by mine inspectors from national offices in Jakarta, and mine operators paid their travel expenses to visit the mine sites, so those expenses often far exceeded any fines that might be levied for violations. Since 2001, when mine inspections for most mines were shifted to local and kabupaten personnel, payment of travel expenses has been less expensive, but fines have not increased appreciably in frequency or amount. Specification of the amount of fines by statute has prevented the Ministry from setting administrative fines at levels that might help deter violations, and illegal mining is rampant in Indonesia. Enforcement of requirements on holders of Mining Authorizations is lax and ineffective under these circumstances. As it stands, the Basic Mining Law prevents effective enforcement of its requirements in Indonesia.

For comparison, in the United States, the Federal Civil Penalties Inflation Adjustment Act of 1990 (*U.S. Code*, Title 28, §2461 note) requires that civil monetary penalties described in national legislation be adjusted for inflation at least once every four years. Indonesia has no comparable legislation. The Office of Surface Mining determines by regulation a schedule of progressively more substantial fines based on the severity of the violation and the previous compliance record of each mining operation, and may reduce the amount of the fines when violations are rapidly corrected by the mine operator.

Intentional violations of permit conditions and failures to comply with enforcement orders may be sanctioned with a fine of not more than US$10 000 or imprisonment for not more than one year or both (*U.S. Code*, Title 30, §1268(e)), and failure to correct a violation within the period specified incurs an additional penalty of not less than $925 for each day a violation continues, for up to 30 days. Consequently, mine operators in the U.S. are more effectively deterred from violating the terms of their permits and regulations, and have substantial incentives to correct violations in a timely manner.

Surface Rights and Prohibited Acts

Holders of Mining Authorizations are required to pay the surface owners and users of land for any damage inflicted on their property. Surface owners are obliged to allow disruption of their land when compensated (Republic of Indonesia, 1969, Articles 25–27). Mining is not allowed in territories which are closed in the interest of the public and in areas near military installations. There are no restrictions in the Act or implementing regulations on mining in national parks, protected forests or other public areas, which may be closed to mining only by the Minister of Energy and Mineral Resources. Implementing regulations state that places declared closed in the public interest by any other agency may be opened to mining by the Minister after consultation with the relevant agency (Republic of Indonesia, 1969, Articles 16, 22).

The Act of 1967 says the working area of mining activities based on a Mining Authorization does not include (1) cemeteries, sacred places, public works, such as public roads, railways, waterlines, electric supply lines, gas lines or (2) buildings, dwellings, factories and surrounding compounds, except with the permission of those concerned (Republic of Indonesia, 1967, Article 16). Such areas would be considered "unsuitable for mining" and mining activities would be prohibited in the United States (*U.S. Code*, Title 30, §1272(e)). However unlike the U.S. legislation, there are no distances specified in the Act of 1967 or in implementing regulations within which mining is prohibited, so it is unclear, for example, whether mining might be allowed right up to the edge of a public roadway or the foundation of a dwelling. Absent specified distances, protection of such facilities from damage due to mining would be difficult to effectively enforce. Moreover, the Act contains authority for "removal" of such amenities or facilities at the expense of the mining operation, in cases of extreme necessity, in the interest of the mining activities based on a Mining Authorization, and after obtaining a permit from the competent authorities (Republic of Indonesia, 1967, Article 16 (4)). Neither the Act nor implementing regulations provide criteria for deciding what constitutes an extreme necessity, so this appears to be a matter of discretion for the Minister or other competent authorities.

National Performance Standards

The only requirement in the Basic Mining Law of 1967 approximating a performance standard for protection of lands subjected to surface mining of coal from undue environmental harm stated that after completion of mining, the holder of the relevant Mining Authorization is obliged to restore the land in such condition as not to evoke any danger of disease or any other danger to the people living in the environment of the mine (Republic of Indonesia, 1967, Article 30). Because mining operations and their closure rarely present any danger of disease other than that associated with human settlements such as villages, the phrase "or any other danger" has more significance for managing environmental impacts.

In the United States, the Surface Mining Control and Reclamation Act of 1977 required that surface mined land be restored to a condition capable of supporting the uses which it was capable of supporting prior to any mining, or higher or better uses of which there is reasonable likelihood (*U.S. Code*, Title 30, §1265). Statutory performance standards for reclamation are detailed and required similarly detailed regulations, including a specific requirement for contemporaneous reclamation during ongoing mining operations; a general requirement and five specific standards for backfilling and grading concerning time and distance requirements, thin and thick overburden, previously mined areas, and steep slopes; a general requirement and three specific standards for revegetation concerning timing, mulching and other soil stabilization practices, and standards for determining success of revegetation efforts (*U.S. Code*, Title 30, §1265; *U.S. Code of Federal Regulations*, Title 30, §816.100–816.116).

Compared to reclamation requirements in U.S. regulatory policy, the language of the Indonesian Basic Mining Law is exceedingly general, and there is no specific requirement that restored land be capable of supporting uses it supported prior to mining. Implementing regulations of 1969 did not require mine operators to restore mined land, displayed little awareness or concern for environmental management of mines, and were substantially oriented toward regulating business practices of mining operations, such as acreage of the various types of Mining Authorizations and ownership of minerals produced.

The only concern for environmental protection evident in these regulations was expressed in the context of determining who owns onsite equipment and facilities when a Mining Authorization is canceled, abandoned or expires, and the focus then was on public safety: "the holder of the Mining Authorization shall take measures to safeguard the objects as well as structures and the condition of the surrounding ground which might jeopardize public safety" (Republic of Indonesia, 1969, Article 46 (4)). Although the Minister was authorized to determine actions required by the mine operator to control the condition of the ground prior to abandoning the former area of Mining Authorization (Republic of Indonesia, 1969, Article 46 (5)), this provided little basis for inspection and enforcement concerning environmental disruption caused by mining operations, and would be of little effect if the mine operator became insolvent.

Contemporaneous reclamation In the Indonesian Basic Mining Law of 1967 there was no requirement for preservation of topsoil, contemporaneous reclamation, or control of onsite or offsite environmental effects such as water pollution during mining operations, and no requirement for posting any bond or financial guarantee sufficient to reclaim disturbed lands if the mine operator should prove unable to do so due to insolvency. Ongoing mine operations provide a cash-flow from sale of production, a portion of which may be tapped periodically to pay for reclamation efforts. After mining is completed, mines which have closed operations generally do not produce a cash-flow. Unless monies have been set aside previously to pay for reclamation costs, there may be insufficient funds to cover the costs – and possibly no responsible party – to restore the land. Consequently, enforcement of a general requirement to safeguard the condition of the surrounding ground may be difficult, and some mines may be abandoned without reclamation.

The specificity of requirements is significant in making them useful during mine inspections. A general requirement that mined land be restored or reclaimed without saying when or in what manner is of little utility to mine inspectors. A requirement that specifies reclamation must be "as contemporaneous as possible" with ongoing mining operations provides a standard for which compliance can be determined during mine inspection. Under a contemporaneous reclamation standard, within a year of a mine commencing production, an inspector visiting a mine knows he or she should see reclamation efforts in process, and subsequent visits should reveal visible expansion of reclamation efforts over larger areas of mined land. If reclamation efforts are not evident, or have not expanded over larger areas since the last inspection, it is easy to determine a violation of the standard.

Except for the broad statement that natural resources are national wealth of the Indonesian people and shall, therefore, be controlled and utilized by the State for the maximum welfare of the people, there was no requirement in the Basic Mining Law of 1967 that production from surface mines be maximized so as to avoid waste of the resource and future remining of the same lands. Likewise, there were no provisions in the Act for protection of the hydrologic balance or water quality, designation of lands unsuitable for mining, or reclamation of abandoned mined lands previously disturbed.

Prevention and Mitigation

Regulation No. 4 of 1977 of the Minister of Mines concerning the Prevention and Handling of Disturbance and Pollution of the Environment Caused by General Mining (Ministry of Mines, 1977) was the first environmental regulation of mining operations in Indonesia under the Act of 1967. Signed on 28 September 1977, little more than a month after the Surface Mining Control and Reclamation Act was signed in the United States, the principal requirements and language of this regulation echo those of the American statute in many respects. American mining firms were actively involved in exploration for minerals in Indonesia at that time, and Indonesian personnel were being educated at American universities, including substantial numbers at Colorado School of Mines. It appears individuals in the Ministry of Mines were aware of this legislation in the U.S. Congress before it was enacted, and moved swiftly to develop similar policies in Indonesia. Similarities between these regulations and the Surface Mining Act constitute evidence of the diffusion of new ideas across national boundaries.

Regulation 4/1977 imposed on mine operators an obligation to prevent disturbances and pollution of the environment where possible, and mitigate any damage caused by mining. Mine operators were obliged to include a working program for prevention and mitigation of disturbance and pollution in the plan for mining activities approved by the Director General of Mining before the Mining Authorization for Exploitation or Contract of Work is approved (Ministry of Mines, 1977, Articles 1, 2). Where disturbance and pollution of the environment occurs, a mine operator was obliged to take action to mitigate the damage at its own expense, including reclamation of land. The Director General was authorized to order such action, conduct investigations, order temporary cessation of mining operations if necessary, and generally implement the regulation. Violations of Regulation 4/1977 were grounds for cancellation of the Mining Authorization or other civil or criminal sanctions specified in the Basic Mining Law of 1967 discussed above (Ministry of Mines, 1977, Article 9).

Reclamation Performance Bond

The Director General was invited to propose to the Minister amounts of guarantee money required to be deposited in state banks by the mine operator, which could only be reimbursed after the mining operation fulfilled its obligations to control the environmental effects of mining and reclaim the land (Ministry of Mines, 1977, Article 8). This was the first mention of a performance bond or reclamation guarantee in Indonesian mining policy, and it replicated the requirement in the Surface Mining Act for performance bonds (*U.S. Code*, Title 30, §1259). However lacking a defensible method for calculating an appropriate amount for such a bond, and a procedure for reimbursement to the mine operator after appropriate performance, the Director General did not propose a bonding system until 1996, as discussed in Chapter 6.

The Director Generals Regulation 4/1977 was only three pages long, written in general language lacking specific requirements for performance. It stated a general goal of preventing unmitigated disturbance and pollution of the environment by mining operations, and imposed a burden for paying the costs of environmental management and reclamation on the mine operator. It did not describe specific performance standards or operational requirements that a mine inspector could compare to circumstances at particular mine sites and use to enforce the law.

Regulation 4/1977 was followed by issuance by the Director General of Mines of Decrees 7/1978 and 9/1978 concerning prevention and mitigation of damage caused by surface mining, mineral processing and refining (Subagyo, 1994, p. 87). Decree No. 7/1978 elaborated on what contents would be required for the working program for environmental management of surface mines described in Regulation 4/1977. It required the location of areas that might be disturbed or polluted by mining activities be shown on maps attached to the application, including mine drainage and settling ponds, surface excavations, waste and overburden disposal sites, and areas planned for topsoil storage. Unfortunately, maps showing areas that might be damaged by mining activities were required on a scale of 1:10 000 (Ministry of Mines and Energy, 1978b, Article 2), consistent with Presidential Regulation No. 32 of 1969, but insufficient to allow effective assessment of environmental impacts or accurate calculation of reclamation performance bonds.

Reclamation and Preservation of Topsoil

Decree No. 7/1978 required a reclamation plan be submitted along with the mining plan before mining operations commence. Although the Decree did not specify the timing for this submission, apparently it must be submitted along with the working program for environmental management which must be approved by the Director General before a Mining Authorization for Exploitation may be issued. The reclamation plan must include a description of the proposed land use after reclamation, methods to preserve topsoil, preservation of water quality, and a schedule for completion of reclamation (Ministry of Mines and Energy, 1978b, Article 3). Required contents of the reclamation plan in this Decree approximate requirements for reclamation plans in the Surface Mining Act (*U.S. Code*, Title 30, §1258) and appear to include the first mention of a postmining land use in Indonesian mining policy.

Decree No. 7/1978 moved in the direction of specifying performance standards for surface mines, requiring mine operators to: preserve topsoil for use in reclamation efforts; prevent erosion of slopes by providing drainage to settling ponds; backfill, level and revegetate mined areas; and begin reclamation no later than six months after mining operations commence. However like Regulation 4/1977, the Decree was only four pages long, stating general goals but not describing specific operational requirements that a mine inspector could compare to circumstances at particular mine sites and use to enforce the law.

Preparation Plants

Similar difficulties are evident in Decree No. 9/1978, which described what contents are required in the working plan for environmental management attached to an application for Mining Authorization for mineral processing and refining facilities. Decree No. 9/1978 required a description of the processing operation; water requirements and their control; the quality of waste materials; locations of product stockpiles; and possible effects on the environment. Some recognition of the need for better maps was suggested in Decree 9/1978 requirements for maps at 1:500 scale for locations of water intakes, tailing disposal, acid water, and settling ponds (Ministry of Mines and Energy, 1978a, Article 2). This was a significant improvement over previous requirements. Facilities using wet processes were required to test tailings disposal areas and process waste water released to public water courses daily, and record the results. Still, Decree 9/1978 did not require regular maintenance or cleaning of settling ponds. Similarly, it did not require location of processing facilities away from wetlands, streams or rivers, or control of storm water discharges from processing facilities.

Not until 1995 were these requirements modified by Decree No. 1211.K/008/ M.PE/1995 of the Minister of Mines and Energy on Prevention and Mitigation of Environmental Damage and Pollution in General Mining Operations, which replaced and expanded Regulation 4/1977, Decrees 7/1978 and 9/1978. Previously, many mining operations in Indonesia were not organized to assign responsibility for environmental management to any distinct subunit of the organization. Having experienced some difficulty getting mining operations to implement their environmental protection responsibilities under previous requirements, Decree 1211/1995 imposed personal accountability on a single employee of each mining operation by requiring appointment of a Mine Technical Manager (*kepala teknik*), with responsibility for prevention of environmental damage and pollution (Ministry of Mines and Energy, 1995b, Articles 4, 7, 8).

Decree 1211/1995 imposed an obligation directly on the Mine Technical Manager to take preventive measures against the possibility of environmental damage and pollution and, if such damage should occur, to take corrective measures immediately. Mine operators were also obliged to submit annual environmental management and monitoring plans to the Chief Mine Inspector of the Ministry's Directorate of Technical Mining (DTPU), and to the head of the appropriate provincial mine inspection agency. These documents replaced the previous working plan for prevention of disturbance required under Regulation 4/1977 (Ministry of Mines and Energy, 1995b, Article 6).

The Mine Technical Manager was required to provide a map of environmental management and monitoring efforts in a form and following a procedure to be determined by the Director General of Mining (Ministry of Mines and Energy, 1995b, Article 14). Provided the Director General required maps of appropriate scale and detail, this would have been a substantial step forward in securing information useful in inspection and enforcement of environmental requirements. Unfortunately, the Director General provided no further guidance on this matter and continued to accept maps of inappropriate scale.

Towards Performance Standards

With Decree 1211/1995, the Ministry took another positive step in the direction of specifying performance standards for surface mines. Blasting activities were prohibited from damaging houses, important buildings, and the environment, but no minimum distances from structures were specified. Underground mining activities were prohibited under important buildings if they might cause subsidence. If subsidence occurs, the land must be restored to its original condition.

Continuous monitoring of toxic tailings was required, with results reported every three months to the Chief Mine Inspector. Transportation activities were prohibited from causing air pollution. Water used for processing minerals was required to use a closed-circuit system (Ministry of Mines and Energy, 1995, Articles 9–24), but this standard was not universally applied to coal washing plants, as discussed below in Chapter 5. Provisions in Decree 4/1977 authorizing a requirement for a reclamation bond were repeated in Decree 1211/1995 (Ministry of Mines and Energy, 1995b, Article 29), without creation of a program to implement such a requirement.

Topsoil was to be used immediately in reclamation efforts or protected from erosion; slopes formed by the mining operation were to be compacted and slope stability monitored. Reclamation and revegetation of mined areas were required to be conducted in accordance with reclamation plans approved by the Director General. Backfilling of mined areas was required. Overburden and other waste materials were to be stored in a safe location. Disturbance of the hydrological balance was to be minimized and ground water protected from pollution. However standards for determining success of these actions were not specified, and the standards were therefore not useful during mine inspections to assess compliance.

Decree 1211/1995 required the mine operator to collect water runoff in channels and divert it to settling ponds constructed on stable land, and that settling ponds be well-maintained and functioning properly. Presumably stable land would not include river banks, wetlands and floodplains subject to periodic inundation. The World Bank recommends settling ponds to catch storm water and reduce suspended solids be provided for all water before discharge from coal mine sites, saying they "can be consistently achieved by well-designed, well-operated and well-maintained pollution control systems" (World Bank, 1995).

Like Regulation 4/1977, the content of Decree 1211/1995 was brief and somewhat lacking in specifics. It stated general goals but did not describe specific operational requirements that a mine inspector could compare to observable circumstances at particular mines and use to enforce a regulation. For example, it did not describe design standards or a maintenance schedule for settling ponds or how one might know if they were functioning properly. Decree No. 1211/1995 did not mandate design of sediment ponds for any particular area of control or size of precipitation event, or require regular clean out of settling ponds so they would continue to function effectively by allowing gravity to remove solid particles from water. Periodic clean out is necessary over the life of the mining operation, to control discharges of sediments into public water bodies. Construction of settling ponds provides little environmental benefit if they are not properly designed, regularly cleaned out and maintained.

By contrast, performance standards for sedimentation ponds in the United States specifically require they be designed to contain or treat the maximum amount of runoff from rainfall likely to fall during a 24-hour period experienced over 10 years (e.g., a 10-year, 24-hour precipitation event), unless more stringent design standards are appropriate. The OSM California mining permit application requires the proponent to submit detailed designs of sedimentation ponds and impoundments certified by a professional engineer. Designs must meet specific technical requirements established in performance standards for impoundments in 30 *Code of Federal Regulations* §816.49 (U.S.DOI/OSM, 1988, p. 126).

National performance standards in the U.S. also require that sediments be removed periodically from settling ponds to maintain adequate volume for the design event, and ponds must provide adequate detention time to allow effluent discharged from them to meet applicable State and national water discharge standards (*U.S. Code of Federal Regulations*, Title 30, §816.46(c)). Using these performance standards, design parameters for ponds observed during inspections may be calculated to determine compliance. Inspectors may compare the results of water sampling to numerical water quality standards as the measure of settling pond performance for purposes of determining compliance.

Decree 1211/1995 did not limit construction of reservoirs in old mine pits or prohibit locating settling ponds on wetlands, banks of rivers, or in stream beds as national performance standards do in the United States. It said water discharges must meet applicable water quality standards but did not say which standards were applicable. Water quality standards issued by the Ministry of Environment for industrial activities did not specify standards applicable to coal mining operations at that time, and required sampling of organic parameters not typically found at mining operations (Ministry of Environment, 1995), so there was confusion about which standards were applicable.

Unlike Decree 7/1978 which it replaced, criteria were not specified in Decree 1211/1995 for determining a safe location for excess overburden disposal, predicting subsidence or ground water pollution, or determining what constitutes an "important," as distinct from an unimportant building. It did not describe what actions might be required after slope stability was monitored, or specify what actions might be required if monitoring of toxic tailings showed they would be harmful to human health or the environment. A troubling provision of Decree 1211/1995 allowed the Chief Mine Inspector to waive the mine operator's obligation to revegetate mined areas in accordance with AMDAL documents (Ministry of Mines and Energy, 1995b, Article 13(2)). Moreover, although it required reclamation be carried out in accordance with a reclamation plan (Ministry of Mines and Energy, 1995b, Article 12(1)), it did not specify reclamation was to begin within six months after mining operations commence, apparently repealing the previous requirement for contemporaneous reclamation. These were steps backwards from previous requirements. Still, requirements that water discharges meet applicable standards, that topsoil be used for reclamation or protected from erosion, that mined areas be backfilled, reclaimed and revegetated, and that process water utilize a closed circuit were improvements over previous regulations. Overall, Decree 1211/1995 represented a tentative step towards more effective implementation of mining environmental policy in Indonesia.

At this point, Indonesian regulatory policy for mines did not provide a process for designation of lands unsuitable for mining if reclamation of the site is not technologically or economically feasible, if mining is incompatible with existing land use plans, would damage fragile, historic, or productive renewable resource lands, or lands in natural hazard areas where such operations might endanger life and property. Moreover, no fund has yet been created in Indonesia to reclaim lands previously mined but abandoned without reclamation. Abandoned coal mine lands often contain serious public health and safety hazards such as highwalls, unstable slopes presenting potential for land slides, open shafts, erosion of soil and acid mine drainage to surface waters. Because the scale of mining was not historically large, Indonesia does not have many large areas of disturbed abandoned mined lands. Their extent will increase in the future as the extent of mining operations continues to expand, unless reclamation and environmental management efforts are improved substantially.

Decentralization of National Regulatory Responsibilities

Regulation of mining operations was decentralized to regional and local governments by Law No. 22 on Regional Administration enacted in 1999 (Republic of Indonesia, 1999b) and a subsequent Presidential Regulation No. 25 of 2000 on the Authority of the Government and the Authority of a Province as an Autonomous Region (Republic of Indonesia, 2000). In this legislation, subnational units were given authority over all activities of government except those specifically enumerated in the Act which were retained by the national government.

The Republic of Indonesia continues as a unitary state, and exclusive authority remains in the national government over matters concerning foreign policy, defense and security, judicial, monetary and fiscal policy, religious and "other areas" (Republic of Indonesia, 1999b, Article 7 (1)). "Other areas" of continuing national authority include policies on national planning and control of national development, funding to provide financial equilibrium, the national administrative system and economic institutions, development of human resources, efficient use of natural resources and strategic high technology, conservation and national standardization (Republic of Indonesia, 1999b, Article 7(2)). Coverage of these areas is broad enough to allow considerable oversight of subnational government action in almost any area of domestic policy.

Continued central government control over national development and efficient use of natural resources respect provisions of the 1945 Constitution giving control over natural resources to the national government, and allow it some continued initiative and oversight of mining activities, including approval of CoW negotiated by regional governments. Presidential Regulation 25/2000 provided long lists in 25 subject-matter areas or sectors identifying activities in which the national government continues to exercise authority (Republic of Indonesia, 2000, Article 2). In the mines and energy area, broad authority was reserved to the national government to set standards for mineral prospecting and management of mineral and energy resources, and to determine criteria for working areas for mining business. Presidential Regulation 25/2000 also provided long lists in 20 subject-matter categories in which provinces may exercise authority. Regulation of mines and energy is not among them.

Authority over funding to provide financial equilibrium allows the national government to determine by legislation the allocation between units of government of mineral royalties and land rents paid by mining operations for development of national resources. Although authority for collection of most taxes and fees for uses of land was shifted to regional governments, those from mining and forestry operations, along with applicable mineral royalties and land rents, continue to be collected by the national government pursuant to previous laws (Republic of Indonesia, 1999b, Article 80), with shares disbursed to regional governments.

Otherwise, administrative authority to regulate mineral developments was shifted from national offices to those of regional governments, defined to include provinces, regencies and municipalities. These subnational units of government were given relative autonomy to regulate and manage the interests of local communities at their own initiative "on the basis of the aspirations of the community" (Republic of Indonesia, 1999b, Articles 2, 4). Regional governments were specifically authorized to manage national resources and maintain environmental sustainability within their territories. Obligations were imposed on them to handle government matters in public works, health, education and culture, agriculture, communications, industry and trade, investment, environment, land affairs, cooperatives and manpower (Republic of Indonesia, 1999b, Article 10). Although the numbers have increased rapidly since 2001, at last count there were 32 provincial administrations, 349 regencies, and 91 municipal kotamadya in Indonesia. Less than 25 percent of these provinces and regencies have jurisdiction over mineral resources, including eight provinces and 73 regencies. However a single coal mining property may involve one province and more than four regencies (Aspinall, 2000, pp. 4–5). The authority of districts and municipalities was made explicitly applicable to mining areas and forest concessions (Republic of Indonesia, 1999b, Article 119).

Some previous CoW and concession documents had given powers of law enforcement and local government to developers (e.g., Kaltim Prima Coal) in frontier areas where there was no existing local government presence. The new authority given regional governments appears to conflict with the terms of some of these CoW and concession documents, producing negative effects on relations between local governments and existing mining firms, and possibly discouraging future foreign investment. For example, assertion of taxation authority by the head of the East Kutai Kabupaten and attempts by the Provincial Governor of East Kalimantan to assert ownership over shares in the mining operations of PT Kaltim Prima Coal without paying for them lead the foreign-owned mining firm to sell its investments in July 2003 to PT Bumi Resources, and pull out of mining in Indonesia (Battle for Kalimantan's Coal, 2002; Taufiqurohman, et al., 2003, p. 56). Kaltim Prima Coal, the second largest coal producer in Indonesia, was owned by British Petroleum (BP, now BP-Amoco) and Rio Tinto (formerly CRA Limited) of Australia. PT Bumi Resources, an Indonesian firm, also owns PT Arutmin, the largest coal producer in Indonesia. This divestiture greatly concentrated ownership in the Indonesian coal mining industry. No foreign investors have opened new coal mines in Indonesia since 1998, and this divestiture is widely believed to have discouraged foreign investment.

Regional governments serve dual roles under this decentralized framework of government. In some policy areas they may function as autonomous regions, while in other policy areas they function as an administrative arm of the national government, receiving some funding from the national government for the latter purpose (Republic of Indonesia, 1999b, Articles 9, 78). This appears to be a mechanism of central control, because provinces are not the only administrative arm of national government in the countryside. Functions of the national government in foreign affairs, defense and security, judicial, religious, monetary and fiscal affairs continue to be implemented by "vertical government agencies," or regional Kanwil offices of national government ministries. Outlying offices of other national ministries such as the Ministry of Energy and Mineral Resources were to become offices of regional governments and their assets transferred to regional government ownership (Republic of Indonesia, 1999b, Article 64).

Although they are described as autonomous, regional governments can be assigned specific tasks by the national government and are obliged to perform them. Regional governments were authorized to issue regulations and make decisions, but they must not contradict national laws and may be canceled by the national government if they do so. The President may remove the head of a regional government from office for poor performance (Republic of Indonesia, 1999b, Articles 13, 46, 69, 70, 72, 114). Violations of regional regulations carried sanctions limited to a maximum of six months imprisonment or five million rupiah (about US$568.00), and were not classified as felonies (Republic of Indonesia, 1999b, Article 71).

Clearly there were limits to how much autonomy the DPR was prepared to grant to autonomous regions. Concern for possible disintegration of the nation-state was evident in repeated references to the necessity of maintaining the unity of the unitary state, applying the government creed of Pancasila and the Constitution of 1945. Heads of regional governments, including provincial governors, regency heads (*bupati*), district heads and mayors must take an oath of office swearing obedience in applying and defending Pancasila, and pledging to uphold the laws of the Republic of Indonesia. They may be summarily removed from office by the President for "treason or other acts which may disintegrate the Unitary State of the Republic of Indonesia" (Republic of Indonesia, 1999b, Articles 42, 52). The meaning of "other acts" is undefined in the statute, suggesting actions considered contrary to the interests of the national government may not be tolerated.

Granting regional autonomy was largely aimed at defusing separatist sentiments in the outlying provinces and ending decades of exploitation by the central government (Ryas Tenders his Resignation, 2001; Turner, et al., 2003). In the countryside, the national government was perceived to be dominated by ethnic Javanese, who were most numerous in its employment, often even in offices located on islands other than Java. The proceeds of natural resource development were perceived to have flowed mostly towards Jakarta, without being disbursed on an equitable basis among provinces that produced this revenue. Local government officials attending a workshop in East Kalimantan in February 2001 were nearly unanimous in questioning the credibility of coal production figures published by the national government, suggesting they underrepresented actual production, thereby depriving local governments of their fair share of royalties. Law 22/1999, signed by President Habibie, contained an enticement authorizing a grant of special autonomy to East Timor if it should choose to remain within the Republic of Indonesia (Republic of Indonesia, 1999b, Article 118). Never used, this provision became moot when the East Timorese on 30 August 1999 voted for independence.

Apparently slow to delineate a role for regional governments in regulation of mining operations, the Director General of General Mining was quoted in the *Jakarta Post* a week after Presidential Regulation 25/2000 was signed, saying a Ministerial Decree was forthcoming, in which provincial administrations would be authorized to issue Mining Authorizations and CoW, and regencies and mayors would be authorized to issue general business permits for mining operations (Government Prepares Provinces, 2000). Later that year, the Ministry issued new regulations that closely tracked Presidential Regulation 25/2000 but bypassed the provinces and authorized regional governments to issue Mining Authorizations and CoW effective 1 January 2001 (Ministry of Energy and Mineral Resources, 2000). Apparently fearing secessionist tendencies in some provinces, the Ministry retained jurisdiction and inspection authority over all previously signed Mining Authorizations and CoW while delegating authority for negotiating new ones to regencies and mayors, subject to approval by the national government. A few were issued to Asian firms, but through 2003, none of them were considered active. Ministry inspectors from Jakarta continue to inspect mining operations conducted under Mining Authorizations, Contracts of Work and Coal Mining Contracts signed prior to 31 December 2000, but as old mines close and new ones open, the role of regional governments in authorizing and inspecting mining operations will increase.

Since 2001, there have been no new CoW signed between regional governments and foreign mining firms, because foreign investors want the national government to be a party to such contracts (Yusgiantoro, 2003). Foreign firms consider agreements with regional governments to be less secure and are concerned banks will be less willing to lend investment capital based on them, preferring a contract with the central government. Moreover, the debacle over ownership of PT Kaltim Prima Coal which lead to British Petroleum and Rio Tinto pulling out of Indonesia has further soured foreign investors on coal mining in Indonesia.

Transfer of Personnel

Law 22/1999 required the transfer of personnel from national to regional and local governments in accordance with the areas of authority and functions transferred (Republic of Indonesia, 1999b, Article 8). This requirement applied to the transfer of about 2.44 million persons by 1 January 2001 (Turner, et al., 2003, pp. 101–102), the largest transfer of government employees between units of national and subnational governments ever attempted as part of a decentralization effort. When authority to grant Mining Authorizations and to negotiate CoW for development of mineral resources was turned over to regional governments in May 2000, there was apparently an expectation in Jakarta that the provinces and kabupaten would quickly take over many of the personnel as well as the buildings housing local offices of the national Ministry of Mines and Energy. This did not happen.

Although the physical infrastructure changed hands in November 2000, in December 2001 the lights were still on and employees of the Ministry of Mines and Energy were still reporting daily to work at Kanwil offices in East Kalimantan, despite the fact their work had mostly been given to kabupaten and kotamadya offices. When asked why they were not working for regional governments, they replied they had received no offers or transfer orders (Whitehouse, 2003). Regional governments were eager to throw off the yoke of Javanese domination, preferring to hire *putera daerah*, "sons of the region" or local people (Turner, et al., 2003, p. 107), even if they lacked appropriate expertise for tasks to be performed. Central government employees were often believed to be corrupt.

Many Kanwil employees sought other employment, and it may be presumed the most capable ones were most successful in this regard. Those remaining were reluctantly absorbed by provincial offices by central government order in January 2002. Mine inspectors in national Ministry offices in Jakarta who previously dominated the inspection and enforcement process have had little to do since most inspections are now conducted by offices of regional governments (Whitehouse, 2003).

Mine Inspections

Previously Indonesian coal mines were required to be inspected only once per year, but a new Ministry Decree 1453/2000 devolving these responsibilities to regional governments requires mine inspections at least once in six months (Ministry of Energy and Mineral Resources, 2000, Article 9). This was a significant improvement, but more frequent inspections are required to provide adequate environmental protection, for reasons discussed below. Until responsibility for mine inspections was shifted to regional governments in 2002, these inspections were almost always performed by personnel of the Directorate of Technical Mining (Direktorat Teknik Pertambangan Umum – DTPU) based in Jakarta. If violations of the law, regulations, Mining Authorizations or CoW were found, a followup inspection might be scheduled about six months later, usually by a different inspector from a Kanwil regional office of the national Ministry. Thus Decree 1453/2000 decentralized the locus of regulatory authority and increased the frequency of inspections but otherwise left the substance of mining environmental policy essentially unchanged.

Since 2002, inspections of new mining operations have been the responsibility of regional governments: kabupaten and kotamadya. Inspections of mines opened under Mining Authorizations and CoW negotiated by the national government prior to 31 December 2000, and where kabupaten have not yet established their own programs, have continued to be conducted by DTPU inspectors. Inspections have been carried out by personnel from kabupaten offices, some of whom were acquired when Kanwil offices of the national Ministry were closed. The expertise of existing kabupaten and kotamadya mine inspectors is extremely uneven, and many of them have not yet received the training required for certification by Ministry of Mines and Energy Decree No. 2555/1993 (Ministry of Mines and Energy, 1993).

Ministry Decree 1453/2000 specifically adopted the performance standards and procedures for mine inspections which were already in force under Decree 1211/1995. An inspection is initiated with a review of office files related to the mining operation. Historically these were maintained by DTPU in Jakarta, and it is unclear whether these files have been successfully replicated in offices of regional governments since 2002. Interviews with DTPU mine inspectors from 1992 to date indicate their coordination and communication with regional government offices prior to 2000 was rare. Consequently, kabupaten and kotamadya had little experience inspecting coal mines. Historically, the review of office files in preparation for an inspection was thorough, covering previous inspection reports, Mining Authorizations or CoW as appropriate, environmental assessments, and whatever updated documents were submitted by the mine operator subsequent to the last inspection. The responsible DTPU inspector might contact the appropriate Kanwil regional office to discuss a planned inspection and identify any findings of recent inspections conducted by Kanwil inspectors. In practice, contact with the Kanwil offices was infrequent at best. The responsible inspector conferred with DTPU colleagues to plan the field phase of the inspection and then contacted the operator to schedule an inspection.

This contact was normally made 30 days in advance of the site visit. The reason for this courtesy was due to the fact that mine operators were expected to purchase airline tickets and provide accommodations for mine inspectors while in the field, either at mine site housing or in local hotels. As a practical matter, air travel away from Jakarta is expensive when compared to government budgets, and the Ministry of Mines and Energy has never had a budget adequate to pay for these expenses, even for one trip per year to 20 mines located on islands so distant as Sumatera, Borneo, Sulawesi or Papua. This practice continues after regional autonomy laws of 1999 shifted mine inspections to regional governments, although travel distances and expenses are considerably less for them.

Of course the result was that site visits by mine inspectors were almost never unannounced. Mine operators generally have about 30 days to clean up or repair the most serious environmental damage and violations of the terms of their authorization to mine. Often key facilities such as coal washing plants would not be operating when inspectors arrived, ostensibly shut down for maintenance. This made it difficult for inspectors to assess water discharges from such facilities. In addition, both the inspecting organization and individual inspectors remain somewhat beholden to the mine operator for paying their travel expenses. It is not possible to determine what impact this may have had on enforcement efforts, which often appeared to be lax.

Mine inspectors usually would take half a day or so to inspect records in the office of the mine manager, and the remainder of that day and perhaps the next day to view all areas of the mine site, from pit to reclamation efforts. Often, but not always they were accompanied by the Mine Technical Manager. The same inspector was responsible for inspecting health and safety conditions as well as environmental issues, unlike the practice in the United States, where separate inspections are conducted by different personnel from separate agencies. Observations of the inspector were usually noted for discussion with the Mine Technical Manager and perhaps the Mine Manager at the end of the second day. Inspections rarely take more than three days, even for large mines having multiple pits and large shipping facilities, such as Kaltim Prima Coal with seven active mine pits and a deep water port 13 km distant. Photos were taken by some inspectors on their own initiative, with permission of the mine operator, but there were no systematic efforts to record environmental violations on video or film.

Findings and recommendations for remedial action were recorded by hand in a Mine Book located and maintained in the mining company offices for this purpose, after discussion and agreement between the Mine Technical Manager and the inspector. Thus the Mine Book constitutes a journal or ledger of mine inspection findings and required remedial actions. After making an exact copy of Mine Book entries for each inspection, often by hand, the mine inspector returns to Jakarta and submits this as the official report to the Chief Mine Inspector. Problems identified during an inspection are divided into two categories, those which can be remediated quickly and problems which will require more time to correct. Recommendations for remedial action typically had a schedule established for implementation, but the schedule might be vaguely worded to require action "as soon as possible." The Mine Technical Manager must send a letter to the Chief Mine Inspector agreeing with or appealing the findings and recommendations recorded in the Mine Book.

On returning to DTPU offices in Jakarta, the inspector would make a formal report and oral presentation to the Chief Mine Inspector concerning a mine inspection. After reviewing this report and any communications from the Mine Technical Manager, the Chief Mine Inspector was required to prepare a memorandum, expressing concurrence or modifying the inspectors findings and recommendations. In interviews, DTPU mine inspectors indicated they were often reluctant to "be too tough" on mining operations they inspected for fear their recommendations would be reversed by their superiors, who had not visited the mine. Copies of the final inspector's report and letters from the Mine Technical Manager regarding findings and recommendations were distributed to both the Jakarta and Kanwil offices, the mine owner and Mine Technical Manager. This action is a final administrative action in that there is no avenue of appeal available within the Ministry other than to the Minister in the form of a complaint. A perception was widely shared among mine inspectors, especially in Kanwil offices, that serious enforcement actions would not be sustained by upper management. Often a significant period of time elapsed before these reports were received by Kanwil offices. Rarely were they sent to regional governments unless there was some open and public controversy involving a mine.

Corrective measures written in the Mine Book may or may not be completed by the operator. A followup inspection might be scheduled about six months after the original inspection, but often did not occur until the next annual inspection. Violations not corrected were usually re-entered on the next inspection without penalty to the operator. Thus the regulatory authority had no way of knowing whether remedial actions requiring less than six months to perform were ever performed. In the United States, minor violations often must be corrected within a few days.

If a followup inspection determined remedial action had not been taken after six months, additional inspections might be scheduled. If repeated inspections revealed continued noncompliance, the mine manager or other senior company officials might be contacted and informed of the lack of compliance. No Indonesian mine has ever been closed down as a result of failure to adhere to environmental requirements (Miller, 1999, p. 325). Prior to 2002, active followup to ensure that corrective measures were taken was limited because of scarce budget resources, which sharply constrained visits by DTPU inspectors to distant mine sites. Mine operators were not always able or willing to promptly arrange travel for inspectors on followup visits. Thus corrective actions might not be taken until after they were observed during a subsequent inspection a year later. Environmental degradation from mining operations might continue indefinitely under this regime.

Mine inspectors must have the financial means to reinspect areas of mines they find in noncompliance. One possibility would be to charge mining companies for reinspections. Companies would have a date certain for reinspection and would be more likely to fix the problem if they knew for sure the inspector would be back to see if they had taken action to comply with requirements noted in a previous inspection. Annual inspections can fail to capture environmental problems in a timely fashion. Mining operations typically move very large volumes of solid material on a daily basis. Mining operations therefore change in appearance and the mine pit may change in location fairly rapidly. Deviations from established plans or requirements can occur quickly, causing environmental degradation which may be expensive and time consuming to mitigate.

Inspection frequency of one or two site visits per year is unquestionably inadequate. Because of the large amounts of coal being mined and large volumes of overburden moved, the physical location of a mine pit and environmental conditions can change rapidly. Coupled with the practice of having different inspectors visit a mine over a period of time, continuity and quality of inspections was hard to sustain. In the United States, at least one partial inspection per month and one complete inspection per calendar quarter of every surface coal mining operation is required. Followup inspections conducted by the same inspectors may occur in less than 30 days for serious violations. Subsequent inspections may be conducted by different inspectors.

In Indonesia, rather lengthy inspection reports are prepared but there is no standard format or field inspection checklist to assure consistency among inspectors. There are no requirements for a standard reporting format or other device which would allow mining or environmental inspection data to be systematically collected and usefully analyzed. The basic record of compliance – entries in a Mine Book – varies widely in scope and specificity. This makes it difficult to aggregate observations and hard to identify trends. Trend analysis which might identify environmental problems is therefore not possible.

The absence of precise national performance standards applicable to all mine operations makes it virtually impossible to apply uniform standards when conducting inspections. This was difficult prior to 2002, and will be more difficult under a decentralized regime of inspections conducted by regional governments. It should be considerably less expensive for regional governments if inspectors travel shorter distances to inspect mines located nearby, perhaps making them less dependent on mine operators for travel funds. It is simply too early to tell if mine inspections may be conducted more frequently than previously, or if they will be unannounced, so inspectors may get a more realistic view of day-to-day operations. To date, there is no indication these practices have changed with mine inspections conducted by regional governments.

Financing decentralization required a second Law No. 25 of 1999 on Financial Equilibrium between the Central Government and Regional Administrations (Republic of Indonesia, 1999a). Among other sources of government revenue, Law 25/1999 specified a division of receipts from development of forestry, mining and fisheries of 20 percent to the central government and 80 percent to the region producing the revenue (Republic of Indonesia, 1999a, Article 6(5)). It did not specify allocations between provincial, kabupaten, and village governments. These funds are collected by the national government and distributed to regional governments.

Water quality control Regulation of water quality was decentralized to provincial and regional governments in Presidential Regulation No. 82 of 2001. National standards were set for discharges of water from industrial activities based on the intended use of receiving waters, and regional governments were authorized to set and enforce more stringent standards (Republic of Indonesia, 2001, Articles 8, 11, 12). Monitoring of water quality is to be done at least six times per month by local governments at each source (Republic of Indonesia, 2001, Article 13). A comparison of Indonesian and U.S. water quality standards is shown in Table 2.1.

Development of water quality standards applicable specifically to mining operations had been discussed within the Ministry of Mines and Energy since 1996. Then-current water quality standards set by the Ministry of Environment were not appropriate for most mining operations, because they required sampling and lab work for parameters which are not typically associated with mining operations (e.g., fecal coliform), but did not specify adequate standards for parameters associated with mining operations (e.g. fine coal particulates).

Table 2.1   Comparison of Indonesian and U.S. Water Discharge Standards for Coal Mining Operations

Parameter	Units	Indonesian Discharge Limits	Indonesian Discharge Limits	U.S. Discharge Limits: New Sources(c)		U.S. Drinking Water Standard (d)	World Bank Effluent Limit(e)
		Regulation 51/1995(a)	Decree 113/2003(b)	1 day max	30 day ave	maximum	maximum
Total Suspended Solids (TSS)							
Mining Activities	mg/L	400	400	70	35	–	50
Processing Activities	mg/L	400	200	70	35	–	50
pH	6–9	5–9	6–9	6–9	6–9	6–9	6–9
Total Iron (Fe)	mg/L	10	7	6	3	0.3	3.5
Total Manganese (Mn)	mg/L	5	4	4	2	0.05	–

Sources:
(a) (Ministry of Environment, 1995).
(b) (Ministry of Environment, 2003).
(c) *U.S. Code of Federal Regulations*, Title 40, §434 (2003).
(d) *U.S. Code of Federal Regulations*, Title 40, §141 (2003).
(e) (World Bank, 1995).

Closer collaboration with the Ministry of Environment was necessary to resolve these issues.

Although they have become more stringent over time, Indonesian water discharge standards are generally not as demanding as U.S. effluent limits. The World Bank recommends a water discharge standard of 50 mg/L for total suspended solids from coal mining operations (World Bank, 1995). A proposal in 2002 to reduce the Indonesian water discharge standard for industrial facilities of 400 mg/L suspended solids to 50 mg/L specifically for coal mining operations, which would have brought it closer to the U.S. one day maximum of 70 mg/L (*U.S. Code of Federal Regulations*, 2000, Title 40, §434), was abandoned in August 2003 (Whitehouse, 2003). New water quality standards in 2003 applied the old standard of 400 mg/L suspended solids specifically to mining activities, and set a new standard of 200 mg/L only for coal processing activities at preparation plants (Ministry of the Environment, 2003). These were small steps towards regulatory improvement of water quality discharged from coal mining operations.

Yet, such concentrations would constitute adequate cause for immediate shut down of a coal mine in the United States, until equipment required to meet the more stringent U.S. standard was installed and operable. U.S. standards are based on impacts to human health. Because many Indonesians who live near coal mines are too poor to drink bottled water, must take their drinking water from rivers and streams, and have no means to purify it other than filtering and boiling it, U.S. drinking water standards are shown for comparison. Boiling water may remove biological impurities, but has little effect on physical or chemical contaminants, and may actually elevate concentrations of them as hot water is lost to steam.

Implementation of Regulatory Policies

What does implementation of mining regulatory policy for control of environmental effects of mining operations look like on the ground in Indonesia, in terms of actual accomplishments? A review of the literature, interviews with officials in the Ministry of Energy and Mineral Resources, provincial and kabupaten governments, and several visits to Indonesian mines with inspection personnel since 1992 provide the following overview and characterization of recent implementation experience with mining regulatory policy in Indonesia, as compared to the United States.

Waste Rock Dumps

Surface mining produces large volumes of overburden that must be moved and placed somewhere during mining to get it out of the way of coal extraction. The most efficient practices, in terms of minimizing expenditures for operating costs for fuel, labor, and equipment maintenance, entail moving as little overburden as possible, as infrequently as possible, for the shortest possible distance. It is always necessary to place some overburden outside an active mine pit, at least temporarily, and many surface mines produce more waste rock than will fit back into a depleted mine pit. Still, after the initial cut is made, the most efficient mining practice often involves placing overburden in mined-out pits near current operations, to the maximum extent possible.

In Indonesia, waste rock dumps are often constructed outside the mine pit, in valleys with active streams, and in ways that appear to maximize disturbance to the environment. Many existing waste rock dumps were constructed without first clearing trees or other vegetation, and without removing the topsoil (U.S.DOI/OSM, 1995d). These practices are not recommended in any civil engineering manual. Buried trees and other vegetation decompose and lead to instability in the fill: slope failures from settling and the creation of slip planes along the soil/waste rock interface. Valuable topsoil needed for future reclamation efforts is buried, instead of being segregated and saved for future use. This is especially significant in Indonesia, where topsoil layers at many mine sites in rain forest are often less than 2 cm deep.

One of the greatest threats to long-term stability of fills is the development of a water table within a fill. This occurs when fills are not compacted sufficiently to shed water or because they are not free draining. Durable rock can be dumped without compaction and will be free draining so long as it makes up more than 80 percent by volume of a fill. Fills with higher amounts of fine, soil like material need underdrains and compaction. Most existing Indonesian coal mine fills were constructed without underdrains or internal drainage systems of

Figure 2.1   Unstable out-of-pit overburden fill pushed into trees with organic material in fill
Source: Alfred E. Whitehouse

any kind. A great deal of the material observed being placed in fills at coal mines appeared to be unconsolidated sediments, soil-like in composition, with little or no durable rock. This material was dumped without compaction in lifts or layers (Whitehouse, 1999, p. 67). The likelihood of future slippage or slumping of such material is therefore great, especially during periodic heavy rainfall.

In Indonesia, some existing coal mine fills were constructed on the sides of steep hollows or valleys above active streams and rivers, and some accumulated water in depressions at the foot of the fill, suggesting the beginnings of a water table (Hamilton, Whitehouse and Tipton, 1992). Slumping of a fill above an active stream or river carries the potential for catastrophic loss of property or life, if it blocks streamflow, creates an earthen reservoir, and then gives way due to the weight of water built up behind it. This hazard potential is greater if communities are located downstream of such fills.

In the United States, national performance standards require preservation of topsoil, compaction of fills, and construction of waste rock dumps designed to ensure permanent stability, prevent spoil erosion and movement. Waste rock dumps such as those observed at Indonesian coal mines either would not receive permits, or would be cited as violations requiring immediate abatement. Those posing an imminent danger to public health or safety or significant harm to the environment would require immediate cessation of operations until the danger was abated.

Surface Water Management

At Indonesian mines, surface water often is not diverted from preparation plants, mining areas or reclamation areas, causing unnecessary pollution and increased mining costs. Visitors to Indonesian coal mines have viewed instances where operations have mined through a stream without diverting it, resulting in excessive water in the mine pit (Hamilton, Whitehouse and Tipton, 1992). Expenditures for pumping water out of the pit and extracting heavy equipment from deep mud decrease productivity of investments. Failure to divert storm water from preparation plants and mining areas increases erosion and transport of contaminants, increases the amount of water which must flow through settlement ponds before leaving the mine site, and therefore increases design and maintenance costs. Failure to divert storm water from reclamation areas not yet completely revegetated often increases erosion, and may cause deep gullies which must be repaired at additional expense. Failure to contain storm water on the mine site through use of perimeter ditches increases environmental disruption from discharge of contaminants to nearby water bodies.

Sediment and erosion control practices are not universally applied, especially during overburden removal, construction of waste rock fills and revegetation of reclaimed land, resulting in deep gullies and large sediment loads leaving the mine area. Outslope drainage control ditches and sediment traps or ponds are generally small, few in number, and poorly maintained. At nearly every coal mine in Indonesia, sediment traps and ponds were not designed for the size of the area contributing water flow or a design storm event, but were constructed to fit the area of land available for their use. Little attention was given to the maintenance of erosion control structures or to removal of accumulated sediment (U.S.DOI/OSM, 1995a). Grasses or annual grains are not routinely planted as cover crops to protect and stabilize soil in reclaimed areas until trees reach sufficient size and percentage of cover to protect against erosion. Mulch such as hay, straw, wood chips, or paper is seldom used to protect bare soil, to keep soil and seeds from being washed away in heavy rains, or to retain moisture during dry seasons (Whitehouse, 1999, p. 67).

Acid Mine Drainage

Low grade coal stockpiles, waste rock dumps, coal storage yards, refuse or tailings disposal sites and processing areas are likely sources of acid mine drainage (AMD) or acid rock drainage (ARD). In Indonesia, this environmental issue was just beginning to be recognized as a potential problem in the mid-1990s, and there was no evidence of preparation for or existing chemical treatment installations to meet Indonesian water quality standards (Whitehouse, 1999, p. 67). For example, visitors to the government-owned PT Bukit Asam Ombilin Mine in West Sumatera during August 1996 observed acid mine drainage discharging from the portal of one of the few underground coal mines in Indonesia. Ombilin is an old mine opened during the Dutch colonial period. No overburden analyses or streamflow studies were available to determine if the mine was being recharged from stream water loss. Neutralization and settling of the AMD before its discharge into the stream was not apparent.

Often acid producing materials in overburden can be segregated, returned to the mine pit, and buried below the water table to isolate them permanently. A prerequisite for such "special handling" of acid or toxic materials is periodic sampling and analysis of drill core samples during exploration and mining. Analysis of overburden for toxic or acid producing materials is required in the United States, but not in Indonesia. Proponents of mines there offer limited quantification (drilling) of the mineral resource prior to and during mining. There are few documented analyses of overburden for acid forming materials or toxics and little or no analysis of overburden to determine its suitability as an alternative growing medium for revegetation of mined areas. Little care was given to burying or covering unmarketable coal or seam partings and binders. This material adversely impacted revegetation efforts and was a source of acid mine drainage at some mines. Outflows of groundwater at seeps and springs were not well managed, diverted or drained from previously constructed waste rock dumps and fills. These can be a primary cause of instability in fills and provide transportation paths for acid mine drainage.

Protection of the surface and groundwater quality and quantity in the vicinity of mining operations both during and after mining and reclamation activities is required by national performance standards in the United States. Special handling is required for all acid-forming, toxic and combustible materials to ensure they are disposed of in a manner that will prevent

Figure 2.2   Reddish-orange acid mine drainage below eroded highwall and carbonaceous refuse
Source: Alfred E. Whitehouse

contamination of surface and groundwaters. Contaminated water cannot be discharged or allowed to flow from the mine site without notice of violation of the standards and payment of administrative penalties. A pattern of repeated violations would justify temporary closure or revocation of the mining permit in the U.S.

Reclamation and Revegetation

There are typically overlapping interests in the land being mined. In some cases forested land has traditionally been used by local people to gather medicinal plants, harvest fruit trees, plant small gardens or hunt small game. In some areas there are ancient burial sites of local families or communities, or locations with religious significance that do not appear on government maps. In nearly every instance where harvestable trees occur, someone has acquired a concession to harvest them from the national government, even if their rights are old and have not been exercised. In many locations, local people would like to use the land for rice or other crops.

These interests collide both during and following mining activities. Thus many decisions about how the land will be used after mining have been avoided or postponed by regulatory agencies, leaving mining companies to debate reclamation options with several conflicting interests. This uncertainty and confusion is often expressed by both private and public mining operations during visits to their mines. Consequently, some mines are very slow to reclaim and rarely have more than a few percent of the total disturbed land reclaimed. Demonstration plots are common and some have excellent results, including invasion of native species. But these successes are not often transferred to large scale reclamation.

Operators at Indonesian mining operations are often proud of their reclaimed areas and careful to point them out to visitors. In most cases, reclaimed areas appear to be essentially test plots. They constitute a very small percentage of land disturbed by Indonesian coal mining operations, and there is often little evidence of recent effort. Reclamation seems to be an action considered appropriate after mining ceases rather than as a part of ongoing mining operations. In several cases, spoil continued to be placed in offsite dumps and fills far from active mining rather than in adjacent portions of the empty pit, thereby increasing the cost of disposal. The lack of a designated postmining land use lies at the root of these difficulties, preventing both effective reclamation and increasing the costs of reclamation if it is done at mine closure. Soil tests to measure fertility and pH of the soil are rarely taken and there is little evidence of the use of chemical fertilizers. Vegetation in many cases appears stressed or stunted. Lack of soil data makes the cause difficult to determine. Revegetation consists mostly of trees and shrubs which do little or nothing to stabilize the land and prevent erosion (Whitehouse, 1999, p. 67). Often trees are planted in "flower pot" fashion on ungraded overburden in which holes are dug by hand, and in which a small amount of topsoil is placed. For such trees, root growth would not reach the drip line before encountering either infertile or possibly toxic overburden.

National performance standards in the United States require all mine excavations be backfilled, graded and seeded as mining proceeds, and all excess spoil material not returned to the mine pit be placed, graded, stabilized and revegetated within the permit area. A diverse and permanent vegetative cover of native species at least equal in coverage to the natural vegetation of the area must be established and maintained for at least five years after the last year of seeding, irrigation or other work, consistent with a pre-approved postmining land use plan. For larger areas of disturbed, unreclaimed land, larger performance bonds are required to be posted. As mined land is reclaimed and disturbed areas shrink in size, portions of the performance bond are refunded to the mine operator.

Coal Processing Facilities

Coal processing and washing facilities, coal stockpiles and refuse disposal areas are often located in the floodplains of major rivers, or in coastal areas near shipping facilities and are active sources of pollution at virtually every coal mine in Indonesia. There do not seem to be any sound environmental or economic reasons to site these facilities in floodplain or coastal locations rather than near the mine pit. To the extent coal processing facilities remove impurities such as rock, clay and slimes that could be disposed of in the mine pit, locating them at distance from the pit increases transportation expenditures and reduces productivity of mining operations in Indonesia.

Although required by Ministerial decree since 1995, coal processing facilities in Indonesia are not zero discharge closed circuit plants as they are in the United States, but active polluters with poorly designed and maintained pollution control facilities. Sediment traps and fine coal catchments at processing facilities are generally undersized and poorly maintained, allowing coal fines, rock and clay sediments to flow into nearby rivers. Fine coal (<2 mm) representing 20 percent or more of production is not being recovered at most operations. A marketable product, this is lost revenue and an unnecessary source of pollution when discharged to nearby water bodies (Hamilton, 1998a).

The market and royalty values of fine particulate coal lost from some Indonesian coal cleaning and loading facilities are substantial, easily sufficient to purchase and operate environmental controls and improve operating efficiencies. Lost production of fine coal particulates from mines to rivers and streams in Indonesia is estimated in Chapter 5 below at over 513 million metric tonnes during the period 1990–2003. The value of lost coal sales in this period is estimated at nearly US$3.08 billion. These figures represent lost profits to mining operations, because production was lost *after* costs of extracting the coal were incurred. The value of unpaid royalties due the Indonesian government on this production is estimated at US$415.6 million. These estimates are quite conservative, because lost production is known to run higher than estimated at some mines. In the United States, coal processing facilities must meet applicable water quality standards for suspended solids, iron, manganese and pH regardless of whether they are located on the mine site or off site. Fine coal recovery systems are routinely used to capture product that is then shipped with larger pieces of coal, thereby substantially increasing productivity of mining operations.

Improving the Mining Regulatory Program

Programmatic and regulatory consistency are important elements under-pinning successful policies and programs of decentralization. Decentralized systems, if they are to achieve effectiveness, must be based on clear operational and regulatory standards. It is important to have clear procedures, consistent regulatory requirements, and substantial technical support capability to successfully implement a decentralized regulatory system.

In Indonesia, there are a number of challenges to successful regulation of mining activities. The enforcement process is not delivering expected results in the field. The reasons for this are numerous and complex, based in part on the terms of national policy, program organization, and differences between Indonesian culture and that of the societies in which these ideas and technologies originated. There are widespread beliefs in regulatory agencies that foreign and multinational mining companies will use the same environ-mental management technologies in Indonesia that are required in more developed countries. Foreign mining firms are urged to conform to more stringent environmental protection standards of their home countries, but unfortunately they rarely do. Adding to the regulatory dilemma and a further decline in environmental protection are what appear to be more relaxed standards and lower expectations for local companies and small scale miners. This apparent double standard sends the wrong signal to large multinational companies who allow their environmental performance to slip (Whitehouse, 1999, p. 61).

Developing and enforcing a few environmental performance standards may provide an opportunity for the regulatory program to improve overall environmental performance by encouraging mining companies to focus on a few important issues. Both mining companies and mine inspectors agree on the benefits of performance standards. They clarify responsibilities, make compliance measurable and apply a single standard of minimum environ-mental performance. While there may be a need to provide special incentives for local or small scale miners, they should take the form of assistance to meet the same minimum environmental protection standards rather than permission to ignore them. Performance standards inspire creativity to lower costs and minimize maintenance while keeping the goal of the standard in clear focus (Whitehouse, 1999, p. 61).

Specific Performance Standards

In Indonesia, there are very few specific environmental standards for mines to meet and little awareness of what should be done when anomalous data are found when sampling. A series of water samples indicating discharges are not in compliance indicates either a design or maintenance problem. Trend analysis can also provide an early warning of acid mine drainage so operational changes can be made to mitigate the problem. Sampling seems to be done to meet government requirements, but little attention is paid to the results or to trends in the data by either mine operators or mine inspectors (Whitehouse, 1999, p. 64). Neither sampling nor inspections are well-related to specific performance standards. When mining industry groups in Indonesia are asked what would make their lives easier, they respond "tell us what you want us to do," so they can know what is required for compliance. When the same question is asked of mine inspectors, they say they want specific standards against which they can measure performance. Both groups are talking about the same thing but strangely, few specific standards have been developed for the mining industry.

In terms of the environmental degradation at existing coal mining operations, the greatest needs are for performance standards for erosion control, prevention and treatment of acid mine drainage, construction of waste rock dumps and fills, coal preparation facilities and revegetation. These standards need to be sufficiently specific that performance to meet them can be measured and enforced, not just guidelines which are advisory or optional. Examples are discussed in Chapter 3. Industry can assist government arrive at performance standards that are reasonable and possible to implement at the mine site. Standardization of the mine authorization and inspection processes will vastly improve the climate for foreign investment by making clear the requirements so they can be planned for, budgeted for, and met (Whitehouse, 1999, p. 70).

Mining Authorizations

The application for mining authorization submitted by the applicant should be in a standard format for ease of compilation by the operator and review by the regulatory authority. Certainly an adequate application for mining authorization must include a complete description of the environmental resources that may be affected including information concerning climate, vegetation, fish and wildlife, soils and land-use baseline data. Maps, plans and cross-sections showing geologic data at a scale specified by the regulatory authority should be required. A detailed reclamation and operation plan must be described in the application, including: the mining operation, blasting, air pollution control, fish and wildlife, reclamation of the mine site, ground water, surface water, postmining land use, ponds, impoundments, diversions, public roads, excess spoil and transportation. An adequate application for mining authorization must show the operation will be in compliance with specific performance standards stated in law and regulations. Information should also be included showing ownership and control of the proposed mining operation is in compliance with applicable regulations and conditions at any other mines they operate. Applications from firms that are not in compliance at other mines should be rejected. Proof of an adequate reclamation bond should be demonstrated.

A regional government should review an application for mining authorization to determine if it addresses each requirement of the regulatory program necessary to initiate review. If the application is not complete, the Dinas Pertambangan (regional mining regulatory agency) should send a letter to the applicant describing any deficiencies. The applicant must be allowed an opportunity to respond to any deficiency letters, modifying the application to address any shortcoming. The applicant's responses and modifications to the application for mining authorization should be reviewed again by the regulatory authority for completeness. This process should continue until all deficiencies are satisfactorily resolved. However if the applicant is unable or unwilling to submit needed information, then the application for mining authorization should be summarily rejected.

After the applicant responds satisfactorily to all deficiencies, an application for mining authorization should be declared complete by the regulatory authority. This is not a final approval of the project, but a preliminary determination of data adequate for decision making. The applicant must ensure that a copy of the application for mining authorization is available for public examination near the proposed mine site, perhaps at a court house or public library. The applicant must publish a notice in local newspapers that briefly describes the proposal, advises the public where they may review a copy of the application and where any comments to the Dinas Pertambangan should be sent. A period of about 60 days should be allowed for the public to review the application for mining authorization, submit written comments and request an informal conference. If requested, the Dinas Pertambangan should chair a public meeting. The public should be given an opportunity to submit written and verbal comments at this meeting. The Dinas Pertambangan should distribute a copy of the application for mining authorization to internal offices and to other appropriate government agencies for review. National and state agencies should be allowed about 60 days to submit written comments on an application. Mine inspectors in the Dinas Pertambangan should review the application for on-the-ground "inspectability" of the proposed mine operation.

Communication and participation between the authorizing and inspection staffs of the Dinas Pertambangan are important. Close coordination will help improve the quality of mining authorizations issued; ensure that inspectors understand the various aspects of the authorization and it is "inspectable;" help to identify any special attributes of the mining operation (e.g., toxic materials) that may require special attention (e.g., toxic material handling plans); and will keep the various staff informed of mutual needs to ensure compliance. To better accomplish these goals, some regional governments might form "mine teams" for large surface coal mines composed of staff from their BAPEDALDA (regional environmental assessment agency) and inspection staff from Dinas Pertambangan. Mine teams comprised of individuals knowledgeable about environmental analysis, inspection and enforcement, and technical disciplines are best suited to achieve these aims and to assure that mines are in compliance with applicable requirements.

National Oversight Role Undeveloped

Although the decentralization laws of 1999 reserved exclusive authority to the national government over matters concerning planning and control of national development, and efficient use of natural resources (Republic of Indonesia, 1999b, Article 7(2)), the Ministry of Energy and Mineral Resources has been slow to define a role in oversight of subnational government actions in the realm of mining environmental policy. Indonesia has moved more quickly on the environmental side of this area, as discussed in Chapter 4 below. Continued central government control over national development and efficient use of natural resources allows the Ministry significant initiative and oversight of mining activities, including approval of CoW negotiated by regional governments. Presidential Regulation 25/2000 reserved broad authority to the national government to set standards for mineral prospecting and management of mineral and energy resources, and to determine criteria for working areas for mining business (Republic of Indonesia, 2000, Article 2). Presidential Regulation 25/2000 also identified subjects in which provinces may exercise authority, but regulation of mines and energy is not among them. Thus regional governments are subject to oversight and supervision by the Ministry of Energy and Mineral Resources, and must continue to abide by policies and directives promulgated by it, whether they be national performance standards for mining operations and facilities, or criteria for approval and supervision of regional government programs implementing mining environmental policies.

The Ministry retains full inspection and enforcement authority over mines existing before 31 December 2000, but has already shifted responsibility for authorization and inspection of new mines to regional governments. Several large older mines are only a few years from closure. Responsibilities and workload of the Ministry for mine inspection and enforcement will decline as old mines close. As new mines are authorized, the role and workload of regional governments in authorizing and inspecting mining operations will increase, eventually displacing the Ministry in these areas. In the absence of a systematic oversight program administered by the Ministry ensuring adherence to uniform national standards, it seems unlikely the quality of mine authorizations and inspections will be comparable across the many kabupaten and kotamadya who will implement them. Environmental degradation from mining operations will increase in some areas more than others. Some mine operators will perceive they are financially disadvantaged by lack of uniform treatment, and be discouraged from making additional investments in Indonesian mining operations. Implementation of mining environmental policies in Indonesia will deteriorate rather than improving due to decentralization of functions.

The Ministry of Energy and Mineral Resources has the authority to establish criteria and a process for evaluating, approving and monitoring mining environmental regulatory programs of regional governments, to ensure the quality of mine authorizations and inspections will be comparable and mining operations will receive uniform treatment across the many kabupaten and kotamadya in which mining occurs. Regional governments should be required to demonstrate to the satisfaction of the Ministry that they have adequate legal authority, qualified personnel and budget to successfully administer a mining regulatory program within their jurisdictions. Appropriate criteria for approval of a regional government regulatory program for mining and reclamation operations would include requirements that the program provide evidence of (1) effective implementation, maintenance and enforcement of a mining authorization, Contract of Work or permit system; (2) sanctions for violations of regional and national government regulations, performance standards, or conditions of authorizations, including civil and criminal sanctions, forfeiture of bonds, suspensions, revocations, and withholding of mining authorizations, and the issuance of cease-and-desist orders by the regulatory authority or its inspectors; (3) a process for designation of areas unsuitable for mining; (4) a process for coordinating review and issuance of mining authorizations with other regional or national authorization processes; (5) rules and regulations consistent with regulations and decrees issued by the national government; and (6) sufficient funding, administrative and technical personnel to effectively regulate mining and reclamation operations.

Regional government programs for mining and reclamation operations which meet these criteria should be approved or disapproved (in whole or in part) by the Ministry by a specified date. Failure of a regional government to submit a fully approvable regulatory program should trigger preparation and implementation of a national program by the Ministry for control of mining within that jurisdiction. Similarly, failure of a regional government to continuously administer or enforce an approved program should trigger either national government take over and enforcement of the regional government program, or adoption of a national program for regulation of mining within the regional jurisdiction. The Ministry should continuously monitor regional government implementation of approved programs, and if necessary take over all or part of a program if the regional government fails to implement it in accordance with Ministerial decrees.

In the United States, the Office of Surface Mining has approval and monitoring authority over subnational state regulatory programs for surface coal mining and reclamation operations in 26 states (*U.S. Code*, Title 30, §1253), and has since 1979 implemented regulations containing criteria for approval of such programs (*U.S. Code of Federal Regulations*, Title 30, §§730.1-736.25). Descriptions of approved state regulatory programs and the cooperative agreements between state and national government agencies that established them are found in the *U.S. Code of Federal Regulations* (Title 30, §§900.1-950.36). The possibility of national government take over of a state regulatory program provided substantial incentives for development and implementation of approvable state regulatory programs in the U.S., as discussed in Chapter 3 below.

Regulation and Development

Does regulation stifle development? There is an undercurrent of fear within the Ministry of Energy and Mineral Resources and in some regional government offices that any environmental requirements will increase costs, limit employment opportunities or cause mine closures. A similar fear was voiced in the United States during consideration of the Surface Mining Control and Reclamation Act of 1977. Experience with regulating mining operations in the United States has proven this fear to be unfounded. After passage of the Act produced some temporary confusion in the mining industry over what the new standards were, how they would be met, and how they would be enforced, coal production increased every year despite stringent requirements. Mining companies simply changed the way they mined. They learned quickly to design and operate mines that were both cost effective and complied with new standards. Many of the technologies required to meet new environmental performance standards applied in Indonesia have been tried and perfected in other countries like the United States. Indonesian mining companies can use these technologies to quickly and profitably adapt to new regulatory requirements. The fear that companies engaging in profitable mining ventures in Indonesia may abandon their investments if the government imposes new regulatory standards and institutes a more vigorous enforcement program are simply not supported by American experience, described in the next chapter.

Note

1   Dollar equivalents in this chapter were calculated at average 2003 exchange rates of 8800 rupiah per dollar.

Chapter 3

Coal Mining Regulatory Policy in the United States

A Federal System of Government

The Constitution of the United States specified the powers of the national government and reserved all other powers to the states or the people. In general, this means the national government can exercise only those powers specifically delegated to it, with a few additional powers determined by subsequent interpretation of the U.S. Supreme Court to be "necessarily implied" in order to exercise powers specifically delegated to it. The Tenth Amendment to the Constitution reserved to the states or the people all powers not explicitly conferred on the national government, or specifically prohibited from exercise by the states. Local governments in the United States, being subdivisions of the states or organized as "home rule" entities under provisions of state constitutions, exercise powers of the state, not the national government.

The reserved powers of the states are often described as "police powers," despite the fact they extend far beyond what are ordinarily thought of as powers exercised by law enforcement authorities. They include the authority to make regulations designed to promote the public convenience and general prosperity as well as those to promote public safety, health, and morals. The police power is not confined to suppression of what is offensive, disorderly, or unsanitary, but extends to what is for the greatest welfare of a state:

> Because the police power of a State is the least limitable of the exercises of government, such limitations as are applicable thereto are not readily definable. Being neither susceptible of circumstantial precision, nor discoverable by any formula, these limitations can be determined only through appropriate regard to the subject matter of the exercise of that power. It is settled that neither the *contract* clause nor the *due process* clause had the effect of overriding the power of the state to establish all regulations that are reasonably necessary to secure the health, safety, good order, comfort, or general welfare of the community; that this power can neither be abdicated nor bargained away, and is inalienable even by express grant; and that all contract and property rights are held subject to its fair exercise. Insofar as the police power is utilized by a State, the means employed to effect its exercise can be neither arbitrary nor oppressive, but must bear a real and substantial relation to an end which is public, specifically, the public health, safety, or public morals, or some other phase of general welfare.

> (U.S. Library of Congress, 1973, p. 23)

71

The general rule is that if a regulation goes too far, it will be recognized as a taking of private property for public purposes, for which fair compensation must be paid (O'Leary et al., 1999, pp. 63–67). Yet where mutual advantage is sufficient compensation, an ulterior public advantage may justify a comparatively insignificant taking of private property. Thus state and local governments in the United States generally have broad authority to regulate the uses of land, including such specific aspects as the location of energy developments, unless preempted by explicit provisions of state or national legislation, or constrained by other constitutional provisions.

As a practical matter, there is no clear division of responsibility in the United States, but substantial duplication and overlap of effort is evident between national, state and local regulations applied to different aspects of the same activity. For example, the national government may regulate safety aspects of a nuclear electric generating station, while a state agency issues permits for discharges of heated water and air emissions, and local government land use and construction permits must be obtained for non-nuclear facilities of the same development (Hamilton and Wengert, 1980). Although roughly one-third of the nation's land is managed by several national government resource agencies, national legislation requires their land use planning processes and development decisions be coordinated with those of state and local authorities and made consistent with them "to the maximum extent" consistent with law (*U.S. Code*, Title 43, §1712(c)(9)). Furthermore, cooperative agreements authorized by national statute allow enforcement of national, state and local regulations by appropriate state and local government officials.

Regulatory Policy for Coal Mining

In the United States, the Office of Surface Mining Reclamation and Enforcement (OSM) has national responsibility for regulating environmental effects of surface mining pursuant to the Surface Mining Control and Reclamation Act of 1977 (*U.S.Code*, Title 30, §1201, et seq). Of 37 states with some coal production, 31 had enacted some form of regulation of surface coal mining for state-owned coal, and 29 of them also applied their regulations to privately-owned coal as of December 1975 (Imhoff, Fritz and LeFevers, 1976). With few exceptions, the record of state control over coal mining activities prior to 1977 was one of ineffective (or nonexistent) planning and regulation. Pressure on state officials to protect the coal industry within their respective states from competitive disadvantage militated against imposition or enforcement of strong reclamation regulations, in the absence of uniform national standards.

About two-thirds of U.S. coal production comes from surface mines. Strip mining of western coal, which contributed only a small portion to U.S. coal production prior to 1970, expanded rapidly in subsequent years and is expected to do so increasingly in the future as the productivity and production of western surface mines with thick seams continue to increase. Coal mined west of the Mississippi River now accounts for about 55 percent of production in the U.S. Destructive environmental and economic impacts of surface and underground coal mining without land reclamation are well documented and generally recognized. These include surface and groundwater pollution and depletion, erosion of land, loss of topsoil and productive subsoil, flooding, land subsidence, blasting damage to property, stream sedimentation and obstruction, aesthetic nuisances and disruption of community life by haphazard coal development (Rochow 1979, p. 560).

Following long after a series of hearings before the Senate Committee on Interior and Insular Affairs in 1968 established the environmental hazards of surface mining, enactment of the Surface Mining Control and Reclamation Act of 1977 (SMCRA) culminated nearly a decade of occasionally bitter debate over the need for national mined-land reclamation standards (Harris 1985; Shover, Clelland and Lynzwiler 1986). The controversy surrounding this debate – which included 52 recorded votes in Congress and two vetoes by President Ford – are reflected in the comments of one participant: "The history of the Act would serve as a textbook for any national legislator desiring to thwart the clear will of the majority of the Congress" (Udall 1979, p. 554). Originally introduced in 1971, by the time it was enacted, SMCRA had garnered attention world wide. Officials in several countries including Indonesia watched the proceedings with interest, and followed suit with their own legislation and regulations.

In enacting the Surface Mining Control and Reclamation Act, the U.S. Congress recognized that:

> surface mining had created widespread disturbance of commerce and the public welfare by irreparably harming the utility of land for commercial, industrial, residential, recreational, agricultural, and forestry uses through shoddy surface mining practices. Erosion, landslides, floods, water pollution, and the creation of virtual wastelands attributable to the coal mining industry had lowered, if not endangered, the quality of life for plants, animals, and humans.
> (Eichbaum and Babcock, 1982, p. 623; *U.S. Code*, Title 30, §1201(c), 2004).

The Act required that coal mine operators: "to the extent possible using the best technology currently available, minimize disturbances and adverse impacts of the operation on fish, wildlife, and related environmental values, and achieve enhancement of such resources where practicable" (*U.S. Code*, Title 30, §1265(b)(24)). A process was provided by the Act for development of state programs to administer national standards for surface mining and reclamation and to set national performance standards for environmental protection and reclamation of lands subjected to surface mining of coal resources. It established procedures for protection of certain lands designated unsuitable for surface mining; required mine operators to post a performance bond adequate to complete reclamation of disturbed areas; established a special fund for reclamation of previously mined and unrestored lands; and provided that surface owners or permit holders and lease holders of land must give written consent before nationally-owned coal beneath their land may be leased by the Department of the Interior for surface mining. The Act did not regulate mining of any other minerals besides coal.

Regulatory Programs

SMCRA established an Office of Surface Mining Reclamation and Enforcement (OSM) in the Department of the Interior and specified the duties of that office, including: implementation of the Act's regulatory and reclamation programs; approval or disapproval of state regulatory programs; establishing standards and regulatory policy for reclamation and enforcement efforts, and providing grants and assistance to state governments. Procedures and criteria were specified in the Act for creation of a national surface mining control and reclamation program, and for OSM approval of state regulatory programs implementing national performance standards.

Interim program   Some states, such as Montana and Pennsylvania, already had surface mining regulatory programs prior to 1977. SMCRA required with some exceptions that permits for new mines issued under existing state programs after January 1978 require compliance with national performance standards concerning: restoration of mined land to approximate original contours; preservation of topsoil; minimization of disturbance to the hydrologic balance; construction of water impoundments from mine spoil; use of explosives; and revegetation. All existing mines with state permits, except small mines producing less than 100 000 tons of coal per year, were required to comply with these standards by April, 1978. Small mines were required to comply nine months later, by January 1979. A national enforcement program was established to require compliance until such time as a state regulatory program could be approved or a national program implemented in lieu of a state program.

Approval of state regulatory programs   State programs were required to be submitted to the Secretary of the Interior and either approved or disapproved (in whole or in part) according to specific criteria stated in the Act by August 1979 (*U.S. Code*, Title 30, §1253). Program approval responsibilities were delegated to the Office of Surface Mining Reclamation and Enforcement by statute. Failure of a state to submit a fully approvable regulatory program triggered preparation and implementation of a national program for control of surface coal mining within that state. Similarly, failure of a state to administer or enforce an approved program may trigger either national government (OSM) enforcement of the state program, or adoption of a national program for regulation of surface coal mining within the state. Either an approved state program or a national agency program was required to be in force for each state by June 1980, with few exceptions. OSM was authorized to monitor state implementation of approved programs, and may take over all or part of a state program if the state fails to implement it in accordance with SMCRA.

In 2004, there were 24 approved State Regulatory Authorities, one in every state having significant coal production except Tennessee and Washington. OSM oversight, regulatory and Abandoned Mined Land responsibilities are implemented through ten Field Offices, seven of which have jurisdiction covering more than one state, including some states with no significant coal production and no State Regulatory Authority. For example, the OSM Knoxville Field Office implements national regulatory program requirements in coal-producing Tennessee, and in Georgia and North Carolina, which have no significant annual production. Twenty-three of the same states and four Indian Tribal Governments have approved Abandoned Mined Land Programs funded by OSM grants. OSM also has three Regional Coordinating Centers in Pennsylvania, Illinois and Colorado which provide technology development and transfer for State Regulatory Authorities and OSM Field Offices. In addition, OSM operates programs to control impacts of acid mine drainage from abandoned mines, encourage reforestation of reclaimed mined land, develop techniques for reclamation of prime farmlands, and publicly recognize outstanding reclamation efforts.

Mining Permits

Eight months after an approved state regulatory program – or a national program for a state not having an approved state program – was established, no surface coal mining could be commenced or continued without a permit from such program. The only exception to this rule concerned an existing mine operating under a permit issued previously by a state regulatory body, when the operator of such a mine had made application for a permit under a new program and no action had been taken on the application. These operations could continue until action was taken by the appropriate regulatory agency. Exploration activities that will remove more than 250 tons of coal require a permit (*U.S. Code of Federal Regulations*, Title 30, §772.12).

Generally, permits may be issued under approved state or national programs for no more than five years. All permits contain a right of renewal within the approved boundaries of the existing permit, provided the mine is in compliance with regulatory requirements at time of renewal. This gives regulatory authorities several opportunities over the 20–40 year lifetime of a mining operation to correct deficiencies or terminate operations. Detailed information concerning ownership of the mining company, surface and subsurface property rights must be submitted to the appropriate regulatory agency in order to obtain a permit. New permits cannot be issued to persons or firms under the ownership or control of individuals who have outstanding violations of the terms of previous permits. Required contents of the permit application and each of these plans are explicitly stated in the Act.

There are usually three stages in processing a permit application for a new coal mine, whether the permitting agency is OSM or a state regulatory authority. They include: administrative completeness review, technical review, and decision document stages (Clark, 1996).

Administrative completeness review The permit application submitted by the applicant must be in a standard format for ease of compilation by the operator and review by the regulatory authority. A permit application must include a complete and accurate description of the environmental resources that may be impacted or affected including information concerning climate, vegetation, fish and wildlife, soils and land-use baseline data. Maps, plans and cross-sections showing geologic data at a scale specified by the regulatory authority are required. A detailed reclamation and operation plan must be described in the permit application, including: the mining operation, blasting, air pollution control, fish and wildlife, reclamation of the mine site, ground water, surface water, postmining land use, ponds, impoundments, diversions, public roads, excess spoil and transportation. The permit application must show the operation will be in compliance with specific performance standards stated in law and regulations. Information must also be included in a permit application showing ownership and control of the proposed mining operation is in compliance with applicable regulations and permit conditions at any other mines they may operate. Proof of an adequate reclamation bond must be demonstrated, and payment of the appropriate permit application fee included.

The regulatory authority reviews a permit application to determine if it addresses each requirement of the regulatory program necessary to initiate processing and public review. If the application is not complete, the regulatory authority sends a letter to the applicant describing any deficiencies. The applicant is allowed an opportunity to respond to any deficiency letters, modifying the permit application to address any shortcoming. The applicant's responses and modifications to the permit application are reviewed by the regulatory authority for completeness. This process continues until all deficiencies are satisfactorily resolved. However if the applicant is unable or unwilling to submit needed information, then the permit application is summarily rejected.

After the applicant responds satisfactorily to all deficiencies, a permit application is declared complete by the regulatory authority. The applicant ensures that a copy of the permit application is available for public examination near the proposed mine site, usually at a court house or public library. The applicant must publish a notice in local newspapers that briefly describes the proposal, advises the public where they may review a copy of the application and where any comments to the regulatory authority should be sent. A period of about 60 days is allowed for the public to review the permit application, submit written comments and request an informal conference. If requested, the regulatory authority chairs a public meeting. The public is given an opportunity to submit written and verbal comments at this public meeting. The regulatory authority distributes a copy of the permit application to internal offices and to other appropriate government agencies for review. National and state agencies are allowed about 60 days to submit written comments on an application. Mine inspectors in the regulatory authority review the permit application for on-the-ground "inspectability" of the proposed mine operation.

Communication and participation between the permitting and inspection staffs of the regulatory authority are important. Close coordination helps improve the quality of permits issued; ensures that inspectors understand the various aspects of the permit; ensures that the permit application is "inspectable;" helps to identify any special attributes of the permit application (e.g., toxic materials) that may require particular attention (e.g., toxic material handling plans); and keeps the various staff informed of mutual needs to ensure compliance. To better accomplish these goals, some offices in OSM formed "mine teams" for large surface coal mines composed of permitting staff from a regional office and inspection staff from field offices. These teams promote closer coordination and enhance communication among OSM units and between OSM, mine operators and the interested public. OSM believes that mine teams comprised of individuals knowledgeable of permitting, inspection and enforcement, and technical disciplines are best suited to achieve these aims and to assure that mines are in compliance (Clark, 1996, p. 7).

Technical review　Technical specialists in the regulatory agency conduct detailed reviews of information in a permit application to determine compliance with the law and regulatory requirements. A mutidisciplinary team of technical specialists assigned to review the permit application includes mining engineers, hydrologists, soil scientists, ecologists, environmental specialists, wildlife biologists, air quality specialists, geologists, archaeologists and project managers. The technical review stage is the most challenging and time-consuming part of the entire mining permit application process. Depending on the environmental setting, analyses of the proposed mining and reclamation activities can be very complex. Determining if the permit application meets performance standards in the regulations often requires application of highly technical skills and a capability to handle large amounts of data. OSM developed a Technical Information Processing System (TIPS) to assist technical specialists in accomplishing these tasks. TIPS is a nationwide net-worked computer system that was jointly developed by OSM and States with primacy under the Surface Mining Act. TIPS provides state-of-the-art automated techniques and support to improve the efficiency and effectiveness of the program. Nothing comparable is currently used by national or regional governments to analyze proposed mining operations in Indonesia.

A site visit is normally conducted by technical specialists. Compliance with other national laws is reviewed by technical specialists, including the Fish and Wildlife Coordination Act, National Historic Preservation Act, Endangered Species Act, Clean Water Act, and Clean Air Act. They prepare reports documenting a permit application's compliance or deficiencies with applicable laws and regulations. These reports become part of the decision document. If any areas of a permit application are not in compliance with the law or regulatory standards, the regulatory authority sends a letter describing the deficiencies to the applicant. Again the applicant is given an opportunity to respond by modifying the permit application as necessary to achieve compliance. Modifications to permit applications are distributed for review by technical specialists. This process continues until deficiencies are satisfactorily resolved. Some deficiencies or issues may be partially resolved, allowing the regulatory authority to approve the permit application with special permit conditions. However if an operator proves unable or unwilling to resolve issues to the satisfaction of the regulatory authority, then a decision document would be prepared denying the application.

EIS or FONSI? If the regulatory authority is OSM, it must meet environmental assessment requirements of the National Environmental Policy Act concurrent with performing the technical review. If the regulatory authority is a state agency, it is not required to meet requirements of NEPA. Most permits are issued in the United States by State regulatory authorities. If appropriate, OSM prepares an Environmental Assessment (EA). The EA is a scoping document used to determine whether to prepare an Environmental Impact Statement (EIS) or a Finding of No Significant Impact (FONSI). Alternatives to the proposed mining operation are analyzed in the EA. If the EA determines that a proposed action does not have a potential for causing significant impacts on the quality of the human environment, OSM prepares a Finding of No Significant Impact. This becomes part of the decision document. OSM may skip preparation of an EA and immediately begin an EIS if it determines the proposed action has a potential for causing significant impacts on the quality of the human environment, as discussed below in Chapter 4.

Decision document A decision document is prepared recommending the permit application be approved, approved with special conditions or denied. Based on the law, regulatory requirements and information contained in the decision document, the regulatory authority determines if the applicant should be awarded a permit. If the regulatory authority issues a decision approving or approving with special conditions the permit application, it must make written findings that the permit application is complete and accurate and the applicant has complied with all requirements of the law and applicable regulations; reclamation can be accomplished as detailed in a reclamation plan; the proposed mine is not within an area designated as unsuitable for mining; and the operation would not affect continued existence of endangered or threatened species or destroy or adversely affect their critical habitats.

Based on assessment of the probable cumulative impacts of all anticipated mining on the hydrologic balance of the area, the regulatory authority must also make written findings that the proposed operation has been designed to prevent material damage to the hydrologic balance outside the permit area. The regulatory authority must also receive bond instruments from the applicant and determine ownership and control of the proposed mining operation is in compliance with applicable regulations and permit conditions at any other mines they may operate before a permit is granted. The permit document is a standard form that includes routine provisions, including required mitigation measures and specific references to applicable performance standards. Special conditions may be attached to a permit to address circumstances for a particular mine site. Special permit conditions are few and are used only when required to ensure an operation will be in compliance with the law and regulatory requirements. The applicant has the right to appeal any special conditions attached to a permit. The public also has the right to appeal the decision to a court of law. If the regulatory authority issues a decision that denies a permit, it must specifically explain the reasons why required findings cannot be made and the application cannot be approved. The applicant has a right to appeal this decision.

Figure 3.1   Wetland postmining land use for wildlife on reclaimed coal mine formerly having acid mine drainage, West Virginia, USA
Source: Alfred E. Whitehouse

Thus mining permits in the United States contain specific requirements each mine must meet, including direct references to applicable air and water pollution standards and national performance standards for coal mining operations. Although national air and water pollution standards for mining operations are set by the U.S. Environmental Protection Agency (U.S.EPA), mining operations are inspected for compliance with these standards and enforcement actions taken by State regulatory authorities and OSM under cooperative agreements between U.S.EPA and OSM. No permit can be issued unless the appropriate regulatory agency finds that the mining operation will conform to national performance standards. Permit renewals are conditioned on compliance with regulatory requirements. In 2002, a total of over 5.8 million acres were under permit for coal mining in the United States (U.S.DOI/OSM, 2002, p. 27) at 1500 mining operations (National Mining Association, 2003).

Inspection and Enforcement

Mine inspections are conducted by both State regulatory authorities and OSM. Each mine must receive a full inspection every calendar quarter, and at least one partial inspection every month by either OSM or the State regulatory agency. OSM is also authorized to carry out oversight inspections in states having approved State regulatory programs to monitor performance of state inspectors (*U.S. Code of Federal Regulations*, Title 30, §§840.11; 842). If an OSM inspector performing an oversight inspection discovers a violation, rather than citing it on the spot it is reported to the State regulatory authority in a Ten Day Notice, giving the State agency 10 days to make its own inspection and decision. Ten Day Notices have become rare as State mine inspectors became more competent over the years.

Inspections are initiated with a review of office files related to the mining operation, covering previous inspection reports, current permit conditions, layout maps, and applicable performance standards. Inspections are normally carried out on an irregular basis, without notice to the mine operator. Inspectors are authorized to issue Notices of Violation (NOV) of the laws, regulations, or any condition of the mining permit and assess administrative fines. They must close a mine down temporarily if they find violations which pose an imminent danger to public health or safety or significant harm to the environment. If a cessation order is issued, it can only be lifted after reinspection reveals conditions that led to the order have been abated. OSM was authorized to suspend or revoke a mining permit if it finds a pattern of violations of any requirement or permit conditions exists due to intentional acts of the permit holder (*U.S. Code of Federal Regulations*, Title 30, §843).

In a Notice of Violation, inspectors describe the nature of the violation, remedial actions required to abate it, and a reasonable period of time for abatement, which may be a few days, weeks, a month, but no more than 90 days, including all extensions. Some violations are cured before the inspector leaves the minesite. Others, such as administrative violations concerning record keeping, may require several months to fix, but are considered less severe than those which result in environmental disruption. Reinspection, usually by the same inspector, occurs after the designated period to determine if violations have been abated. Extensions of the designated period are routine if work is in progress when reinspection occurs and there are credible reasons why it is not yet completed. In practice, a mine inspector may be on the property nearly every week, performing reinspections of violations cited during previous inspections to determine if they have been abated in the allowed period of time. This is possible in the United States because mine inspectors have duty stations near enough to the mines that they can often stop off for a few minutes to check on abatement while on their way to another mine for inspection.

During 2002, over 35 700 complete and nearly 53 000 partial inspections of coal mines, were conducted by State regulatory authorities and OSM (acting in states lacking a state program) but only 68 cessation orders were issued based on imminent harm, and 3961 Notices of Violation were issued (U.S.DOI/OSM, 2002, p. 27). OSM also conducted nearly 2300 oversight inspections, issuing only 19 Notices of Violation, four cessation orders for failure to abate previous violations, and no cessation orders for imminent harm. Low numbers of cessation orders issued by OSM indicate State regulatory authorities were doing a credible job of enforcing the law. Low numbers of Notices of Violation, compared to the number of inspections conducted, indicate most mining operations were usually in compliance with regulatory requirements and conditions of their permits. There were no suspensions or revocations of mining permits due to patterns of violations by permit holders in 2002.

OSM is authorized to assess civil penalties not exceeding US$5500 per violation, and:

> Each day of continuing violation may be deemed a separate violation for purposes of penalty assessments. In determining the amount of the penalty, consideration shall be given to the permittee's history of previous violations at the particular surface coal mining operation; the seriousness of the violation, including any irreparable harm to the environment and any hazard to the health or safety of the public; whether the permittee was negligent; and the demonstrated good faith of the permittee charged in attempting to achieve rapid compliance after notification of the violation.
>
> (*U.S. Code*, Title 30, §1268(a))

Civil penalties of up to $5500 per day may be assessed for violations of regulations, national performance standards or the terms of permits. In addition, any person who knowingly violates a condition of a permit, or misrepresents information in a permit application, can be fined up to $10 000, imprisoned for one year or both. Primary enforcement authority resides in the states, under national supervision by OSM.

To provide guidance for mine inspectors in determining civil penalties, and provide some semblance of consistency in the treatment of violations by different inspectors at different mining operations nationwide, OSM developed a point system for assessment of administrative fines. Under this system, minor violations may not incur an administrative fine, if they are cured in a timely manner, but greater numbers of more serious violations receive more severe penalties. In determining the amount of fine for each violation, the mine inspector is required to take into consideration three factors: the history of violations at a single mining operation during the previous year, the seriousness of the current violation, and whether it occurred through negligence, intent or recklessness of the operator. Thus mine inspectors are allowed meaningful discretion to determine the severity of the penalty imposed.

The inspector must assign up to 30 points based on the history of violations, one point for each previous violation. Another 30 points may be assigned based on the seriousness of the violation, up to 15 points of which is based on the probability of occurrence of the event a violated standard is designed to prevent (insignificant = 1–4 points, unlikely = 5–9 points, likely = 10–14 points, and occurred = 15 points), and up to 15 points of which is based on the extent of potential or actual damage (on site damage = 0–7 points, off site damage = 8–15 points). If a violation is of an administrative requirement, such as a requirement to keep records, the inspector can only assign up to 15 points total for seriousness, based on the extent to which enforcement is obstructed by the violation. Finally, the inspector must assign up to 25 points based on the degree of fault of the operator (no negligence, intent, or recklessness = 0 points, negligence = 1–12 points, and reckless, knowing or intentional conduct = 13–25 points). Rapid compliance with an order to abate a violation may be rewarded with subtraction of 1–10 points per violation, provided it is actually abated before the time set for abatement. Normal compliance within the time given for abatement receives no reduction in points assessed (*U.S. Code of Federal Regulations*, Title 30, §845.13).

Thus each violation may be assigned up to a total of 85 points (or 70 points for administrative violations). All violations assigned 31 points or more result in a fine. A violation assigned 30 points or less may or may not be fined, at the discretion of the regulatory agency. Consequently, rapid abatement of a violation assigned 31–41 points may erase a fine. This is significant, because points translate directly into dollars. OSM regulations provide a graduated scale of civil penalties based on the total number of points per violation. Violations are assessed at US$22/point for up to 25 points, and thereafter at US$110/point up to $5500 per violation. Any violation which continues for two or more days and is assigned more than 70 points must be assessed a penalty for a minimum of two days. Moreover each day of a violation may be assessed a separate civil penalty from the date of the notice of violation to the date set for abatement, especially if the operator gains financially as a result of a failure to comply. Whenever a violation is not abated within the abatement period given, an additional civil penalty of not less than US$925 must be assessed for each day the violation continues, for no more than 30 days (*U.S. Code of Federal Regulations*, Title 30, §§845.14; 845.15). These penalties are higher than they were before November 2001, when they were adjusted pursuant to the Federal Civil Penalties Inflation Adjustment Act of 1990 (*U.S. Code*, Title 28, §2461 note; *Federal Register*, Vol. 66, p. 58647, 2001), which requires that civil monetary penalties be adjusted for inflation at least once every four years. Indonesia has no comparable legislation.

Thus the point system for assessment of administrative fines contains several financial incentives for mine operators, first, to avoid violations, and second, after receiving a Notice of Violation, to quickly comply with applicable regulations and standards. A history of compliance with requirements results in lesser fines for violations; minor violations may not result in fines; rapid abatement of some violations may reduce or eliminate a fine; and normal compliance within the period allowed for abatement avoids assessment of larger fines. More damaging violations are penalized more severely than less serious ones. Mine operators with a history of violations are penalized more severely than those with good records of compliance. Intentional, knowing and reckless violations are penalized more severely than unintentional or ignorant violations. Rapid compliance is rewarded. Moreover the Surface Mining Act provides explicit criteria for determining standing to sue and procedural rules allowing citizen lawsuits to enforce its provisions. Suits are specifically allowed to be filed against the U.S. or other government agencies, or against any person for violation of rules, regulations, orders or permits issued under the Act.

National Performance Standards

The Surface Mining Act for the first time set national performance standards for protection of lands subjected to surface mining of coal from undue environmental harm. These standards required that coal production from surface mines be maximized so as to minimize future disruption of the land due to further mining, and that surface mined land be restored: "to a condition capable of supporting the uses which it was capable of supporting prior to any mining, or higher or better uses of which there is reasonable likelihood" (*U.S. Code*, Title 30, §1265). Performance standards are specified in 53 requirements subject to inspection by mine inspectors employed either by OSM or by state regulatory authorities (*U.S. Code of Federal Regulations*, Title 30, §§773.11; 773.17; 778.13; 800.60; 816–817; 842.11(e); 870.15).

These include, in general, backfilling, grading and compaction of mined lands for restoration of the "approximate original contour" of the land where feasible; or at a minimum, restoration of the "lowest practicable grade" necessary to stabilize the land "in order to achieve an ecologically sound land use compatible with the surrounding region" (*U.S. Code*, Title 30, §1265(a)(3)). This provision reflects concern for prevention of landslides, erosion and water pollution, and preparation of a surface suitable for revegetation and some productive postmining land use. All surface areas, including spoil piles and tailings, must be stabilized for similar purposes. All highwalls must be backfilled and completely covered with spoil material. An exception to requirements for restoration of "approximate original contours" is contained in provisions governing "mountaintop removal" for purposes of removing an entire seam or seams running through the upper fraction of a mountain, ridge or hill (*U.S. Code*, Title 30, §1265(c)). In this case, a level plateau or gently rolling contour suitable for an approved postmining land use must be provided as mining of each area of the mine is completed.

Variances may be granted from requirements for restoration of "approximate original contours" where other contours are deemed desirable to permit a "better economic or public" postmining land use. These "better" uses may be industrial, commercial, residential or public (including recreational) in nature and must be approved in advance by the regulating body. Nowhere in the Act are any particular uses specified or ranked. Determination of what constitutes a "better economic or public use" is therefore a matter of discretion for the appropriate regulatory authority. Examples of postmining land uses in the United States include cropland agriculture, commercial forestry, recreation and tourism; public works such as airfields, roads, housing developments, and industrial sites; fish and wildlife conservation on shrub lands, woodlands and wetlands, for hunting and fishing, and to conserve biological diversity.

Preservation of topsoil All topsoil (or best available subsoil, if better than the topsoil) removed must be segregated from other mine spoil and saved for reuse and revegetation during reclamation of the mine site. It must be protected from wind and water erosion and contamination by acidic or toxic materials, and must be revegetated in the interim if necessary to achieve these aims. Special requirements are imposed on surface mining of prime agricultural lands to preserve and restore both topsoil and root zone of the natural soil.

Hydrologic balance and water quality Surface mining must be done in a manner which will minimize disturbance to the prevailing hydrologic balance and protect surface and groundwater quality and quantity in the vicinity both during and after mining and reclamation activities. All auger holes must be sealed with an impervious material after use. Special attention was given to preserving "essential hydrologic functions of alluvial valley floors in arid and semiarid areas" such as are found in much of the western United States. All waste piles and excess spoil dumps must be located on the mine site, designed to ensure permanent stability, prevent spoil erosion and movement. Offsite areas must be protected from slides or other damage from coal mining and reclamation operations. All debris, acid-forming, toxic and combustible materials must be disposed of in a manner which will prevent contamination of surface and ground waters, and sustained combustion above or below ground. Access roads must be constructed so as to control erosion, siltation, water pollution and damage to fish and wildlife and both public and private property; they must not alter the normal flow of water in stream beds or drainage channels.

Limited reservoir construction   Reservoirs may be created as reclamation measures *only* where water impoundments are necessary for effective revegetation, the water level of the reservoir is relatively stable, water quality is suitable for the postmining land use proposed and safe access is provided for proposed water users. No diminution of quality or quantity of water used by surrounding land owners will be permitted in such instances. Merely allowing an open pit mine to fill with acidic or toxic water, as was often done in the past, is not permitted under the Surface Mining Act. In addition, all temporary and permanent water impoundments associated with surface mining operations and reclamation must be designed, built and operated in accordance with standards (including safety specifications) promulgated by the Secretary of the Interior with written concurrence of the Corps of Engineers.

Blasting and safety   Mine operators are prohibited from surface mining coal within 500 feet of active or abandoned underground mines unless approved mitigation measures are taken to prevent breakthroughs and protect the health and safety of miners. Public notice must be given of blasting schedules, which must be carried out by specially trained and certified explosives personnel in accordance with all existing state and national laws.

Contemporaneous reclamation and revegetation   Under the Surface Mining Act performance standards, reclamation of surface mined lands must proceed as "contemporaneously as practicable" with surface mining operations, except under specified conditions when a variance may be granted which specifies a reclamation timetable. All excess spoil material or solid waste piles not returned to the mine excavation must be placed, graded, stabilized and revegetated within the permit area in a manner compatible with the natural surroundings. The mine operator must establish a "diverse, effective and permanent vegetative cover" of native species at least equal in coverage to the natural vegetation of the area, unless introduced species are necessary and approved in advance in a postmining land use plan. The mine operator must also assume responsibility for successful revegetation for five years after the last year of seeding, fertilizing, irrigation or other work, except in areas receiving 26

Figure 3.2   Trout hatchery on coal mine property receives all its water from Mettiki Coal Company acid mine drainage facility following chemical treatment and aeration, before water is discharged into receiving stream, Maryland, USA
Source: Alfred E. Whitehouse.

inches or less annual average precipitation (including most of the Western United States), where the term is extended to ten years. All of the requirements above apply to surface effects of underground coal mining operations, and land subsidence causing material damage to surface features must be prevented (*U.S. Code*, Title 30, §1266).

Lands Unsuitable for Mining

The Act specifies several ways in which lands may be designated unsuitable for certain mining activities. Different criteria are specified for surface coal mining and non-coal mining operations. For a state regulatory program to be approved, it must establish "a planning process enabling objective decisions based upon competent and scientifically sound data and information as to which, if any, land areas of a state are unsuitable for all or certain types of surface coal mining operations ..." (*U.S. Code*, Title 30, §1272). Standards governing such designations are statutorily specified. These standards include: (1) mandatory designation of unsuitability if reclamation is not technologically and economically feasible; and (2) discretionary designation if surface mining operations would be incompatible with existing land use plans, would damage fragile, historic, or productive renewable resource lands, or lands in natural hazard (e.g., flood, earthquake) areas where such operations could substantially endanger life and property.

Where a national program has been promulgated in the absence of an appropriate state program, the Secretary of the Interior must establish a process for designation of private lands within the state as unsuitable for surface coal mining in accordance with the above standards. The Secretary must also review all nationally managed public lands to determine whether, pursuant to these standards, there are areas which should be designated unsuitable for all or certain types of surface coal mining. Public lands so designated may be withdrawn from the public domain or otherwise rendered ineligible for surface coal mining operations.

Non-coal mining unsuitability   The Secretary was authorized (and if requested by a state governor, required) to assess any area of public lands within a state for designation as unsuitable for mining minerals other than coal, in accordance with specified criteria (*U.S. Code*, Title 30, §1281). Such a designation may be made, and the lands withdrawn from mineral leasing or entry, if they are predominantly urban or suburban or if a mining operation would adversely affect lands used primarily for residential purposes. Any person with a valid legal interest (e.g., property) which might be adversely affected by mining may petition for exclusion of such an area from mining activities. Mines existing in 1977 when the statute was enacted were exempted from operation of this provision.

Abandoned Mine Reclamation Fund

The Surface Mining Act established within the U.S. Treasury a special, self-supporting Abandoned Mine Reclamation Fund (*U.S. Code*, Title 30, §§1231–1242) for acquisition and restoration of unreclaimed lands ravaged by uncontrolled surface and underground mining operations of the past. The Fund is supported by a reclamation fee assessed on each ton of coal mined in the U.S., and by monies derived from the sale, lease or use of land reclaimed through the Fund. Production fees of 35 cents per ton of surface mined coal, 15 cents/ton of coal mined underground and 10 cents/ton of lignite are collected from all active coal mining operations. States having approved regulatory programs may submit reclamation plans for abandoned mine sites and receive funding for their implementation, including grants of up to 90 percent of acquisition costs for lands to be reclaimed. Nearly US$7 billion was collected from January 1978 to September 2003 (U.S.DOI, 2004).

Surface Rights

Because mineral rights may be severed from other property rights in land in most of the United States, nationally-owned coal deposits located beneath lands owned by individuals were sometimes leased and mined without the consent of persons holding title to surface lands. The Surface Mining Act defined the "surface owner" of land as one who has held legal title to the land surface or has lived on the land or received directly a "significant portion" of their income from farming or ranching operations on the land for at least three years (*U.S. Code*, Title 30, §1304(e)). Under the Act, the consent of these persons must be obtained before such nationally-owned coal deposits may be leased for surface mining. Where nationally-owned coal lies beneath lands subject to an existing national government lease or permit, either written consent of the surface lease or permit holder must be obtained, or a bond must be posted by the mine operator to cover any damages or losses imposed on the surface lease or permit holder by surface mining operations.

Prohibited Acts

By statute, surface coal mining is specifically prohibited (1) within 100 feet of any public road, (2) within 300 feet of any occupied dwelling, unless permission has been granted by the owner, (3) within 300 feet of any public building, school, church, community or institutional building, or public park, and (4) within 100 feet of a cemetery. It is also prohibited on any lands within the National Parks, National Wildlife Refuges, Wilderness Areas, the Wild and Scenic Rivers System, the National System of Trails, and National Recreational Areas. Surface mining may be permitted on lands in the National Forest System only if the Secretary finds there are no significant recreational, timber, economic or other values that are incompatible with surface coal mining. Similar restrictions apply to surface coal mining operations that might adversely affect any public lands listed on the National Register of Historic Sites.

In spirit, the Surface Mining Act lies in the mainstream of traditional American conservationism. It provides for the wise use of coal-bearing lands in accordance with known scientific standards of hydrology, agronomy and geology, while protecting only lands with special characteristics from surface mining activities. However in detail and specificity of its substantive and procedural features, as well as in its approach to state-national relations, the Surface Mining Act is decidedly modern. It is original, landmark legislation in the realm of mined land restoration, with broad coverage and far-reaching ramifications. Its language is hardly of a constitutional nature – the detail, specificity and number of exemptions to its requirements closely limit the amount of discretion which may be exercised by administrators attempting to implement this national policy.

Coal production in the United States increased from about 697 000 short tons per year in 1977 to over 1.1 billion tons per year in 2001, and nearly that much in 2002 (U.S. Energy Information Administration, 2003, p. 89). It is evident the imposition of mining environmental regulations pursuant to the Surface Mining Act was not an excessive burden and did not prevent expansion of the coal mining industry in the United States. Fluctuations in annual coal production are influenced more by economic activity than by regulatory activity.

Chapter 4

Environmental Assessment Policy in Two Countries

NEPA in the United States

There is a popular myth in the United States, and perhaps in some quarters throughout the world, that the National Environmental Policy Act (NEPA) of 1969 requires regulatory decision making to accomplish protection of the environment. This mistaken view is widely held both by proponents of development, who often demonize NEPA as preventing economic growth, and by environmentalists who glorify the statute as the one best hope for environmental preservation, commonly viewing it as a means of stopping environmentally disruptive development projects. Substantial portions of the general public probably accept both views. Nothing could be further from reality. A close reading of the statute indicates NEPA is not an environmental regulatory statute. That is, the Act does not prohibit disruption of the environment. Environmentally disruptive actions continued to be made after its enactment, and can still be made today.

On the few occasions when NEPA speaks in terms of preservation or protection, the language is so heavily qualified as to make it unenforceable. For example, the national government is given the responsibility to "preserve *important* historic, cultural, and natural aspects of our national heritage, and maintain, *wherever possible*, an environment which supports diversity and variety of individual choice" (*U.S. Code*, Title 42, §4331(b)(4)). Who is to say what aspects of our national heritage are not important enough to preserve? Under what circumstances is it not possible to maintain the environment simply by refraining from action that harms it?

Nowhere in the statute are definitions or criteria specified that would guide an official decision maker in determining what is an "important" aspect of our national heritage, as distinct from one which is not deserving of preservation. No indication is given about when it may or may not be possible to maintain such an environment, or how to decide. Moreover there are no sanctions for failure to preserve and maintain the environment. In the absence of criteria for decision and sanctions for noncompliance, such provisions remain lofty statements of goals which are largely unenforceable.

Reform of Administrative Procedure

Nonetheless, examination of the literature on its implementation suggests that NEPA stimulated a learning process which improved the quality of decision

89

making in national government agencies, in turn resulting in less environmental disruption than previously. NEPA accomplished this goal through a far-reaching reform of administrative decision making, not by directly requiring environmental protection *per se*. Improved results were achieved in the United States neither automatically nor immediately, but only after a learning period of many years during which substantial procedural refinements, guidelines, rules and other clarifications of the policy were gradually imposed by the national courts, administrative agencies, and by the President's Council on Environmental Quality.

The vehicle for this accomplishment was a series of procedural requirements which effectively broadened the scope of information that must be considered by government officials before they make important decisions affecting the environment. That is, NEPA imposed significant planning and information gathering requirements on national government decision makers before they can make a final decision in an official capacity. NEPA did not prescribe the substance of what decisions may or may not be taken, but established a new procedure for making decisions. In this respect, NEPA is similar to and elaborated on the Administrative Procedures Act of 1946, (*U.S. Code*, Title 5, §551 et seq) which set forth for all national government agencies a set of procedures for administrative decision making that must be followed to implement the due process clause of the U.S. Constitution.

Substantive policy Environmentally damaging decisions may be made, but only after correctly following prescribed information gathering and analytical procedures. The substantive policy of NEPA was simply to require that previously neglected environmental values be taken into consideration along with economic and engineering values during decision making by national government agencies in the United States. There was no explicit ranking of these value sets raising any above the others, only a new requirement that they all be considered. So, how do we understand what NEPA actually requires? One avenue, not often taken, is to read the statute itself, or the myriad of law texts, journal articles and books that have been published about it since 1970. This can be a bit tedious for persons lacking legal training. An only slightly less heinous alternative is to examine publications of the President's Council on Environmental Quality (CEQ) where a substantial body of well-informed interpretation may be found.

Council on Environmental Quality

CEQ was created by NEPA in the Executive Office of the President to advise the President on national and international policies "to promote the improvement of the quality of the environment"(*U.S. Code*, Title 42, §4342), and monitor national government implementation of NEPA. CEQ functions in an advisory capacity and is not a regulatory agency, those functions having been delegated by the Congress to the U.S. Environmental Protection Agency, which was created by separate legislation in 1970 (*Statutes at Large*, Vol. 84, §2086; *U.S. Code*, Title 5, appendix). Because of this, and due to the fact it first served President Nixon, who had not favored its creation, CEQ in its early years suffered from considerable role ambiguity and did not initially take a strong leadership stance. It published guidelines in 1970 and 1973 which were unfortunately limited to nonbinding suggestions for agency preparation of environmental impact statements (U.S.CEQ, 1973), and did not interpret the entire Act. Consequently, those guidelines were largely ignored.

It was not until 1978, during the term of an environmentally somewhat more sympathetic President Carter, that Executive Order 11991 (*U.S. Code of Federal Regulations*, Title 2, §123) directed CEQ to publish binding regulations interpreting NEPA and imposing requirements for preparation of environmental impact statements on all national government agencies (U.S.CEQ, 1978). These regulations drew upon and codified a substantial body of case adjudication by the national courts which accumulated in the eight years after NEPA was enacted.

NEPA Requirements

CEQ regulations state:

> NEPA procedures must insure that environmental information is available to public officials and citizens before decisions are made and before actions are taken. The information must be of high quality. Accurate scientific analysis, expert agency comments, and public scrutiny are essential to implementing NEPA. Most important, NEPA documents must concentrate on the issues that are truly significant to the action in question, rather than amassing needless detail.
>
> ***
>
> Ultimately, of course it is not better documents but better decisions that count. NEPA's purpose is not to generate paperwork – even excellent paperwork – but to foster excellent action.
>
> (*U.S. Code*, Title 40, §1500.1(b), (c))

Integration   CEQ regulations repeatedly emphasized the need to integrate NEPA with other planning and review procedures (*U.S. Code*, Title 40, §§1500.2(c); 1500.4(g); 1500.5(d); 1501.1(a)), to insure that planning and decisions reflect environmental values, to avoid delays later in the process, and to head off potential conflicts (*U.S. Code*, Title 40, §1501.2). Special attention was given to coordinating preparation of environmental impact statements with related surveys and studies required by the Fish and Wildlife Coordination Act, the National Historic Preservation Act, and the Endangered Species Act (*U.S. Code*, Title 40, §1502.25) so information obtained about these resources would be used in decision making.

Applicability   The requirements of NEPA apply directly only to important decisions taken by officials of the national government, not state or local governments. NEPA requires all agencies of the national government to prepare a detailed statement about the environmental impacts of every major action "significantly affecting the quality of the human environment" (*U.S. Code*, Title 42, §4332(C)). Others may gather or provide information, but it is the responsibility of a national government official to verify the accuracy of all information included in the statement.

This is the well-known "environmental impact statement" (EIS) about which so much has been written since 1970. There was no indication in the statute concerning the scale of activities or harm that might be considered significant. Subsequent court decisions and CEQ regulations have defined "significantly" as referring both to intensity or severity of impact, and the context in which an impact occurs, suggesting it varies with the setting of the proposed action (*U.S. Code*, Title 40, §1508.27). For example, national government decisions concerning the granting of permits for nuclear electric generating plants, transmission line rights of way across national public lands, and the financing of projects such as major new impoundments of water may significantly affect the quality of the environment. Because proponents of such projects often are not national government agencies, NEPA applies indirectly to many development proposals advanced by state and local governments and firms in the private sector, insofar as permits or approval decisions are required for them from officials of the national government. National government agencies were required to use a scoping procedure to identify and focus on the most significant environmental issues deserving of study while de-emphasizing insignificant issues and narrowing the scope of the process accordingly, thereby reducing paperwork and delays (*U.S. Code*, Title 40, §§1500.4; 1500.5; 1501.7).

Implementation Litigation

In instances where the procedural requirements of NEPA were not met, based on intentional omissions of required information or analysis, or other deficiencies in information provided in the EIS document, national courts in the United States issued injunctions delaying development projects until the procedural shortcomings were corrected – usually through preparation of supplements to the original documents. Such cases were numerous in the first five years following enactment of NEPA, but declined thereafter as agencies learned how to meet the procedural requirements (Anderson, 1973; Liroff, 1976).

When procedural requirements are met, the courts are reluctant to intervene to stop a project, even temporarily. The reason for this is obvious: NEPA confers no authority on any national government agency to prohibit an environmentally disruptive development project, and therefore confers no authority on the courts to ban permanently a disruptive action. This may only be done when a particular action is prohibited by some other policy or statute besides NEPA. Some development interests would prefer the general public believe otherwise. Their obvious interest in organizing public support for large-scale developments often leads into criticism of NEPA and other environmental policies that restrain their freedom to do whatever they please. A few projects have been abandoned by proponents after being delayed temporarily by the courts, but no project has ever been terminated permanently by court order in the United States on the basis of requirements contained in NEPA.

Preparation of a supplemental EIS, depending on the deficiencies identified and the nature of information required to cure them, may take a year or more – during which interest payments may be required on substantial loans. Required information gathering and analysis may cost additional millions of dollars. Proposals may be abandoned at any time economic conditions or the expected total investments for a project change. Similarly, only in cases where a national government official has discretionary authority under some other statute has a project ever been denied approval for environmental reasons. NEPA does not authorize denial of official action on substantive environmental grounds, even if the projected environmental disruption would be substantial.

A survey of 70 national government agencies preparing EIS for their own development projects, and reviewing applications for approval of those sponsored by others, revealed numerous instances where designs were modified, locations of facilities were changed, the scope of alternatives broadened, and operating regulations improved. Actions sponsored by the agency preparing an EIS were sometimes canceled during internal reviews, if locations proposed were determined to be environmentally unsuitable for activities proposed. Requests for approval were sometimes withdrawn on the basis of information compiled in EIS documents. Overall, the EIS was viewed as an important aid to planning and decision making by national government agencies which had substantially improved government decisions after enactment of NEPA (U.S.CEQ, 1976).

Systematic Interdisciplinary Approach

Although it has become the most well-known provision, preparation of an EIS was not the only requirement imposed on decision makers of the national government. NEPA requires that the policies, regulations, and public laws of the United States be interpreted and administered in accordance with requirements of the Act, and that all agencies of the national government utilize a systematic interdisciplinary approach in decision making that may have an impact on the environment. It specifically directed that all agencies of the national government utilize ecological information during planning and development of resource-oriented projects, and develop methods to insure that "unquantified environmental amenities and values" be given appropriate consideration in decision making "along with economic and technical considerations" (*U.S. Code*, Title 40, §4332).

In imposing these requirements, the U.S. Congress was mindful of a well-documented tendency among some agencies of the national government – particularly the Atomic Energy Commission, the Army Corps of Engineers, and the Bureau of Reclamation – to place heavy reliance on criteria of economic and engineering feasibility during decision making for construction of nuclear electric generating plants and major new water impoundments, to the exclusion of information about environmental values which might be provided by the natural and social sciences and the environmental design arts.

The U.S. Congress required that this information and analysis be described in the environmental impact statement prepared by the responsible official for disruptive projects.

Contents of the EIS

Subsequent court interpretations of NEPA determined the EIS must contain descriptions of: the environment in the vicinity of each site proposed for development before construction begins; the proposed development and facilities required to implement it; the probable impacts or changes in environmental conditions induced by implementation of the proposal; any unavoidable adverse effects on the environment; any irreversible or irretrievable commitments of resources and discussion of any actions proposed to mitigate or reduce adverse environmental effects (Liroff, 1976). Logically, assessment of change induced in environmental conditions dictates there be some baseline data about the environment as it exists prior to implementation of development actions. This baseline data must be based on actual onsite observations by persons with relevant expertise in the field, not just on references to previously published literature about nearby locations. Projects producing probable impacts on air and water resources thus would require a full year of air and water quality monitoring data to describe seasonal climatological fluctuations, especially concerning wind direction and velocity, rainfall, and runoff during storm events. Information about the development activity must be of sufficient specificity to facilitate analysis of probable effects on the environment. This requires actual data about construction techniques, locations, and operating capabilities of equipment to be used throughout the term of the development, related to the physical and climatological context of the site proposed.

Description of unavoidable adverse environmental effects of a proposed development requires some analysis of which specific effects might be avoided. Description of any irreversible or irretrievable commitments of resources invites discussion of the financial, labor and materials requirements for the proposed development, and any unavoidable destruction of natural, biological, cultural, or historic resources. This further ensures that consideration of damage to environmental amenities will be factored into the costs of proceeding with a project, even if they are unquantifiable. Discussion of actions proposed to mitigate or reduce adverse environmental effects requires identification and assessment of opportunities and technologies to minimize environmental harm of otherwise disruptive development projects.

The EIS document must also contain descriptions of: environmental impacts of alternatives to the proposed action, including the alternative of doing nothing; a discussion of the relationship between short-term uses of the environment and the maintenance and enhancement of long-term productivity of resources; and discussion of any national policies which justify causing unavoidable adverse effects (*U.S. Code*, Title 40, §1502.16). Predetermined geological locations of mineral resources limit somewhat the consideration of alternative locations for mineral development projects, but not consideration of alternative mining technologies (e.g., underground mining) that would allow avoidance of surface mining in areas where surface resources may be more valuable than minerals, or reclamation may not be feasible. Consideration of alternative technologies and locations for some mining activities and facilities such as settlement ponds, topsoil storage piles, excess overburden disposal areas, coal washing and crushing plants, offices, maintenance garages, housing and transportation load-out facilities may provide opportunities to reduce environmental disruption of mining operations. For example, locating coal washing and crushing plants near the mine pit and away from rivers and streams may avoid siting them on wetlands valuable for wildlife, or in flood plains prone to periodic inundation. It also increases efficiency of operations by locating them near an easy and safe disposal site for process waste materials – the mine pit.

Discussion of short-term uses and long-term productivity of resources was intended to stimulate assessment, especially where proposals concerning uses of national public lands are concerned, of whether the proposed development is temporary or permanent, and whether surface resources are renewable and therefore can be harvested "without impairment of the productivity of the land" (*U.S. Code*, Title 43, §1732(a)) as required by the Multiple-Use Sustained-Yield Act of 1960 (*U.S. Code*, Title 16, §528 et seq). In the United States, mining of a nonrenewable resource is considered a temporary use of land, which must be reclaimed for other postmining uses, thereby insuring its continued productivity.

Depth of Analysis

Environmental effects and values must be identified in adequate detail so they can be compared to economic and technical analyses, and so significant issues may be analyzed in depth in the environmental impact statement, with only brief discussion of other than significant issues. The purpose and need for action may be "briefly" described, and the environment of the area affected must be "succinctly" described (*U.S. Code*, Title 40, §§1501.2(b)-1502.15). However all reasonable alternatives including the proposed action must be rigorously explored and objectively evaluated, with "each alternative considered in detail" (*U.S. Code*, Title 40, §1502.14), and appropriate mitigation measures specified. Environmental consequences of each alternative must be discussed, including both direct and indirect impacts and possible conflicts with land use plans in the area concerned (*U.S. Code*, Title 40, §1502.16). Clearly some detailed discussion, not brevity, is expected in these areas of the EIS.

Timing: Feasibility Study

In the United States, preparation of an EIS must begin early – as close as possible to the time the agency is developing or is presented with a proposal – so it can be completed in time for the final statement to be included in any recommendation or report on the proposal. For projects undertaken by a national government agency, the EIS must be prepared at the feasibility study stage and may be supplemented later if necessary. For applications to an agency, environmental assessments or EIS must be begun no later than immediately after an application is received (*U.S. Code*, Title 40, §1502.5). Supplements may be prepared to either draft or final EIS if significant new circumstances or information relevant to environmental concerns arises (*U.S. Code*, Title 40, §1502.9(c)(1)). The time, effort or expense necessary to prepare environmental assessments at such an early stage does not excuse poor quality information. When an agency discovers incomplete or unavailable information is essential to a reasoned choice among alternatives, and the overall costs of obtaining it are not exorbitant, CEQ regulations require the agency to gather and analyze the information and include it in the EIS (*U.S. Code*, Title 40, §1502.22). This includes information about potentially severe impacts even if their probability of occurrence is low.

The agency preparing an environmental assessment or EIS is responsible for the accuracy of information presented in these documents, even if the information was prepared by others such as contractors. It must independently evaluate environmental information prepared by others and make its own evaluation of the environmental issues and take responsibility for the scope and content of the documents. If information is prepared by a contractor, the contractor must be selected by the agency and must execute a disclosure statement specifying that they have no financial or other interest in the outcome of the project (*U.S. Code*, Title 40, §1506.5). Methodologies used in preparation of EIS must be identified with explicit references by footnote to scientific and other sources relied upon for conclusions, and any discussion of methodology should be placed in an appendix (*U.S. Code*, Title 40, §1502.24).

Public Disclosure

Before finishing an EIS, the responsible official must consult with and obtain comments from other national government agencies having either jurisdiction by law or special subject-matter expertise relating to a probable environmental impact (*U.S. Code*, Title 42, §4333(2)(C)). For example, if a proposed coal-fired power plant financed by the U.S. Rural Electrification Administration (REA) will likely disturb habitat of an endangered species under the jurisdiction and expertise of the U.S. Fish and Wildlife Service, that agency must be consulted by the REA during preparation of an EIS. National, state and local agencies authorized to develop and enforce environmental standards must also be consulted (*U.S. Code*, Title 40, §1502.19). Joint preparation of an EIS by two or more agencies is authorized by regulation (*U.S. Code*, Title 40, §1505.2) and ensures interagency coordination where jurisdictions overlap.

Comments from such agencies must accompany the proposal throughout the agency decision making process, must be included in the EIS document, published and made available to the general public pursuant to the Administrative Procedures Act (*U.S. Code*, Title 5, §552). Full disclosure of the EIS was intended to invite scrutiny by both experts and nonexpert members of the general public as a check on the appropriateness of national government administrative decision making.

However according to court decisions and CEQ regulations: "An environmental impact statement is more than a disclosure document. It shall be used by Federal officials in conjunction with other relevant material to plan actions and make decisions" (*Calvert Cliffs Coordinating Committee v. Atomic Energy Commission* 1972; *U.S. Code*, Title 40, §1502.1). Thus NEPA requires an environmental impact statement be prepared as an integral part of the planning process for major development decisions, before a decision is made to proceed on a single alternative among several considered. The EIS was conceived and expected to serve the function of an actual decision document, a useful aid to decision making. It was never intended to be a pro-forma or after-the-fact paper shuffling exercise divorced from analysis of relevant information, compiled merely to justify decisions already made on other grounds.

Time Limits?

In the United States, prescribed universal time limits for the entire NEPA process were determined to be too inflexible to accommodate projects of greatly varying size, complexity, and severity of environmental impact. Instead, agencies were encouraged to set time limits "appropriate to individual actions," and required to set time limits if an applicant requests them, provided the time limits "are consistent with the purposes of NEPA" (*U.S. Code*, Title 40, §1501.8). Agencies were authorized to set time limits for each constituent part of the NEPA process, taking into consideration the distinctive characteristics of each project, including: the potential for environmental harm; size of the proposed action, number of persons and agencies affected; degree of public need for the proposed action, including the consequences of delay; availability of relevant information and the amount of time required to obtain it; state of the art of analytic techniques; and the degree of controversy attending the proposal.

Regulations normally allow a minimum of 90 days for public comment after publication of a draft EIS before final decision is made by the government official but in special circumstances this may be reduced to 45 days. The 90 day period may be routinely extended for 30 days by EPA, and has sometimes been extended for several months by the lead agency for complex and controversial projects. Moreover no action may be taken for a period of 30 days after a final EIS is published, to allow concerned persons an opportunity to challenge it in court before any damage to the environment begins (*U.S. Code*, Title 40, §1506.10).

An early survey of national government agencies preparing and reviewing EIS found they did not want a specific time limit to finish a final EIS after receiving comments on a draft. Many found 45 days was not enough time for review and comment on a draft EIS, especially when their technical expertise was decentralized in field offices. EPA had difficulties completing review of final EIS in 30 days (U.S.CEQ, 1976, pp. 31, 40, 42). In the United States, there is no requirement that the agency preparing an EIS make a final decision within a specified period of time. There is no automatic approval if an agency fails to make a final decision within any period of time. The most significant requirement concerning time limits is that they be "consistent with the purposes of NEPA," which requires environmental values actually be taken into full consideration with other economic and technical values.

Mitigation of Impacts

The record of decision in the final EIS must state which alternatives were considered environmentally preferable. It must also state whether all practicable means to avoid or minimize environmental harm from the alternative selected have been adopted, and if not, why they were not adopted. Mitigation and other conditions established in the EIS or during its review and committed as part of the decision must be implemented by including them in grants, permits or other approvals, and government funding of any actions must be conditioned on mitigation. Upon request, the results of relevant monitoring efforts must be made available to the public (*U.S. Code*, Title 40, §1505).

Organizational Capability

Each agency must maintain adequate resources and personnel with appropriate expertise to prepare environmental impact statements and evaluate those prepared by others in areas where the agency has jurisdiction by law or special expertise, or is authorized to develop and enforce environmental standards. These personnel must be capable of using a systematic interdisciplinary approach in planning and decision making which may have an impact on the human environment. CEQ regulations appear to require the employment of persons with training in ecology who are capable of initiating and utilizing ecological information in the planning and development of resource-oriented projects (*U.S. Code*, Title 40, §1507.2). Meeting these requirements would appear to necessitate assembly of multidisciplinary teams of personnel with training in the natural and social sciences and the environmental design arts, not reliance on persons with training only in engineering or economics.

A Constitutional Law for the Environment?

In the United States, usually legislation is applicable to one or a few programs or agencies. NEPA has been described as a statute of "Constitutional scope" due to its general language and broad coverage of all national government agencies – even decisions granting foreign aid for construction projects in other countries (but see Caldwell, 1989). NEPA was not self-enforcing, but required the issuance of guidelines and regulations to flesh out required procedures, and considerable early interpretation by the courts (Anderson, 1973).

Although NEPA is not a regulatory statute, it forms a common link between other national environmental statutes enacted since 1970 concerning air and water quality and the uses of land, most of which require use of a systematic interdisciplinary approach to decision making advanced by it. Thus NEPA provides a single general orientation towards disparate policies concerning environmental decision making which ties them all together, in addition to a unifying mission orientation for the Environmental Protection Agency and the President's Council on Environmental Quality.

Global Reach of NEPA

Moreover the National Environmental Policy Act is global in its scope. In NEPA, the U.S. Congress directed that:

> all agencies of the Federal government shall ... recognize the worldwide and long-range character of environmental problems and, where consistent with the foreign policy of the United States, lend appropriate support to initiatives, resolutions, and programs designed to maximize international cooperation in anticipating and preventing a decline in the quality of mankind's world environment.
>
> (*U.S. Code*, Title 42, §4332(2)(F))

This provision formed the basis for OSM participation in science and technology exchanges with foreign governments, and provision of technical assistance to them (Bohlen, 1993), discussed below in Chapters 5–9.

In 1979, President Jimmy Carter issued Executive Order 12114 (*Code of Federal Regulations*, Title 3, §356) requiring environmental analysis and documentation for actions affecting the global commons; actions affecting the environment of a foreign nation not participating with the U. S. in an action (the "innocent bystander" situation); and actions providing a product which in the U.S. is prohibited or strictly regulated to protect the environment against toxic or radioactive substances (Bear, 1989, p. 10067). This order imposed environmental assessment requirements on expenditures outside the United States by many foreign aid programs and projects of the U.S. Agency for International Development and the U.S. State Department.

Sustainable Development Embraced

A crucial policy issue in development is to balance actions which support continued economic growth with environmental quality, to achieve sustainable development. To achieve balanced economic growth, it is necessary to develop mining environmental policy which will encourage environmentally sound, efficient mining practices while maintaining or improving water quality, air quality and reclamation of disturbed land for other productive postmining uses. Eighteen years before the Brundtland Commission described sustainable development as "development that meets the needs of the present without compromising the ability of future generations to meet their own needs," (World Commission on Environment and Development, 1987), NEPA effectively embraced this concept when it articulated a goal of creating and maintaining "conditions under which man and nature can exist in productive harmony, and fulfill the social, economic, and other requirements of present and future generations" (*U.S. Code*, Title 42, §4331(a)).

The Act declared it is the responsibility of the national government of the United States to "use all practicable means, consistent with other considerations of national policy," to "fulfill the responsibilities of each generation as trustee of the environment for succeeding generations" (*U.S. Code*, Title 42, §4331(b)(1)). Because it applied to all environmental resources, not just those which are renewable (*U.S. Code*, Title 43, §1732(a)), this was a much broader statement in support of sustainable development than the endorsement of sustained-yield in the Multiple-Use Sustained-Yield Act of 1960 (*U.S. Code*, Title 16, §528 et seq).

Yet the U.S. Congress stopped short of granting what might be considered a Constitutional right to a clean environment when it enacted NEPA. The version of this legislation initially approved by the Senate and sent to Conference Committee with the House of Representatives stated that "each person has a fundamental and inalienable right to a healthful environment" (U.S. Cong. Senate Comm. on Interior and Insular Affairs, 1969, p. 2). Apparently there was concern among some members of the Conference Committee about the broad legal scope of the Senate version and litigation it might engender (U.S. Cong., 1969, p. 8). Granting such a right by statute might give legal grounds for citizen suits for compensation or injunction if the environment in which they live was damaged sufficiently to be considered harmful to their health. Moreover it might prove difficult to abrogate such a right after it was granted without payment of just compensation to every person harmed in the United States. Consequently, before enacting NEPA, the Conference Committee changed this provision to say "each person should enjoy a healthful environment" (Congressional Quarterly, Inc., 1969, p. 527; *U.S. Code*, Title 42, §4332(c)).

AMDAL in Indonesia

In Indonesia, a Ministry of Environment and Population [now Ministry of Environment] was established in 1978 to implement policies concerning environmental protection (Prasodjo, 1996). An Environmental Impact Management Agency (Badan Pengendalian Dampak Lingkungan – BAPE-DAL) with jurisdiction over environmental impact assessments was established as a non-departmental agency reporting directly to the President, separate from the Ministry of Environment in 1990 (Republic of Indonesia, 1990b). Regional BAPEDAL were established in each province in 1994. BAPEDAL was consolidated into the Ministry in 2002, and now reports to the Minister of Environment. The role of the Ministry of Environment historically was primarily concerned with formulation of environmental legislation and policy, and BAPEDAL was established as an environmental enforcement agency (Warren and Elston, 1994, p. 19). In practice, there was considerable overlap between their respective functions concerning regulation of environmental impacts and pollution control, which was resolved by consolidating them.

A Right to a Clean Environment

The initial Indonesian statement of national environmental policy, the Act on Basic Provisions for Management of the Living Environment of 1982 said every person has a right to a good and healthy environment (Republic of Indonesia, 1982, Article 5(1)), providing a sharp contrast with NEPA. This provision was retained when the Act of 1982 was substantially amended in 1997 (Republic of Indonesia, 1997, Article 5(1)). Furthermore, the Act of 1982 imposed upon every person an obligation to prevent and abate environmental damage and pollution. Those whose rights are violated are entitled to compensation, and liability was imposed on those who damage or pollute the environment, who must also compensate the government for costs of restoration (Republic of Indonesia, 1982, Article 20).

Indonesian society is considerably less litigious than the United States, and its Constitution lacks provisions requiring payment of compensation for abrogation of the rights of persons. Consequently, the foregoing provisions may be viewed as strengthening the hand of those seeking compensation during proceedings for grant of business licenses, but they have not produced successful citizen litigation seeking compensation after environmental damage has occurred (Petrich, 1994, p. 8; Lucas, 1998, pp. 181–208). Where mining is concerned, these provisions would appear to require control of environmental impacts during mining operations, and reclamation of disturbed land for postmining land uses.

Sanctions   The Act on the Living Environment of 1982 provided penalties for intentional damage to the environment of imprisonment not exceeding 10 years and/or a fine not exceeding 100 million rupiah (about US$11 367.00).[1] Persons damaging or polluting the environment through negligent acts are subject to detention not exceeding one year and/or a fine not exceeding one million rupiah (about US$113.64). Although these penalties were more substantial than those provided in the 1967 Mining Law, no individual or firm was prosecuted for failure to adhere to requirements of the Act on the Living Environment of 1982 (Government Urged, 1992).

Nonetheless, the list of environmental crimes was expanded in Act No. 23 on Environmental Management of 1997, and some sanctions made more severe. The maximum fine for intentional pollution was increased to 500 million rupiah (about US$56 835.00), and for negligent damage to the environment, 100 million rupiah (about US$11 367.00) with the maximum period of imprisonment for negligent damage increased from one to three years. More serious sanctions of a maximum of 15 years in prison or fine of 750 million rupiah (about US$85 252.00) were specified for intentional environmental crimes resulting in death or serious injury of a person (Republic of Indonesia, 1997, Article 41). New environmental crimes were established for importing, exporting, trading, transporting or storing toxic or hazardous materials, operating a dangerous installation, or providing false information, damaging, destroying or concealing information needed to investigate such actions (Republic of Indonesia, 1997, Article 43). These are significant penalties, if enforced.

Balance of Sustainable Development

The Indonesian Act on the Living Environment of 1982, adopted 13 years after NEPA and five years before the Brundtland Commission report, echoed goal statements in NEPA concerning achievement of "conditions under which man and nature can exist in productive harmony, and fulfill the social, economic, and other requirements of present and future generations" (Republic of Indonesia, 1982, Article 4a, d). The Act of 1982 also related environmental management directly to a concept of sustainable development: "The management of the living environment is based upon the sustenance of the capability of the harmonious and balanced environment to support continued development for the improvement of human welfare" (Republic of Indonesia, 1982, Article 3).

Control of Natural Resources

Provisions in the Constitution of 1945 and the Mining Act of 1967 which say Indonesian natural resources are controlled by the government and utilized for the maximum welfare of the people were reiterated in the Act of 1982. The Act elaborated on this power by asserting the government has the authority to "regulate the allocation, development, use, reuse, recycling, provision, management and supervision of resources [and] regulate legal actions and legal relations between persons and/or other legal subjects pertaining to resources" (Republic of Indonesia, 1982, Article 10).

The first environmental assessments on major projects in Indonesia actually pre-dated the 1982 Act: for a cement plant in 1974, a resettlement site in 1976, and an oil-related project approved by the Ministry of Mines and Energy in 1976 (Petrich, 1994, p. 2). The Act of 1982 said every plan which is considered likely to have a significant impact on the environment must be accompanied with an analysis of environmental impact, carried out according to government regulations (Republic of Indonesia, 1982, Article 16). This is the source of a formal Indonesian requirement for environmental assessment of major development projects. A focus on significant environmental impacts in the Act of 1982 and the lack of any definition of what might be "significant" parallels NEPA. Yet there is no description in the Act of 1982 of what procedure must be followed or what information the analysis must contain, leaving these details for later regulations.

AMDAL Commissions

Authorization was provided in the Act of 1982 for creation of new national, regional, and sectoral institutions to implement its terms (Republic of Indonesia, 1982, Article 18). The exact nature of these institutions was left for later legislation and regulations to specify. Presidential Regulation 29 of 1986 described these organizations and specified a procedure for preparation and review of environmental assessments (Republic of Indonesia, 1986), which is popularly referred to by its acronym, AMDAL (Analisis Mengenai Dampak

Lingkungan). Subsequently, a Central AMDAL Commission was established in BAPEDAL, and regional AMDAL Commissions in each of 27 provinces. AMDAL Commissions were also established in 14 sectoral Ministries and agencies of the national government with functional jurisdiction over projects within their areas of responsibility, including the Ministry of Mines and Energy. The Ministry of Environment and Population issued decrees in 1987 providing guidance on determination of significant impacts, preparation of environmental assessments, establishment of time limits for their review, and specifying the composition of AMDAL Commissions (1987a; 1987b; 1987c; 1987d; 1987e).

Presidential Regulation 29/1986 included provisions applying environmental impact assessment requirements to existing development facilities. This created considerable difficulties and confusion concerning how to apply such provisions to projects for which decisions and substantial investments had already been made (Dick and Bailey, 1992, p. 69). Similar difficulties were evident in the early years of NEPA implementation concerning projects already proposed but not yet constructed (Anderson, 1973). Consequently, the AMDAL process was substantially revised and simplified by Presidential Regulation 51 in 1993 (Republic of Indonesia, 1993), which abolished requirements for environmental assessments on existing facilities. Subsequently, the Ministry of Environment in 1994 canceled its previous decrees and issued new guidance for preparation of environmental assessments, clarifying which activities would be required to prepare them (1994a; 1994b; 1994c; 1994d; 1994e; 1994f). Most of these decrees were later revoked or amended by Presidential Regulation 27/1999 or by Ministerial decrees, as discussed below.

Law No. 23 of 1997 on Environmental Management replaced and expanded the Basic Law on the Living Environment of 1982, retaining the right of persons to a clean environment, state control of natural resources, the policy of sustainable development, and the requirement for an environmental analysis of every business activity likely to produce a significant impact on the environment. The Act of 1997 banned imports of hazardous or toxic waste and authorized a regulatory program for control of the generation and use of hazardous and toxic materials in Indonesia. It expanded inspection and enforcement powers of the Ministry of Environment, authorized the Ministry to conduct environmental audits of businesses, placed additional emphasis on enforcement by regional and local governments, authorized enforcement actions by citizens and nongovernmental organizations, specified procedures for compensation for environmental damage, significantly expanded the list of environmental crimes and increased the sanctions applied to them (Republic of Indonesia, 1997).

In the United States, citizen complaints have successfully identified violations that might otherwise have gone undetected, and have detected violations occurring between governmental inspections. Local citizens tend to be good observers of changes in off-site conditions, such as changes in surface or groundwater quality. Similarly, it is reasonable to expect that Indonesian citizens would be good observers of off-site effects of violations, such as fish kills, and operations outside lease boundaries. A procedure that has worked well in the United States is for an inspector to meet with citizens filing a complaint, and then allow them to accompany the inspector on an inspection of the mining operation. Unfortunately, in many instances, Indonesian inspectors of foreign-owned mining operations may continue to be stationed a great distance from the mine site, and there may be a lack of transportation and other infrastructure in the area to facilitate a prompt response to citizen complaints. Thus it is a difficult management problem to decide how to take best advantage of information from citizen complaints, and prevent mining operators from committing violations between inspections.

Advocacy by non-governmental organizations (NGO's) has often been successful in the United States in promoting compliance with the laws and regulations. NGO's in the United States have developed the legal and technical expertise to screen out complaints that are insignificant or not related to the mining regulatory program. They present the more significant issues cogently, which saves staff time of the regulatory authority. They develop long-term, programmatic interest and insight into the practices of government and the mining industry. They also help to insulate citizen informants from retribution by mining companies.

On the other hand, NGO's tend to promote their own agendas, which may differ from priorities set by government agencies. In the United States, successful citizen complainants may have their attorney's fees paid by the violator or by the government, and NGO's have sometimes appeared to plan their agendas to obtain the maximum payment of attorney's fees. On the whole, NGO's tend to be useful sources of information, and effective vehicles for relatively unsophisticated citizens to communicate with government agencies and to participate in regulatory processes. It is often worthwhile to solicit comments or participation of NGO's in various phases of the regulatory process. However local and regional governments may find that NGO's may not share their concern with balanced, efficient implementation of a regulatory program.

Administrative Procedure

The AMDAL procedure requires project proponents to prepare a series of documents, each of which are commonly referred to by their acronyms, including:

> Kerangka Acuan (KA), or Terms of Reference, produced after preliminary screening of a project to determine whether it requires a full-scale environmental assessment, this constitutes a contract describing the scope of a proposed development and what information must be included in a more detailed assessment;

> Analisis Dampak Lingkungan (ANDAL), or Environmental Impact Analysis, a more detailed presentation of pre-development baseline data on local environmental conditions, description of the project, and identification of its probable environmental impacts;

Rencana Pengelolaan Linkungan (RKL), or Environmental Management Plan, describes mitigation measures and what steps will be taken to manage anticipated environmental impacts; and

Rencana Pemantauan Linkungan (RPL), or Environmental Monitoring Plan, describes how actual environmental impacts will be monitored during operations.
(Republic of Indonesia, 1993, Articles 1, 8;
Republic of Indonesia, 1999c, Articles 15, 17)

The KA is roughly comparable to scoping documents prepared pursuant to NEPA in the United States prior to preparation of a detailed environmental impact assessment. Taken together, the ANDAL, RKL and RPL comprise what would normally be encompassed in a final Environmental Impact Statement in the United States. Every application for a KA is accompanied by an application fee, portions of which contribute to an informal "mystery budget" which may provide honoraria for technical teams and decision makers involved in evaluating and approving the application.

Activities Requiring Environmental Analysis

Not content to allow the types of projects required to prepare environmental assessments to be determined by the courts as in the United States, Presidential Regulation 51/1993 described them in general terms and authorized the Minister of Environment and Population to determine which would be required to prepare assessments, after consultation with sectoral departments. Business activities for which AMDAL requirements might apply included those involving modification of land forms and the natural landscape, exploitation of renewable and non-renewable natural resources, and using processes with potential to cause waste, damage, and a decline in natural resource utilization (Republic of Indonesia, 1993, Article 2). Clearly this included most mining operations.

Significant impacts A level of significance of impact sufficient to require an assessment was to be determined by the number of people and extent of the area affected, the duration and intensity of impacts, the number of environmental components affected, cumulative impacts of multiple projects, and the reversibility or irreversibility of impacts (Republic of Indonesia, 1993, Article 3). These criteria were reiterated in guidelines for determination of significant impacts issued by BAPEDAL (1994). Influence of the language of NEPA was evident in use of the terminology of irreversibility of impacts. Types of business activities requiring AMDAL and criteria for determining significance of impacts were retained when Presidential Regulation No. 27 of 1999 replaced Presidential Regulation 51/1993 (Republic of Indonesia, 1999c, Article 3).

Application The Ministry of the Environment issued Decree 11/1994 on Activities Required to Prepare Environmental Assessments specifying threshold levels for each activity (1994b). Project size in each sector was the principal criterion for determining when AMDAL procedures are required. Mining operations larger than 200 hectares or producing more than 200 000 tonnes/year of coal, 60 000 tonnes/year of primary ores, or 100 000 tonnes/year of secondary ores are required to prepare AMDAL documents (Ministry of Environment, 1994b, Appendix I). Operations with 50 or more hectares open for mining were added to this list in 2001, when the other criteria were relaxed to apply to operations producing more than 250 000 tonnes/year of coal, 200 000 tonnes/year of primary ores, or 150 000 tonnes/year of secondary ores (Ministry of Environment, 2001, Attachment, 2.J.1–2.J.2). New provisions required AMDAL for smaller projects of any size if there is a reasonable expectation of significant environmental impacts.

Prior to 2001, any development project proposed in a specified protected area was required to prepare AMDAL documents, including those proposed in: forest protection areas, coastal edges, river edges, nature conservation areas, national parks and nature parks (Ministry of Environment, 1994b, Appendix II). These included areas of habitat occupied by rare and endangered species of plants and animals, and other areas identified in the Act on Conservation of Living Resources and their Ecosystems No. 5 of 1990 and Presidential Regulation on Management of Protected Areas No. 32 of 1990 (Republic of Indonesia, 1990a; 1990c). Criteria issued in 2001 for types of activities required to prepare AMDAL did not mention protected areas.

AMDAL Commissions within the sectoral Ministries were responsible for preparing technical guidelines for the environmental impact assessment process, and for evaluating AMDAL documents (KA, ANDAL, RPL and RKL). They were intended to assist in the decision making process of the Ministry or agency, and were advised by a technical team during review of AMDAL documents (Republic of Indonesia, 1999c, Article 8). Technical teams in the Ministry of Mines and Energy for coal mine AMDAL documents were usually staffed by mine inspectors from DTPU on temporary assignment. Staff provided by DTPU generally have good field experience but are all mining engineers and lack any credible technical skills in geology, geochemistry, surface and groundwater hydrology, soils, civil engineering, forestry, biology, agronomy, archaeology or cultural and historic resource protection. Consequently, possible impacts in these technical subject areas were poorly evaluated, missed, or ignored during ANDAL review. Provincial AMDAL Commissions appointed by Governors were authorized to evaluate AMDAL documents at the regional level. If AMDAL documents failed to provide the required information, the proponent was obliged to revise them as directed by the reviewing AMDAL Commission (Republic of Indonesia, 1999c, Article 21).

Timing of Environmental Analysis

The AMDAL process has been criticized for failing to shape development of environmentally sensitive policies, plans and resource programs because it was prepared after key decisions were made or provisional development permissions were granted (Welles, 1995, p. 111). Presidential Regulation 51/1993 added a provision, apparently intended to strengthen enforcement (Warren and Elston, 1994, p. 23), that said a final operating permit (*izin usaha tetap*) can be granted by the authorized government agency only after the RPL and RKL are implemented (Republic of Indonesia, 1993, Article 5).

However after Presidential Regulation 51/1993 required AMDAL documents be prepared along with the feasibility study, they continued to be prepared after engineering and economic feasibility studies were completed and a decision had been made to go ahead with a project (Soemarwoto, 1996, p. 79). More recent regulations require the ANDAL, RPL and RKL be prepared simultaneously with the feasibility study and submitted at the same time to the authorizing agency (Republic of Indonesia, 1999c, Article 2). These regulations also require environmental management and monitoring activities be included in operating permits (Republic of Indonesia, 1999c, Article 3(5)). Although applicable guidelines recognized the AMDAL process constitutes an integral part of the feasibility study, (Ministry of Environment, 1994e, Appendix I.6; BAPEDAL, 2000b, Attachment I.A.6) there was no specific requirement in the Acts of 1982, the Act of 1997 or subsequent regulations that costs of environmental controls or reclamation be included. Consequently, environmental compliance costs have usually not been included in feasibility studies (Soemarwoto, 1996, pp. 79–80), and mine operators have later argued some environmental costs are therefor not economically feasible for their operations (Whitehouse, 2003).

Inclusion of Environmental Values

Guidelines established by the Ministry of Environment for preparation of environmental assessments state that the purpose of an ANDAL study is to integrate environmental considerations into the detailed planning stages of a business or activity, to assist decision makers in selecting the best environmental alternatives, and to serve as a basis for monitoring and management of environmental effects of development projects (Ministry of Environment, 1994e, Appendix I.B.1.2; BAPEDAL, 2000b, Attachment I.B.1.2). Statements of purpose such as this in Indonesian regulations are typically strong, but when it comes to stating operational requirements, those are often weak.

Contents of AMDAL Documents

Prior to 1993, documents produced by the AMDAL process were often deemed to be of unsatisfactory quality. Inadequate technical competence of preparers and reviewers, institutional reliance on part-time members of AMDAL Commissions, and a complex and time-consuming process were seen as factors which produced poor quality results (Welles, 1995, p. 111). Several subsequent modifications of AMDAL procedures were designed to produce higher quality documents.

Guidelines issued in 1994 governing data collection and analysis during preparation of ANDAL appeared to call for field studies by referring to data collection sites and use of observation/analysis (Ministry of Environment,

1994e, Appendix II.B.II.3b, c), but did not explicitly do so. More recent guidelines issued in 2000 are an improvement, suggesting field observation should be used for identification of significant impacts during scoping, and that description of baseline data should rely upon actual data from field studies as much as possible (BAPEDAL, 2000b, Attachment I.A.8.1.1; I.B.2.2.a). However these guidelines do not require field studies of any particular data within any specific period of time before construction, not even to secure baseline data on air and water quality or flora and fauna at the project site.

Quality of Data

Baseline data acquired through field studies has been rare in ANDAL, which tend to rely heavily on published literature, often about locations distant from project sites. For example, the ANDAL for PT Kaltim Prima Coal, the second largest coal mine in Indonesia, did not contain baseline water quality data obtained from field sampling, because no water quality sampling was conducted until 30 days after mine construction had started (Irving, 1992). The single exception where onsite field data is commonly used in ANDAL concerns geological data acquired during exploration, describing the mineral resource.

Similarly, guidelines for preparation of ANDAL published in 1994 required proponents to describe plans for reclamation and rehabilitation of land, and plans for return of the site to other uses after operations are finished (Ministry of Environment, 1994e, Appendix II.B.3.4.e.3, 4), but did not require identification of a specific use for mined land after operations cease. In the case of mining, it is not possible to maximize efficiency and effectiveness of operations during movement of large volumes of soil and overburden unless the mine operator knows in advance the specific postmining land use for a site. Simply knowing the postmining land use tells the mine operator where to put the rock and dirt so most of it doesn't have to be moved a second time, at additional expense.

For example, a visit to PT Bukit Asam's Ombilin Mine in West Sumatera in July 1996 afforded an opportunity to discuss site specific environmental issues and compliance with AMDAL requirements with both a mine inspector and an AMDAL reviewer. The BLT-OSM Project Director was accompanied by a mine inspector from the Kanwil regional office of West Sumatera and the Head of the BLT Division of Environmental Management and Spatial Zoning. Acid mine drainage was observed discharging from the portal of this state-owned underground mine, but no overburden analyses or streamflow studies were available to determine if the mine was being recharged from stream water loss. Neutralization and settling of acid mine drainage before its discharge into a stream was not apparent. This was both an enforcement and a permitting issue. This information had not been included in AMDAL documents for the mine. For inspection and enforcement to be effective, standards for performance must be included in the ANDAL, Environmental Management Plan (RKL) and the Environmental Monitoring Plan (RPL), but they were not.

A large surface coal mine nearby run by PT Allied Indo Coal was also visited in July 1996 by this group. Here they found reclamation was not considered part of the mining process and was being done only on a small scale long after mining was completed. This increased costs and made attaining a reasonable postmining land use difficult. Erosion and sediment control practices were poorly designed or built to fit the available space rather than sized for surface runoff from any expected rainfall event. No serious effort to manage acid mine drainage was evident and there was little understanding of the geochemistry of the mine site apparent on the part of mining personnel. As with many mines in Indonesia, the ratio of disturbed land to reclaimed and unmined land at PT Allied Indo Coal was very high. Conditions at this mine illustrated the need for a postmining land use to be specified in either AMDAL documents or the Mining Authorization. Without a specific postmining land use, there was no reclamation goal. A specific postmining land use coupled with a reclamation plan would assist mine inspectors in evaluating a mine's environmental performance and would make the calculation of a realistic reclamation guarantee possible.

Supplementary guidelines issued in 2000 say the ANDAL should include a description of reclamation plans and planned land use after project activities are completed (BAPEDAL, 2000b, Attachment II.B.4.3.e.4(b), (c)). They also require discussion of plans for reclamation during continuing operations, which appears implicitly to require contemporaneous reclamation of disturbed land during mining (BAPEDAL, 2000b, Attachment II.B.4.3.e.3(b)). If implemented, these would constitute significant improvements in the contents of ANDAL. There are no parallels in Indonesian mining regulations. BAPEDAL guidelines for preparation of AMDAL documents provide much more detailed guidance than previously on what baseline information must be included in ANDAL documents (BAPEDAL, 2000b, Attachment II.B.5.3), comparable in many respects to what must be required in the United States. Information on climate, air quality, soils and geology, hydrology, topography, existing land uses, protected areas, biological and cultural features are identified with sufficient specificity to provide real guidance for both preparers and reviewers of AMDAL documents. Water quality data must be described in terms of applicable water quality standards, and the biological data required could only be obtained by field studies of the project site. These were significant improvements in the AMDAL process.

Water quality standards set by the Ministry of Environment for industrial activities were not well-related to mining operations but provided useful reference points for assessment of environmental impacts, inspection and enforcement activities. The Indonesian water discharge standard of 400mg/L suspended solids from mine sites, and 200mg/L from preparation plants is considerably less stringent than the U.S. one day maximum of 70mg/L (Ministry of Environment, 2003; *U.S. Code of Federal Regulations*, Title 40, §Part 434). BAPEDAL guidelines for preparation of AMDAL might be improved by requiring information be collected and presented about impacts going beyond a concern for preventing diseases to data concerning the physical and chemical composition of water discharges that might impact human health (e.g., pH, concentrations of iron and heavy metals like lead). Likewise, requirements for presentation of biological baseline data would be more useful if they explicitly required collection and analysis of information which might indicate the presence of unusually high concentrations of biodiversity in the study area.

Depth of Analysis

Guidelines established by the Ministry of Environment for preparation of the terms of reference for ANDAL did not require or encourage production of high quality documents, but opened the door for "quick and dirty" environmental assessments which provide little of use to decision makers and enforcement personnel. The scope of AMDAL studies was frequently not well defined, and the usefulness of resulting ANDAL in the decision making process has been questioned (Soemarwoto, 1996, p. 80). Statements suggesting there may be limitations on resources, time, funds and personnel for preparation of environmental assessments by project proponents were included in guidelines for preparation of KA-ANDAL, and justified an official policy of adjusting the objectives and outcomes expected in the ANDAL (Ministry of Environment, 1994e, Appendix I.4.b; BAPEDAL, 2000b, Attachment I.A.4.2), rather than a requirement proponents acquire and present adequate data for analysis. These statements are significant because they determine the content of a key document, the KA-ANDAL, which later determines the content of the ANDAL, Environmental Management Plan (RKL) and Environmental Monitoring Plan (RPL).

The scope of the ANDAL study area is to be determined in the KA-ANDAL by the totality of project boundaries within which a business will carry out activities; plus ecological boundaries within which environmental impacts are expected; plus social boundaries within which social interactions of community groups will undergo fundamental change; plus administrative boundaries of affected governmental units, concessions, mining authorizations or CoW. However consideration of administrative boundaries may be influenced by "technical constraints" concerning availability of funds, time and manpower (BAPEDAL, 2000b, Attachment I.A.8.2.4). This provision may allow the scope of environmental assessment to be limited to the boundaries of a mine concession, ignoring downstream water quality impacts on human health if it is expected to be expensive to study them.

Although this provision appears to allow consideration of offsite air and water quality issues, the guidelines say determination of the study area ultimately will be decided on the basis of the usual capabilities of the developer, based on resources available such as time, money, manpower, technology and research methods (BAPEDAL, 2000b, Attachment I.A.8.2.5). Ministry of Environment guidelines issued in 2000 for evaluation of KA-ANDAL unfortunately repeat the concern expressed in BAPEDAL guidelines for "technical constraints" concerning availability of funds, time and manpower in determining the scope of environmental assessment, thereby excusing inadequacies of data or analysis that might be found (2000a, Attachment

II.B.2.e). These statements allow the scope of the ANDAL study area and the quality of information developed for analysis to be determined by how much project proponents are willing to spend. Under these requirements, a mine operator – even a well-financed multinational corporation – need only cry poverty to avoid presenting adequate data for rational decision making.

Authorizing discretionary spending by the developer in the terms of reference for preparation of ANDAL may go a long way in explaining the widely-acknowledged poor quality of information actually presented in most AMDAL documents. For example, downstream degradation of water quality from black water discharges of fine coal particulates have not been described in any ANDAL prepared for a coal mining operation in Indonesia, despite their visible presence on stream and river banks downstream from most previously existing operations.

Mitigation Plans

Guidelines for preparation of the Environmental Management Plan (RKL) concerning mitigation of environmental impacts also suffer from the defect of too much deference for "technical constraints" concerning availability of funds, time and manpower (BAPEDAL, 2000b, Attachment III.B.3.6). In addition they state that the RKL is only required to give the basic direction, principles or requirements for prevention, mitigation and control of impacts because in the feasibility study phase, information on the proposed project is still of a relatively general nature (BAPEDAL, 2000b, Attachment III.A.2). This invites preparation of RKL lacking sufficient information to be useful to decision makers and inspectors, resulting in weak enforcement of environmental requirements. RKL and RPL documents prepared before 1993 were criticized as inadequate (Welles, 1995, p. 111). The 1994 guidelines appeared to require specification of applicable environmental performance standards and specific plans of the proponent for meeting these standards (Ministry of Environment, 1994e, Appendix III.B.II.3, 4). These guidelines were strengthened in 2000 by a specific requirement that impact criteria be described based on environmental quality standards in existing regulations (BAPEDAL, 2000b, Attachment III.B.3.2 and 3).

A requirement for identification of planned locations of any environmental management activities with maps, sketches, and drawings "as far as possible" (BAPEDAL, 2000b, Attachment III.B.3.5) appeared to apply to facilities designed to keep surface water quality impacts on site, including locations of perimeter ditches, settlement ponds, topsoil storage areas, and preparation plants for coal mining operations. These requirements were rarely (if ever) fulfilled in approved RKL documents. Maps, sketches and drawings provided in RKL have typically been of such large scale as to be virtually worthless in determining the adequacy of facilities to control environmental effects of mining operations, and were therefore of little enforcement utility to mine inspectors. More recent guidelines for preparation of environmental assessments have required presentation of maps of appropriate scale including alternative project locations, layout of buildings or facilities and production process technology, without specifying what scale they might be in any case (BAPEDAL, 2000b, Attachments II.B.4.3; III.B.3.5). This is certainly more appropriate than requirements for larger-scale maps provided in applications for Mining Authorizations under the Basic Mining Law of 1967 and its implementing regulations, discussed previously, but it remains to be seen whether these provisions will be interpreted by the various AMDAL Commissions in a manner that actually produces maps useful for evaluating and managing environmental impacts. Some authority must specify what "appropriate scale" means, and this has not yet reached specification of a requirement for maps at the necessary scale of 1:500.

Any developer who can afford to capitalize a multi-million dollar coal mine development can afford to provide detailed information about the environmental costs of a project in an environmental assessment, and specific plans and commitments about how to mitigate those impacts in an RKL prepared at the same time. Moreover any developer who cannot estimate the environmental costs of mitigation measures and land reclamation when completing a feasibility study to determine the technical and financial viability of a coal mining operation does not have adequate information to make a sound investment decision. These costs are predictable and not difficult to calculate when the operating capabilities of expected equipment such as for coal washing plants are known. Operating parameters of equipment determine the cost of such equipment, which in turn affects the financial feasibility of planned operations. Consequently, this information must be known by developers during studies of financial feasibility for proposed development projects if they are to make sound investment decisions. Environmental costs are routinely estimated in the United States during calculation of reclamation performance bonds before permits authorizing coal mining operations are issued.

It appears Indonesian policy makers bought the specious argument often advanced by developers in the United States that they cannot afford to acquire or generate this information so early in the development process. In all probability, either they have the necessary information and don't wish to share it, or they simply plan to avoid spending much money on environmental mitigation and reclamation, and do not wish to reveal it is financially feasible to do so. The risk of making a losing investment decision compels responsible developers of major projects such as coal mines to acquire this information before they commit major capital resources to a project.

Guidelines for preparation of the Environmental Monitoring Plan (RPL) compounded these defects by specifying that only significant impacts identified in the ANDAL for which management plans have been described in the Environmental Management Plan (RKL) will be monitored, and stating that environmental monitoring shall be economically feasible (Ministry of Environment, 1994e, Appendix IV.A.2.b, d). These guidelines contained no provision for amending the RPL or monitoring unexpected but significant impacts discovered only after operations begin (e.g., saline ground water, or toxic overburden strata which may generate acid mine drainage), which – because unexpected – were not included in the ANDAL. Although the latter statement about economic feasibility seems to state only the obvious, in this context it invokes previous statements about "technical constraints" of funding, time and manpower on efforts by the developer and therefore allows the developer discretion to determine what is economically feasible.

Thus in guidelines governing the AMDAL process, a sequence of requirements appear to work to the advantage of any developer who might wish to avoid expenditures on environmental measures, and to the disadvantage of government efforts to control environmental effects of large development projects. The ANDAL, RKL and RPL must all be prepared by the proponent on the basis of what is described in the KA-ANDAL (Republic of Indonesia, 1999c, Article 17). Lax requirements concerning what must be included in the KA-ANDAL result in poor data and analysis in the ANDAL, which produce vague plans for environmental management in the RKL, and no figures for environmental costs in the feasibility study. Lack of specificity in the RKL provides little or no basis for monitoring in the RPL, or for enforcement of environmental requirements during inspections. When demands are made by AMDAL Commissions for a mine operator to spend funds controlling environmental impacts, lack of cost estimates for environmental management and reclamation in the feasibility study allow the mine operator to claim it cannot afford them because such expenditures were not taken into consideration during preparation of the feasibility study, were not considered during the initial investment decision, and will wreck the financial viability of the mining operation. These claims may be false, but through poor policy and development of regulations, the government has placed itself in a weak position to refute them, or to enforce environmental requirements on mining operations.

BAPEDAL guidelines for preparation of RKL and RPL also contain troubling provisions requiring identification of organizations that will be responsible for implementation and funding of environmental management for a project (BAPEDAL, 2000b, Attachment III.B.3.8). These provisions suggest the developer may be allowed to assign these responsibilities to another organization, but provide no clear statement that the developer remains ultimately responsible for their performance. Some clarification of this responsibility seems desirable.

Reporting Monitoring Results

In guidelines for preparation of RPL concerning monitoring, there are no requirements setting the frequency with which samples must be taken or inspections must occur (BAPEDAL, 2000b, Attachment IV.B.2.5.c). Provisions for reporting the results of environmental monitoring likewise fail to specify any required reporting periods (Republic of Indonesia, 1999c, Article 32). Historically, the frequency of environmental monitoring and reporting in Indonesia has been irregular at best. Inspections of mines have been conducted once or twice per year, and monitoring data received less frequently than that. Consequently, government environmental agencies with jurisdiction over mining operations conducted in frontier conditions often have been unaware of significant violations of their standards for considerable periods.

Who Prepares Analysis?

Unlike NEPA, which requires environmental assessments be prepared by the responsible government official, regulations issued pursuant to the Acts of 1982 and 1997 require the KA, ANDAL, RPL and RKL all be prepared by the project developer at their own expense (whether government agency or private firm), for review and approval or rejection by the appropriate AMDAL Commission (Republic of Indonesia, 1999c, Article 8). Because of this, the AMDAL process has been described as "basically a self-assessment procedure" (Warren and Elston, 1994, p. 23), producing documents which must be reviewed by government agencies, but for which the agency is not directly responsible.

Environmental Feasibility

Also unlike NEPA, development proposals can be rejected by an AMDAL Commission as not environmentally feasible if they appear to be excessively disruptive to the environment and the impacts cannot be mitigated, or if mitigation costs are higher than benefits (Republic of Indonesia, 1999c, Article 22). Previously, if a project was rejected, that decision might be appealed to an authority superior to the authorized government agency (Republic of Indonesia, 1993, Article 11 (2)), who could approve the proposal. Because the decision of the authorized government agency was generally made by a sectoral Minister, the only superior authority to whom an appeal might be taken was the President. This opened the appeal to political influence at the highest level, in a manner which might be less likely to occur if the only appeal of an administrative decision was to the courts on purely legal grounds. Presidential Regulation 27/1999, which replaced previous regulations on this issue did not include this or any explicit avenue of appeal, thereby strengthening the regulatory authority of AMDAL Commissions.

Environmental assessments are prepared by a government agency in the U.S., but by the proponent of development in Indonesia. Such documents are reviewed by citizens and other agencies in the U.S., but by an AMDAL Commission in Indonesia. Because AMDAL documents are prepared by the project proponent for review and approval or rejection by a government agency in Indonesia, the Acts of 1982 and 1997 were designed as regulatory statutes, not as attempts at administrative reform of government decision making, as they were in the United States.

Time Limits and Automatic Approval

Presidential Regulation 51/1993 reduced the time allotted for review of AMDAL documents by sectoral Ministries from 90 days to 45 days (Warren and Elston, 1994, p. 25). This 45 day time limit included the time it took to distribute the document to members of the AMDAL Commission and any technical review they conducted. The review period for Commission members effectively became 21 days because their comments must be received and consolidated before the Commission could meet for discussion and decision. This short time seriously limited their substantive review. The pro-development bias of this policy was further evident in the fact failure to meet this deadline resulted in automatic approval of the documents (Republic of Indonesia, 1993, Article 10 (3)). Presidential Regulation 27/1999 increased the period of agency review to 75 days prior to decision, an improvement, but retained automatic approval if no decision is made within that period (Republic of Indonesia, 1999c, Article 20).

Moreover Presidential Regulation 27/1999 added a provision for automatic approval of the KA-ANDAL, which must be prepared by the proponent and presented to an AMDAL Commission for review. If the terms of reference are not approved within 75 days, they are assumed accepted (Republic of Indonesia, 1999c, Article 16). These provisions further weaken the utility of the KA-ANDAL and reduce AMDAL Commission review of AMDAL documents from an analytical function to a clerical function, accentuating difficulties described previously. More recent regulations required AMDAL Commissions and their technical teams be given a minimum of ten days to review KA-ANDAL and other AMDAL documents (Ministry of Environment, 2000b, Articles 11, 12, 15, 16). Ten days is not adequate to allow a site visit or complete any thoughtful review by persons who have other obligations, and participate on these teams and commissions as temporary duty of an *ad hoc* nature.

The setting of time limits for review of complex, environmentally disruptive projects, and automatic approval of a proposal at the end of some period regardless of whether an adequate review of environmental impacts has been completed, appears to be inconsistent with the policy that environmental values be taken into consideration along with other values. These practices allow environmentally destructive projects to be approved without complete evaluation or specification of mitigation measures, and put the responsible agency under great pressure to approve every project, regardless of the severity of its impact on the environment. To be certain environmental values are given adequate consideration along with other values, the terms of reference for preparation of environmental impact analysis documents should be based on stated requirements determined by the government agency, not negotiated on a case-by-case basis with proponents. Any negotiations should concern how stated requirements will be met by a proponent, not what the requirements will be. An AMDAL Commission should be authorized to reject the KA-ANDAL of a proponent as unacceptable for failing to meet stated requirements. As it stands, a proponent need only refuse to agree to changes in its proposed terms of reference and wait the required 75 days for automatic approval.

Decentralization of AMDAL Responsibilities

Provincial and local governments received greatly increased responsibility for environmental management pursuant to Law No. 22 on Regional Administration enacted in 1999 (Republic of Indonesia, 1999b, Article 7) and

Presidential Regulation No. 25 of 2000 on the Authority of the Government and the Authority of a Province as an Autonomous Region (Republic of Indonesia, 2000). General authority was given to regional governments for control over the environment (Republic of Indonesia, 2000, Article 3(2)). In the environmental sector, provinces were specifically authorized to exercise control over activities affecting the environment crossing regency and municipal boundaries, to supervise environmental impact analysis of activities with locations encompassing more than one regency or municipality, and development of environmental quality standards based on national standards (Republic of Indonesia, 2000, Article 3(5) 16). Regencies and municipalities were authorized to regulate activities affecting the environment solely within their boundaries.

Broad authority was reserved to the national government to set environmental quality standards; guidelines for environmental pollution control; guidelines for conservation of natural resources; and to supervise environmental impact analysis of activities affecting the nation as a whole, development locations encompassing portions of more than one province, and those in the territorial seas within 12 miles of land (Republic of Indonesia, 2000, Article 2(3) 18). Thus the national government was to set standards and guidelines, which were to be implemented by regional governments.

On the same day these decentralization laws were enacted by the Indonesian House of Representatives, President Habibie signed a new Presidential Regulation No. 27/1999 on Environmental Impact Analysis which dramatically changed the AMDAL review process. AMDAL Commissions in the various sectoral departments such as the Ministry of Mines and Energy were abolished and their responsibilities decentralized to regional government AMDAL Commissions. Review of national projects, those with locations encompassing more than one province, and those in the territorial seas within 12 miles of land transferred to a single national AMDAL Commission located in BAPEDAL, the national environmental impact management agency (Republic of Indonesia, 1999c).

Public Disclosure

Presidential Regulation 27/1999 increased transparency of the AMDAL process and required citizen and community input early in the scoping procedure for development of terms of reference for preparation of ANDAL, as well as during evaluation of the ANDAL, RPL, and RKL. Local knowledge of cultural and environmental conditions at a proposed minesite can be valuable in assessing environmental impacts and identifying an appropriate post-mining land use, but the quality of AMDAL documents in these areas was often poor. Previous regulations did not provide for direct public participation in the AMDAL process and were criticized as inadequate (Welles, 1995, p. 111). Presidential Regulation 27/1999 provided for representation of environmental organizations and affected communities on the various AMDAL Commissions, membership of which now routinely exceeds 20 persons (Republic of Indonesia, 1999c, Articles 9, 10, 33, 34). As is the case with NEPA, all

AMDAL documents and decisions must be disclosed to the public and citizen input is required to be invited before decision.

Presidential Regulation 27/1999 and the two decentralization laws of 1999 necessitated substantial revision and restatement of previous guidance issued by the Ministry of Environment and BAPEDAL concerning preparation and review of AMDAL documents. In February 2000, the Ministry of Environment issued several decrees providing guidelines for evaluation of AMDAL documents (2000a), establishment of regional and municipal AMDAL Commissions (2000b), membership composition and work procedures for the various AMDAL Commissions (2000c) and technical teams set up to assist them (2000d). On the same day, BAPEDAL issued decrees providing new procedures for community involvement in all portions of the AMDAL process (BAPEDAL, 2000a), and new guidelines for preparation of AMDAL documents (BAPEDAL, 2000b) which supplement but do not replace those issued by the Minister of Environment in 1994 (Ministry of Environment, 1994e). Previous decrees and regulations of BAPEDAL and the Ministry of Environment did not require community involvement or public participation in the AMDAL process, so this was a significant improvement. Later that year the Ministry of Energy and Mineral Resources issued a decree formally decentralizing approval of AMDAL documents for mining operations to regional governments (Ministry of Energy and Mineral Resources, 2000, Articles 4, 5).

Guidelines specifying the membership composition and work procedures for regional and municipal AMDAL Commissions provide detailed procedural rules for meetings, some including specified font size for announcements published in print media. However neither they nor guidance concerning preparation and review of AMDAL documents requires that meetings be held near a location proposed for development, and none of them specifically provide travel expenses for representatives of affected communities or nongovernmental organizations to attend such meetings. In many cases, whether such representatives could afford to pay their own expenses is questionable. Public notice and comment on the Terms of Reference for mining ANDAL should provide the proponent and the regulatory authority with valuable information, and give citizens most affected by mining operations a sense that the process is responsive to their views. Public involvement during preparation of Terms of Reference should help resolve technical and policy issues before mining begins.

Effective public participation requires public knowledge of the proposed Terms of Reference. In the United States, an applicant for a permit must submit a copy of the complete application to a local government clerk, and publish a summary of the application in a local newspaper several times. Any person may submit written objections to the regulatory authority, or request an informal conference. Similar provisions might be useful in the Indonesian regulatory program, particularly for citizens residing in towns or cities. However publication of written notices as required in the United States might not be sufficient to inform citizens in remote Indonesian villages. It might be more effective to hold meetings in each village that would be likely to be significantly affected by a mining operation. At such meetings, representatives of the proponent, the regulatory authority, and local or regional government officials could explain the project, ask for the citizens' opinions, and answer questions. The regulatory authority could require the proponent to address all significant comments, either at meetings or in subsequent written submissions.

Implementation of Environmental Assessment Policy

Conceptually, the AMDAL process is sound, but it has encountered numerous difficulties during implementation. Preparation of AMDAL documents continues to follow other approvals, and this applies pressure on review staff to approve the mine operator's environmental assessment. It is rare that field visits are made to proposed mining sites before mining actually commences to determine the accuracy of data submitted in AMDAL documents. The lack of accountability in payment of honoraria to technical teams and decision makers may have facilitated approval of some grossly inadequate KA-ANDAL and ANDAL.

Usefulness to decision makers   The AMDAL process is not a strong decision making tool used by the Ministry to allow decision makers to fully consider environmental factors in making a decision on mining activity or determining whether a proposed mining activity should or should not take place. There are a number of reasons for this inadequacy. Review of several AMDAL documents for coal mining operations found them lacking in detail and specificity in the areas of baseline data on flora and fauna, mine design, cut sequence, design of waste rock disposal dumps, water and sediment control, overburden chemistry, special handling of potentially acid forming materials, facilities locations, cleaning plant technology and design features to assure maximizing production of coal resources and revenues while protecting water quality and the environment (e.g., see PT Arutmin Indonesia, 1994). This information is critical if an AMDAL Commission is to analyze the probable environmental effects of a mining operation and make an informed decision about its ability to meet environmental performance standards.

The feasibility study, which frames the initial decision on whether to proceed with mining continues to be prepared before the ANDAL, RKL and RPL. It does not include reclamation costs or costs of environmental protection practices needed by the AMDAL Commission to determine environmental feasibility. The terms of reference (KA) for an AMDAL study are proposed by the proponent rather than being required in a standard format by the appropriate AMDAL Commission. The government has 75 days to review, comment, and request revision or the KA-ANDAL is approved by default. After a KA-ANDAL is approved it is binding, and the AMDAL Commission is generally reluctant to ask for additional information.

The current ANDAL, RKL and RPL contain too little information for a competent technical review to be made of the environmental effects of a proposal. Maps are of too large a scale to be useful for adequate environmental assessments and are not always tied to established benchmarks of longitude

Figure 4.1 Extensive erosion of unvegetated, unconsolidated mine spoil in unstable out-of-pit dump, with active pit in background, South Sumatera
Source: Alfred E. Whitehouse

and latitude. Measurable performance standards for environmental controls or reclamation during mining are not specifically described in AMDAL documents. Postmining land uses which utilize infrastructure constructed by mining companies and which are sensitive to overlapping interests in the land (e.g., forestry, agriculture) need a premining description to insure reclamation is part of the mining process and not an expensive after-thought. A postmining land use sets environmental performance and reclamation objectives against which performance can be measured during mining. Yet there are no requirements for description of a specific postmining land use or mine closure plan in AMDAL documents. Often they are not specified before mining operations actually begin, but are presented to the Ministry in an annual update of the RKL a year or two before the closure of a mine.

Evaluating Environmental Decision-making Procedures

Before evaluating existing regulatory processes it is necessary to specify a set of criteria for examining the quality of procedures used by decision makers in selecting a course of action. The literature on effective decision making provides criteria that can be used to determine whether decision making procedures are of high quality. The decision maker, to the best of ones ability and within ones information processing capabilities:

(1) thoroughly canvasses a wide range of alternative courses of action;
(2) surveys the full range of objectives to be fulfilled and the values implicated by the choice;
(3) carefully weighs what everyone knows about the costs and risks of negative consequences, as well as the positive consequences, that could flow from each alternative;
(4) intensively searches for new information relevant to further evaluation of the alternatives;
(5) correctly assimilates and takes account of any new information or expert judgment to which one is exposed, even when the information or judgment does not support the course of action one initially prefers;
(6) reexamines the positive and negative consequences of all known alternatives, including those originally regarded as unacceptable, before making a final choice;
(7) makes detailed provisions for implementing or executing the chosen course of action, with special attention to contingency plans that might be required if various known risks were to materialize (Janis and Mann, 1977, p. 11).

A decision making procedure that meets all seven criteria is characterized as vigilant information processing. Failure to meet any of the seven criteria when making a fundamental decision (e.g., one with major consequences for attaining or failing to attain important values) constitutes a defect in the decision making process.

Vigilant information processing is not absolute, but a matter of degree. Each of the seven criteria may be conceived as forming a scale with ratings varying from zero to, say, ten. A decision making procedure characterized as lacking vigilant information processing would be one rated low on several of the seven criteria by an objective observer. Fairly high vigilance would be characteristic of a procedure scoring high on all the criteria but one.

Use of these criteria assumes that the more adequately each criterion is met, the lower the probability that a decision maker will make serious miscalculations that jeopardize immediate objectives and long-term values (Janis and Mann, 1977, p. 11). This does not suggest each and every decision should be preceded by striving for the highest possible score on all seven criteria. Obsessional mulling over the uncertainties of a major decision, and preoccupation with the search for an ideal choice may result in information overload, where the decision maker fails to appreciate the most important factors that need to be taken into account (Janis and Mann, 1977, pp. 11–13; Miller and Starr, 1967, p. 62).

Vigilant information processing does not require a process of optimization. Selecting an optimal course of action would involve a systematic comparison of every viable alternative decision in terms of all expected benefits and costs (Young, 1966, pp. 138–47). This would entail several steps: (1) identification and definition of the stimulus or problem requiring decision; (2) identification of all relevant objectives or values to be served; (3) identification of all possible alternative decisions or means to attain the desired objectives; (4) identification and weighing of all the benefits and costs of each alternative; and (5) a systematic comparison of the alternatives, assessing the probabilities for success and costs for each, resulting in selection of the single alternative with the optimum combination of pros and cons, in terms of attaining desired objectives.

Such a comprehensive information search and appraisal has been characterized as an ideal type of decision making which might take place only in a hypothetical decision making context which bears little resemblance to the reality facing most decision makers (Simon, 1960, pp. 64–181). Meticulous performance of an optimizing strategy would impose a tremendous burden for information gathering and analysis on the decision maker: to survey the universe of objectives to be served, values to be considered, alternative courses of action which might be chosen, and all of their respective anticipated benefits and costs (Braybrooke and Lindblom, 1963, pp. 43–50; Lindblom, 1965, pp. 137–39; Etzioni, 1967, pp. 385–86). Where complex issues arise, rarely does a decision maker have the opportunity or capability to adequately clarify objectives and values which may be many and conflicting, ambiguously stated, or undetermined (Lindblom, 1959, pp. 81–82).

Collecting, organizing and comparatively assessing the information required in turn would require substantial expenditure of time, effort and money – resources which are often in short supply. The pressure of deadlines for final decisions may limit the search for, or analysis of information. Funding for needed staff assistance, consultation or computer time may be limited, thereby reducing the scope of information search or rigor of analysis. While an optimizing strategy might approach a theoretical ideal for vigilant information processing, as a practical matter it would generally be too time consuming. Consequently, decisions are often of a suboptimal nature, maximizing some benefits at the expense of others (Young, 1966, p. 144).

For complex decisions, persons attempting to optimize may exceed limitations of the human mind for perceiving and processing information. Determining all the potentially favorable and unfavorable consequences of all feasible courses of action would require a decision maker to process so much information that impossible demands would be made on resources and mental capabilities (Janis and Mann, 1977, p. 22). Decision makers rarely optimize "because they have not the wits to maximize" (Simon, 1976, p. xxxvii). This is not to say human beings are inherently stupid, but that the complexity and time dimensions of some decisions exceed the capacity of the human mind to process information. The human mind cannot at a single moment grasp the consequences in their entirety. Instead, attention shifts from one value to another with consequent shifts in preference (Simon, 1976, p. 83). Information overload and fatigue decrease the adequacy of information processing when complexity of a decision exceeds the limits of cognitive abilities (Janis and Mann, 1977, p. 17).

Decision makers may not optimize but instead "satisfice," or look for a course of action that is "good enough" (Simon, 1976, p. xxix). When such a course of action is detected, identification of alternatives and the search for new information terminates. Only a small number of alternatives may be identified.

But even if a large number of alternatives are eventually considered prior to decision, they are examined sequentially as they arise, with no attempt to make a systematic comparison of benefits and costs of each to the others. Reliance on a simple decision rule (e.g., "good enough") risks the mistake of premature decision, which may overlook negative consequences that are not obvious (George, 1974). One result of satisficing is a more superficial information search and analysis, and therefore a lack of vigilant information processing, in terms of criteria #3 through #6 above. Not all consequences of each alternative are carefully weighed. The search for new information may be terminated before all alternatives are identified or their consequences known or understood. Consequences of all alternatives are not reexamined in light of new information before a final decision is made.

A series of satisficing policy choices might result in incremental improvements, or slow progress towards an optimal course of action (Miller and Starr, 1967, p. 51). This result is not inevitable, but is possible. However, policy change is usually incremental because decision makers in large organizations tend to consider only a few decision alternatives, which differ only slightly from existing policy (Lindblom, 1965, p. 146). Decision makers may be attracted to the incremental approach because it allows them to avoid difficult cognitive tasks: Examination of business decision making and governmental policy making suggests that, whenever possible, decision makers avoid uncertainty and the necessity of weighing and combining information or trading-off conflicting values (Slovic, 1971, p. 41). Cognitive complexity in issues requiring decision generates psychological tension or stress, especially under conditions of inadequate knowledge and conflicting information (George, 1974, p. 187). Uncertainty and the fear of making a serious error may trigger defensive avoidance behavior to alleviate stress (Janis and Mann, 1977, p. 560; Benveniste, 1977, pp. xvi–xix) – behavior which is often dysfunctional for the organization and the individual.

Information search and assessment stages of incremental decision making are simplified by limiting alternatives to those similar to what is familiar (e.g., to what has been done in the past), and weighing them successively rather than simultaneously. Identification of alternatives is focused to respond to particular stimuli – specific short-comings in existing policy – and no attempt is made to select the best course of action from a comprehensive set of alternatives. Overall or long term goals may be ignored in favor of short-term or single administrative unit objectives (or putting out brush fires), in decision making through "successive limited comparisons" of alternatives without systematic comparison of the benefits and costs of each to all others (Lindblom, 1959, p. 81).

Incremental decision making scores low in terms of criteria #1–5 above for vigilant information processing. Not all possible alternative decisions may be identified, and some will receive less consideration than others. Less than the full range of objectives and values to be fulfilled may be considered. More reliance may be placed on what is already known to have worked in the past, than on searching for new information. Consequences of some alternatives may therefore remain unanticipated or poorly understood. Decisions made on the basis of simple decision rules, which minimize cognitive complexity and alleviate stress induced by fear of error, may overlook nonobvious negative consequences of the chosen decision alternative and thereby result in error.

Thus incremental decision making is a pragmatic adaptation to the shifting compromises and coalitions of bureaucratic politics in a pluralistic society, serving to sustain the organization and facilitate its efforts toward policy implementation. Members of organizational coalitions in government include administrators, workers, appointed and elected executives, legislators, judges, clientele, interest group leaders and other influential individuals (Cyert and March, 1963, p. 27). Thus, agency coalitions include internal constituencies comprised of other agency employees – subordinates, superiors and colleagues – as well as external groups which make claims on the decision maker. The decision maker is obliged to produce a decision which will be acceptable to ones superiors in the agency, to legal authorities, and to those persons who will share in its implementation – and to avoid stirring up employee or constituent resistance to the decision (Allison, 1971).

Some incremental changes in policy may therefore be made primarily to keep some politically influential groups in an organization sufficiently satisfied that they will not complain or obstruct movement in a new policy direction (Halperin, 1974, pp. 99–157). Such compromises are based largely on the criterion of agreement, rather than on the actual values embodied in the issue at controversy. There seems to be little room in the concept of incrementalism for leadership initiative. Central issues may therefore get lost in the search for sufficient agreement to support a decision which will maintain the organizational coalition, and existing relationships between the agency and its "negotiated environment" (Cyert and March, 1963, pp. 119–20).

Incremental decision making may serve democratic values in a pluralistic society by avoiding some of the pitfalls of centralized decision making, like forced, drastic social changes (Braybrooke and Lindblom, 1963). But this approach may also stifle policy innovation and needed social change (Dror, 1964, p. 155), reflects primarily the interests of the most powerful groups, and under represents those of underprivileged and politically weak or unorganized people (Etzioni, 1967, p. 387). On the one hand, incremental decision making based on a succession of satisficing choices can have functional value for decision makers who wish to avoid the risks of drastic social change, but on the other hand, it may slow needed social change, and "there is the danger that it can prove to be a zigzag passage to unanticipated disaster" (Janis and Mann, 1977, p. 34).

In a synthesis of optimizing and incremental decision making strategies called mixed scanning, fundamental policy decisions which set basic programmatic directions require information search and analysis similar to but less demanding than optimization, while minor decisions made within this policy context require only an incremental strategy based on simple forms of satisficing (Etzioni, 1968, p. 294). Resulting incremental changes may then lay the groundwork for new fundamental policy decisions through active monitoring and review of any information acquired by the decision maker while implementing policy.

The term "scanning" refers to information search, gathering and analysis, which are the principal cognitive activities required in vigilant information processing. Each choice between alternatives requires a judgment by the decision maker as to how much resources (e.g., time, energy, money) one is willing to allocate to scanning activities, based on the following rules:

> a. On strategic occasions ... (i) list all relevant alternatives that come to mind, that the staff raises, and that advisors advocate (including alternatives not usually considered feasible). (ii) Examine briefly the alternatives ... and reject those that reveal a "crippling objection." These include (a) utilitarian objections to alternatives which require means that are not available, (b) normative objections to alternatives which violate the basic values of the decision makers, and (c) political objections to alternatives which violate the basic values or interests of other actors whose support seems essential for making the decision and/or implementing it. (iii) For all alternatives not rejected under (ii), repeat (ii) in greater though not full detail ... (iv) For those alternatives remaining after (iii), repeat (ii) in still fuller detail ... Continue until only one alternative is left ...
> b. Before implementation (in order to prepare for subsequent "incrementing"): (i) when possible, fragment the implementation into several sequential steps ... (ii) When possible, divide the commitment of assets into several serial steps ... (iii) When possible, divide the commitment of assets into several serial steps and maintain a strategic reserve ... (iv) Arrange implementation in such a way that, if possible, costly and less reversible decisions will appear later in the process than those which are more reversible and less costly. (v) Provide a time schedule for the additional collection and processing of information ...
> c. Review while implementing: (i) Scan on a semi-encompassing level after the first sub-set of increments is implemented. If they "work," continue to scan on a semi-encompassing level after longer intervals and in full, over-all review, still less frequently. (ii) Scan more encompassingly whenever a series of increments, although each one seems a step in the right direction, results in deeper difficulties. (iii) Be sure to scan at set intervals in full, over-all review even if everything seems all right ...
> Formulate a rule for the allocation of assets and time among the various levels of scanning ... (Etzioni, 1968, pp. 296–98).

If we assume a progressively more stringent standard of assessment and adequate, competent staff, mixed scanning scores high on all seven of the criteria for vigilant information processing. It explicitly entails a thorough scanning of alternatives, taking into account utilitarian, normative and political values, as well as alternatives not usually considered feasible. An intensive search is made for new information and detailed provisions are made to monitor implementation and make adaptive changes as needed. In order to accomplish these activities conscientiously, the decision maker would find it necessary to carefully weigh the consequences of each alternative, thoroughly assimilate new information, and reexamine consequences before making a final decision.

Of primary concern here is specification of a set of criteria for examining the quality of procedures used by decision makers in selecting a course of action as related particularly to surface mining activities. Particular attention is directed to formal statutory procedures or rules providing guidance to decision makers concerning the scope and detail of information search and appraisal activities required during mining environmental policy implementation. A set of decision criteria has been specified and four approaches to decision making from the literature have been characterized in terms of the criteria. The criteria are realistic and practical. It is not necessary to approximate an ideal decision making context or to adopt an optimizing strategy for a procedure to rate fairly high in terms of vigilant information processing. Satisficing or incremental decision making strategies are found to be defective procedures, rating low in terms of several of the specified criteria. Mixed scanning scores high on all seven criteria, but is by no means the only procedure which could conceivably do so. A brief application of these criteria to procedures required by the National Environmental Policy Act is provided below, for illustrative purposes.

NEPA and Vigilant Information Processing

The National Environmental Policy Act requires that officials in federal agencies: "utilize a systematic, interdisciplinary approach which will insure the integrated use of the natural and social sciences and the environmental design areas in planning and decision making which may have an impact on man's environment" (42 U.S. Code, §4332(2), 1978). The statute requires federal decision makers to survey a broad range of objectives and values prior to selecting a final decision, to insure that environmental values will be given due consideration alongside economic and technical values.

A nonexhaustive list of values and objectives to be considered by federal decision makers may be extracted from the statute and its judicial interpretation:

- productive and enjoyable harmony between man and his environment;
- safe, healthful, and esthetically and culturally pleasing surroundings;
- prevention or elimination of damage to the environment;
- understanding of ecological systems and natural resources important to the Nation;
- fulfill the social, economic and other requirements of present and future generations;
- improve and coordinate national government plans, functions, programs and resources;
- attain the widest range of beneficial uses of the environment without degradation, risk to health or safety, or other undesirable and unintended consequences;
- preserve important historic, cultural and natural aspects of our national heritage;
- an environment which supports diversity and variety of individual choice;
- achieve a balance between population and resource use which will permit high standards of living and a wide sharing of life's amenities;
- enhance the quality of renewable resources;

- maximum attainable recycling of depletable resources;
- other essential considerations of national policy (42 *U.S. Code*, §§4331, 4335, 1978).

Some of these are values and objectives are so general and ambiguous as to elicit near universal support, and little action. Many of them overlap or shade into others. Some of them are quite concrete. Cumulatively, the requirement that they be considered in agency decision making substantially expanded the scope of choices implied in organic mission statements of most national government agencies, while preserving traditional missions intact. This requirement – when observed – has increased the likelihood that decision makers will survey a fuller range of objectives to be fulfilled and values implicated by the decision than simply those expressed in basic mission statutes. Therefore, the decision making procedure specified in the National Environmental Policy Act rates high on criteria #2 for vigilant information processing.

NEPA and implementing regulations issued by the Council on Environmental Quality distinguish between major and minor decisions, requiring preparation of a detailed statement by the responsible official only for each "major federal action significantly affecting the quality of the human environment" (42 *U.S. Code*, §4332(2) (C), 1978). Preparation of this document entails gathering information necessary to complete an assessment of environmental effects of the proposed action and *bona fide* alternative actions prior to deciding which action (if any) to take. NEPA therefore rates high on criteria #1: canvassing a wide range of alternatives; criteria #3: assessing the costs and risks of negative consequences of each alternative; and criteria #4: intensive search for new information relevant to evaluation of alternatives.

The responsible government official must obtain comments from other agencies with special expertise and include them – with appropriate agency responses – in the environmental statement. (40 *U.S. Code of Federal Regulations*, §1502.9(b)), must evaluate any new alternatives thus identified (40 *U.S. Code of Federal Regulations*, §1501.7(c)), and may modify the proposed action as appropriate before making a final decision (40 *U.S. Code of Federal Regulations*, §1503.4). Criteria #6 for vigilant information processing, requiring reexamination of the positive and negative consequences of all known alternatives before final decision, is served by this provision of NEPA as well as the foregoing ones.

Regulations issued pursuant to the National Environmental Policy Act, both by the Council on Environmental Quality and by individual agencies, provide that measures to mitigate adverse environmental effects of a decision must be discussed in the statement, and appropriate monitoring and enforcement programs must be established (40 *U.S. Code of Federal Regulations*, §1505.2(c)). These may include attachment of conditions to permits or other approvals requiring implementation of mitigation plans (40 *U.S. Code of Federal Regulations*, §1505.3). Such requirements serve criteria #7: making detailed provisions for implementing the chosen course of action, with special attention to contingency plans that might be required if various known risks materialize.

In the first few years after its enactment, language of the statute and initial, non-binding guidelines for its implementation issued by the Council on Environmental Quality (U.S. CEQ, 1973) were often, perhaps erroneously interpreted as imposing an impossible burden of information gathering and analysis on federal agencies. One result was considerable litigation over the timing and content of environmental impact statements (Anderson, 1973). But only a few agencies were involved in a large number of court suits. After several years of experience, court interpretation, study and issuance of binding regulations for preparation of environmental statements by the Council on Environmental Quality, (40 *U.S. Code of Federal Regulations*, §1500 et seq.) it is apparent this statute does not require an ideal or optimizing decision making strategy, as is often supposed, but a strategy similar in many respects to mixed scanning.

This is most evident in the regulations, which require an early environmental assessment to determine whether it is necessary to expend the resources to prepare a full environmental statement; a scoping process to determine which issues are significant enough for inclusion in the statement and which can be excluded from detailed study; and a standard format for environmental statements which includes requirements for page limitations, consistent organization, and plain, nontechnical language (Bass and Warner, 1979). The standard format is intended to assist agencies in reducing resource commitments to preparation of documents, while providing adequate treatment of salient issues, and facilitating review and comment by other agencies and citizens. Some comprehensiveness is sacrificed for the sake of significance and usefulness of the final document, and savings in time, effort and money. But this is compensated by provision of mitigation programs to ameliorate negative consequences, and monitoring programs to provide opportunities for further corrective action if it becomes necessary. Provision for adaptive corrections after a decision is made and consequences arise is the principal distinguishing feature of the mixed scanning strategy.

Thus while it certainly does not incorporate vigilant information processing by direct reference, the National Environmental Policy Act does include procedural requirements and general references to goals and objectives sufficient to substantially meet the criteria for vigilant information processing. It may therefore be characterized as specifying a decision making procedure of high quality.

AMDAL and Vigilant Information Processing

The Act on Basic Provisions for Management of the Living Environment of 1982 as amended by the Act on Environmental Management of 1997 says every person has both a right to a good and healthy environment, and a duty to prevent and abate environmental damage and pollution, in order to achieve environmentally sustainable development (Republic of Indonesia, 1982, Article 5(1); 1997, Article 3; Article 5(1)). Government officials are required to perform environmental assessments "in an integrated manner with spatial management, protection of non-biological natural resources, protection of artificial resources, conservation of biological natural resources and their ecosystems, cultural preservation, biodiversity and climate change (Republic of Indonesia, 1997, Article 9(3)). Accomplishing these aims would appear to require utilization of a "systematic, interdisciplinary approach" comparable to that required in the United States (42 *U.S. Code*, §4332(2), 1978). The Act requires government decision makers to survey a broad range of objectives and values prior to final decision, to insure that environmental and human development values will be given due consideration alongside economic and technical values.

A nonexhaustive list of values and objectives to be considered by federal decision makers may be extracted from the Act:

- achievement of harmony and balance between humans and the environment;
- formation of the Indonesian person as an environmental being disposed toward and acting to protect and foster the environment;
- preservation of environmental functions;
- guaranteeing the interests of present and future generations;
- prudent control of the exploitation of resources;
- protection of the Unitary Indonesian Republic against impacts of business or activities outside the country which cause environmental pollution or damage (Republic of Indonesia, 1997, Article 4).

As in the United States, these values and objectives seem laudable and so ambiguous as to elicit near universal support. Yet some are quite concrete, and there are subtle differences between the two decision making processes. While NEPA sought to increase awareness of the environment, the AMDAL Act of 1997 would go further in the direction of human development by fostering positive and protective attitudes towards the environment, perhaps because it conferred a right to a healthy environment. Where NEPA called for reform of administrative decision making, the 1997 Act called for direct regulation of resource development. And where NEPA sought to maximize international cooperation in anticipating and preventing a decline in the quality of mankind's world environment (42 *U.S. Code* 4332 (2)(F)) almost as an afterthought, the 1997 Act called for protection of the nation from international sources of environmental harm, specifically including conservation of biodiversity, climate change and international trade in hazardous waste (Republic of Indonesia, 1997, Article 21).

Cumulatively, the requirement that these objectives be considered in government decision making substantially expanded the scope of choices implied in more limited mission statements of most government agencies to include environmental values, while preserving traditional missions intact. Despite the fact a clear cultural bias in favor of growth and development is apparent in both the U.S. and Indonesia, this requirement – when observed –

has increased the likelihood that decision makers will survey a fuller range of objectives to be fulfilled and values implicated by the decision than merely those expressed in basic mission statutes. Therefore, the decision making procedure specified in the Act of 1997 rates high on criteria #2 for vigilant information processing.

The 1997 Act and implementing regulations issued by BAPEDAL distinguish between major and minor decisions, requiring preparation of an environmental assessment for activities having "a large and important impact on the environment" (Republic of Indonesia, 1997, Article 15) . Although preparation of an ANDAL entails gathering information necessary to complete an assessment of environmental effects of the proposed action prior to approving it, consideration of alternative actions is not required. Moreover, detailed guidance concerning what information must be presented, requirements for field surveys of biological and cultural resources, maps of adequate scale, and standardized formats for the KA, ANDAL, RKL and RPL documents are all lacking. "Technical constraints" concerning availability of funds, time and manpower may determine the scope of environmental assessment and excuse inadequacies of data or analysis. Combined with specification of short time limits before automatic approval of proposals, high workloads and lack of expertise of staff preparing and reviewing AMDAL documents, this results in inadequate information and insufficient time and resources for decision makers to effectively evaluate the environmental impacts of development proposals in Indonesia. Therefore the Act of 1997 does not rate as high as NEPA on criteria #1: canvassing a wide range of alternatives; criteria #3: assessing the costs and risks of negative consequences of each alternative; criteria #4: intensive search for new information relevant to evaluation of alternatives; and criteria #5: correctly assimilating any new information or expert judgement. As a result, the AMDAL Act of 1997 and BAPEDAL procedure for mining operations must be rated as seriously deficient in terms of all four criteria. Therefore the AMDAL process seems unlikely to be capable of preventing serious mistakes in decision making which may result in failure to attain important values and objectives stated in the Act of 1997, as discussed earlier in this Chapter.

Agencies with relevant expertise are assumed to be represented on the appropriate AMDAL Commission charged with evaluating documents for a specific development proposal, but there is no specific requirement that all such agencies be represented (BAPEDAL 2000a). The door is propped open but there is no right of entry without invitation. Neither the applicant nor the AMDAL Commission is obliged to obtain comments from other agencies with special expertise or jurisdiction and include them – with appropriate responses – in the ANDAL or the approval decision. The AMDAL Commission is not required to evaluate any new alternatives or mitigation measures identified during review of the ANDAL, and is not authorized to modify the proposed action before making a final decision. Criteria #6 for vigilant information processing, requiring reexamination of the positive and negative consequences of all known alternatives before final decision therefore is not well served by the Act of 1997 and BAPEDAL guidelines.

Regulations issued pursuant to the Act of 1997, both by BAPEDAL and by individual agencies, provide that measures to mitigate adverse environmental effects of a decision must be discussed in the RPL, and appropriate monitoring programs must be specified in the RKL. Because guidelines for preparation of RPL and RKL suffer from too much deference for "technical constraints" concerning availability of funds, time and manpower, and because they state the RKL is only required to give the basic direction, principles or requirements for prevention, mitigation and control of impacts, these documents often lack sufficient information to be useful to decision makers and inspectors. There is no provision for attachment of conditions to Mining Authorizations, CoW or other approvals requiring implementation of mitigation plans. Such lax requirements poorly serve criteria #7: making detailed provisions for implementing the chosen course of action.

Guidelines for preparation and review of AMDAL documents have been revised many times since 1982. Still the Act of 1997 and implementing guidelines do not specify adequate procedural requirements or sufficiently detailed guidance to substantially meet the criteria for vigilant information processing. The AMDAL decision making procedure may therefore be characterized as defective, one with laudable goals and objectives, but deficient because it does not provide adequate information for decision making. These deficiencies might be cured by implementing recommendations described in this Chapter.

Integration of Mining Environmental Policy in Indonesia

The Basic Provision for Management of the Living Environment Act was enacted in 1982, and implementing regulations were issued in 1986, establishing an environmental impact assessment process. However the Act and regulation were general in nature and did not specify requirements for particular sectors of development such as mining. A monitoring plan and reclamation plan were required, but were not prepared until after approval to mine was issued, reducing their value in determining conditions and requirements to be included in the Mining Authorization.

Thus Ministry of Mines and Energy regulations providing environmental protection from the effects of mining activities were issued in 1977, before the Basic Provisions for Management of the Living Environment Act was enacted in 1982. In the early 1990s, it was recognized by officials in the Ministry of Mines and Energy that their regulations had not been, but needed to be adjusted to bring them into conformance with the Act of 1982 (Kuntjoro, 1993). Poor compliance with environmental requirements by mine operators, deviations from environmental standards, and enforcement problems were seen as threats to sustainable development (Kuntjoro, 1994, p. 22). Moreover Ministry officials recognized they did not have sufficient numbers of technical staff and mine inspectors with knowledge and experience in the fields of environmental management to inspect the rapidly increasing number of mines, design effective regulations, or implement mining environmental policies (Subagyo, 1994, p. 88).

Presidential Regulations were issued on environmental impact assessment in 1986, 1993 and 1999; on mining in 1969; and on water quality and pollution control in 2001. The Ministry of Environment issued guidelines for preparation of environmental assessments in 1987, 1991, 1994, and 2000; guidance for determining environmental quality standards in 1978 and 1991; and quality standards for waste water in 1991 and 1995. BAPEDAL issued guidelines for determination of significant environmental impacts in 1994 and for preparation of environmental impact assessments in 2000. The Ministry of Mines and Energy issued guidelines for environmental impact assessment in 1988, 1989 and 1996 (1988a, 1988b, 1988c, 1989, 1996a). The Ministry of Health issued regulations for water pollution control relating to public health (1977), establishing parameters for contaminants such as fecal coliform, which are relevant to water pollution from human settlements, but were not well related to chemical and mineral pollutants from mining operations. Thus several Ministries and agencies were involved over time in issuing many versions of regulations and guidance for environmental impact assessment, control of mining operations, and water quality standards, leaving many gaps as well as some areas of overlap and duplication. Consequently, a lack of integration is apparent in mining environmental policy in Indonesia.

Mining Feasibility Study and the AMDAL Process

The feasibility study for a Mining Authorization and the AMDAL process are separated by time and organization and would benefit greatly from integration. The feasibility study is a presentation to a mining regulatory authority on the economic and technical aspects of a proposed mine. Can the company open and operate the proposed mine at a profit? The appropriate AMDAL Commission is the interface between many facets of government, the public, and the mining industry. However the mining regulatory authority does not use the AMDAL process with its broad based inputs as a decision making tool to determine whether mining should or should not take place as proposed. That decision is made *a priori* during the feasibility study process.

At the feasibility stage, there is some discussion of equipment selection and the beginnings of a mining plan with just enough detail to suggest financial feasibility of a mining operation. Unfortunately, the feasibility study does not account for costs associated with environmental requirements. That puts pressure on the AMDAL Commission to use the feasibility study as the economic baseline for expenditure even if environmental costs are not included and environmental standards can not be met. It is a fundamental axiom of mineral development that if the mine operator can avoid enough costs, almost any resource is feasible to mine.

Despite requirements they be prepared at the same time, the feasibility study is a separate document and is not presented as part of the ANDAL, so it cannot be re-evaluated to determine whether current mining and environmental protection technologies will be applied or whether adjustments could be made to accommodate both economic and environmental interests. Mining Authorizations and Contracts of Work with foreign partners are dependent on approval of environmental assessment documents. It may be difficult for AMDAL Commissions to deny approval because important contracts depend on it.

Environmental impact assessment policy is in itself poorly integrated. Guidelines for review and *evaluation* of AMDAL documents issued by the Ministry of Environment are considerably less detailed than guidelines for *preparation* of those documents issued by BAPEDAL, and the two sets of guidance are not well integrated or explicitly related. That is, in guidelines for *review* of KA-ANDAL, ANDAL, RKL and RPL there are no requirements that the substance of these documents actually be consistent with detailed guidelines for their *preparation*. Thus there is a significant disconnect between guidelines for preparation and guidelines for evaluation of AMDAL documents.

Evaluation guidelines do not specify evaluation criteria, but merely provide lists of things that must be included in the documents. These lists do not require evidence of field surveys for acquisition of baseline data, and appear to emphasize air and water data to the neglect of flora and fauna. Although layout maps of facilities are required, evaluation criteria do not specify any scale for maps or require inclusion of topographical maps essential to environmental assessment of mining operations. Guidelines for preparation and review of AMDAL documents sometimes look more like instructions for preparation of scholarly papers and journal articles than guidance for preparation of decision documents for government officials. Instructions for review of AMDAL documents at several points focus attention of reviewers on study methods, credentials of those preparing the documents, and clarity of the narrative, but do not emphasize assessing the adequacy of information presented as a basis for making an informed decision. This tends to reduce the review function from an analytical task to a clerical task.

Moreover guidelines for evaluation of AMDAL documents are not internally well-related. For example, although they require that applicable environmental standards be identified, instructions for review of the ANDAL evaluation of significant impacts do not require comparison of impacts to applicable standards (Ministry of Environment, 2000a, Attachments III.B.1.a; III.B.7). More important, there is no mention of applicable environmental quality standards in guidelines for evaluation of Environmental Management Plans (RKL) or Environmental Monitoring Plans (RPL). Under these guidelines, it appears such plans are not required to identify or use applicable standards. Consequently, guidelines for evaluation of AMDAL documents are weak, especially compared to guidelines for preparation of these documents. Proponents of development can figure this out. Therefore existing guidelines for evaluation of AMDAL documents do not provide much incentive for preparation of sound environmental impact analysis documents, because there appears to be no sanction for failing to do so.

Linkages between the AMDAL process and project operating authorizations have been criticized as inadequate (Welles, 1995, p. 111). Failure to comply with conditions identified during the AMDAL process does not trigger enforcement actions or revocation of operating permits. Environmental impact assessment procedures and information requirements are not well-integrated with procedures for granting Mining Authorizations or Contracts of Work. In part this is due to the fact the Basic Mining Law has not been updated since its enactment in 1967, in a period when there has been considerable development and improvement of environmental policy. Although a point in the process for granting Mining Authorizations has been identified for preparation of AMDAL documents, the two processes are not well integrated. The AMDAL process does not produce information useful for mine inspections. Monitoring data acquired during mine inspections concerning compliance with the terms of ANDAL, RKL or RPL is not reported to appropriate AMDAL Commissions or BAPEDALDA.

Usefulness to mine inspectors Because the ANDAL is broad and conceptual in nature, it generally does not contain specific requirements for mitigation measures or other operational requirements which a mine inspector can use to determine compliance during inspections of the mine site. Premining inspections are rarely conducted, and there is little communication between mine inspectors in the Dinas Pertambangan and AMDAL Commissions of regional governments after mining begins. Violations of AMDAL requirements and conditions are not systematically or regularly enforced. Mine Book entries and requirements for abatement of violations of AMDAL conditions are not reinspected to judge whether action was taken to bring them into compliance. A lack of coordination is also evident between Dinas Pertambangan, the AMDAL Commissions, and the national Ministry of Energy and Mineral Resources concerning implementation of mining regulatory policy, national performance standards, and environmental assessment policy.

The absence of a detailed mine plan in ANDAL or RKL severely limits the ability of an inspector to evaluate a mine in progress by comparing what is seen on the ground it to a mine plan approved in the AMDAL process. An *ad hoc* approach to mining is characteristic of Indonesian coal mines. There is very little documented planning for a progression of mining operations, even though five-year plans were mentioned in interviews with mine inspection officials. Often it does not appear such plans are followed, and mine operators are unable or unwilling to produce them on visits to mining operations. At several mines visited in 1995 there were an extensive number of active pits and unreclaimed areas along the contour, and spoil was generally disposed of outside of the pit. These inefficient practices were attributed by mine operators to a need to meet customer coal quality specifications or pressing production requirements (U.S.DOI/OSM, 1995d). There are no enforceable requirements to follow mining plans approved in the AMDAL process.

Revision of documents Most disturbing to inspectors is that after approval, changes may be made to AMDAL documents and Mining Authorizations (and therefore in the proposed method and schedule of operation of mines) by the operator in annual environmental reports to the mining regulatory authority, without approval by the AMDAL Commission, its technical advisors, and

Figure 4.2   Unrepaired breach of narrow dike by Ombilin River in background created a lake in coal mine pit 800 m wide by 150 m deep and 1.5 km long in foreground, flooding mining equipment and small scale illegal mine workings, West Sumatera
Source: Alfred E. Whitehouse

usually without advance knowledge of the inspector. Significant changes are made without review to ensure that environmental mitigation measures are appropriate. Often inspectors first learn about such changes when they arrive at mine sites for annual inspections. These unannounced changes further complicate a difficult inspection process because each minesite has different environmental requirements imposed by the AMDAL Commission. This results in inconsistent inspections and uneven application of general standards by inspectors. Because standards are expressed in general terms, some inspectors may be more stringent in their interpretation of requirements than others.

A formal process for revision of ANDAL, RKL, RPL, and the Mining Authorization is needed to accommodate appropriate changes during mining, after review by the appropriate AMDAL Commission. Revisions could be proposed by the mine operator due to modifications desired in day-to-day operations or discovery of unforeseen circumstances, or ordered by the appropriate mining regulatory authority to bring a mine into compliance with regulatory requirements or with revised performance standards. Any time a mine operator wishes to deviate from an approved mining plan, a request for revision should be submitted – with appropriate details – and approval obtained from the AMDAL Commission and the mining regulatory authority prior to implementation. The mining regulatory authority would order revisions when it determines there is a problem with implementation of the RKL, RPL, or there is a rule change requiring some new mining practice.

Applications for revision of ANDAL, RKL, RPL and Mining Authorizations might propose significant or minor changes to approved documents. Significant revision applications would include those that propose: major changes in production or recoverability of the coal resource; major changes in environmental effects of the operation; additional lands to be mined; changes that may intensify public interest; or activities that could result in significant adverse impacts on fish or wildlife, endangered species, or cultural resources. If a proposal is determined to be significant, then it would be processed like a new application for Mining Authorization, requiring preparations of supplements to the original AMDAL documents, with public disclosure and participation.

Minor revision applications are those that are not significant revision applications. Examples of minor revision applications would include: changes to locations or design of roads and facilities; new locations of topsoil stockpiles; changes in the revegetative seed mix; adjustments to bond calculation; responses to special permit conditions; changes to the ground or surface water monitoring plan; modifications to sedimentation pond designs; or adjustments to approved water, air, and revegetation standards. If a revision proposal is determined to be a minor change to the approved documents, then it could be processed in a two stage process involving a technical review and a decision document. Copies of the revision proposal would be distributed to all appropriate agencies for review. The public need not be advised of minor revisions, but if interested, they may visit appropriate offices to review public files.

National and regional government agencies would have a period of time (about 60 days) to submit written comments on the proposal. Mine inspectors in the appropriate regulatory agency should review the application for on-the-ground "inspectability" of the proposal. Appropriate technical specialists could conduct detailed reviews of the application to determine compliance with applicable laws and regulatory standards. For a minor revision this normally would not require all available technical specialists, but only those with expertise necessary for review of a specific proposal. For example, an application to modify a groundwater monitoring program would only require review by a hydrologist. Technical specialists should prepare reports that document the application's compliance or deficiencies with the law and regulatory standards. These reports should become part of the decision document.

If any areas of a revision proposal are not in compliance with the law or regulatory standards, the AMDAL Commission should inform the appropriate regulatory authority, who should send a letter describing the deficiencies to the applicant. The applicant should be given an opportunity to respond by modifying the revision application as necessary to achieve compliance. Modifications to the application could also be distributed for review by technical specialists. This process could continue until all deficiencies are satisfactorily resolved. However if the operator proves unable or unwilling to resolve the issues, then a decision document would be prepared denying the application.

Finally, a decision document would be prepared recommending that a revision proposal be approved, approved with special conditions or denied. Based on the law, regulatory standards and information contained in the decision document, the appropriate regulatory authority would determine if an application is approved. If the appropriate regulatory authority issues a decision approving a revision application, the public should have the right to appeal the decision to the courts. The appropriate regulatory authority would update official records and insert the appropriate pages and maps of the revision application into the Mining Authorization. If the appropriate regulatory authority denies an application for revision, the denial must specifically explain the reasons why the required findings cannot be made and the proposed revision cannot be approved. The operator should have the right to appeal this decision to the courts.

Integration of Mining Environmental Policy in the United States

In contrast to the sequence of policy enactments in Indonesia, a national policy on the environment was enacted in the United States before the current regulatory program was created for surface mining coal. The National Environmental Policy Act establishing requirements for environmental assessments was enacted in 1969, before the Surface Mining Control and Reclamation Act (SMCRA) was enacted in 1977. There were specific provisions in the Surface Mining Act saying it did not amend NEPA (*U.S. Code*, Title 30, §1292), and the language and policy of the Surface Mining Act was well-informed by NEPA.

In substantive policy, The Surface Mining Act went considerably beyond the requirement in NEPA that previously neglected environmental values must be included during planning and decision making by national government agencies in the United States. The Surface Mining Act embraced the prior NEPA reform of administrative procedures, and stated its purpose to "establish a nationwide program to protect society and the environment from the adverse effects of surface coal mining operations" and "assure that surface coal mining operations are so conducted as to protect the environment" (*U.S. Code*, Title 30, §1202(a), (d)). Thus the Surface Mining Act established a substantive policy of environmental protection, stated in much stronger terms than NEPA. Where NEPA required inclusion of environmental values, the Surface Mining Act required protecting them from surface coal mining.

Integration with other policies   Coordination of regulatory and inspection activities required by the Clean Air Act concerning air pollution, and the Clean Water Act concerning water pollution were specifically required by the Surface Mining Act (*U.S. Code*, Title 30, §1303). Permit applications were required to include a description of steps to be taken to comply with applicable air and water quality laws and regulations (*U.S. Code*, Title 30, §1258(a)(9)). A national performance standard requiring use of best technology currently available to minimize disturbances from surface mining operations on fish and wildlife, and prohibiting mining activities likely to jeopardize endangered or threatened species or their habitat (*U.S. Code of Federal Regulations*, Title 30, §816.97) was designed to integrate the Surface Mining Act with the Fish and Wildlife Coordination Act and the Endangered Species Act. Regulations issued by OSM concerning the contents of applications for surface mining permits requiring maps showing locations of any cultural or historic resources and known archeological sites made it clear the surface mining regulatory program was intended to be integrated with requirements of the National Historic Preservation Act (*U.S. Code of Federal Regulations*, Title 30, §779.25).

Focus on significant impacts   NEPA required a scoping process to identify and focus environmental assessments on the most significant impacts of a proposed project. The Surface Mining Act went a step further by identifying the most significant impacts of surface coal mining (*U.S. Code*, Title 30, §1201(c)) and requiring detailed information about them be presented in permit applications so an assessment could be made by the regulatory authority concerning severity of probable impacts and the adequacy of proponents plans to control or mitigate them (*U.S. Code*, Title 30, §§1257–1258; 1265).

Systematic interdisciplinary approach   Use of a systematic interdisciplinary approach was required by NEPA in decision making that may have an impact on the environment (*U.S. Code*, Title 42, §4332). The Surface Mining Act required presentation of information in mine permit applications requiring analysis by specialists in a variety of disciplines, including hydrology, mapping, vegetation, fish and wildlife resources, air quality, climate, soils, geology, land use, and mining engineering (U.S.DOI/OSM, 1979; 1988) before a permit can be granted, thus requiring use of a systematic interdisciplinary approach in decision making.

Information required for decision   Contents of an environmental impact statement required by NEPA concerning baseline data, a description of the proposed action, probable impacts, and mitigation measures (*U.S. Code*, Title 42, §4332) are virtually identical to contents of a mining permit application required by the Surface Mining Act (U.S.DOI/OSM, 1988). And like NEPA, the Surface Mining Act requires full disclosure of information presented for decision, and public review and comment before a decision is made. NEPA required identification of unavoidable adverse impacts as a means of encouraging avoidance of some adverse impacts and creation of mitigation measures for others. The Surface Mining Act went further by specifying regulatory performance standards to control unavoidable impacts of surface coal mining within specified parameters, and establishing inspection and enforcement programs to ensure compliance with them.

Identification of irreversible and irretrievable commitments of resources to complete a proposed action was required by NEPA as a means of identifying the environmental costs of a development. The Surface Mining Act went further by requiring damages to air, water and land resources be minimized during mining, and that they be restored as nearly as possible to their premining condition after mining. Enforcement of national performance standards concerning reclamation of mined lands, special handling of toxic and acid-forming materials, protection of the hydrologic balance, surface water quality, and fish and wildlife were central components of the policy to minimize irreversible commitments of resources and repair environmental damage during mining.

Discussion of the short-term uses and long-term productivity of resources resulting from developing a proposed project was required by NEPA to stimulate assessment of trade-offs between current and future users of a resource. Long-term productivity was assumed to be desirable. The Surface Mining Act embraced this concept, treating surface resources such as land and vegetation as renewable resources which in most environmental conditions (e.g., all but desert and alpine tundra) can be managed to allow one-time extraction of coal, a nonrenewable resource. Coal mining is considered a temporary use of land. Temporary impairment of land for some uses is tolerated as a matter of policy to allow extraction of a valuable mineral commodity. Continued productivity of the land is assured by preventing its permanent destruction during mining (e.g., through loss of topsoil, permanent exposure of toxic overburden), and requiring it be reclaimed for a specific postmining land use, determined before mining begins. Thus long-term productivity of the land is assured as a policy goal, after temporary disruption is allowed. Premining designation of a specific postmining land use towards which reclamation efforts are oriented is central to effective implementation of this policy.

Sustainable development   Development of coal resources in a manner that meets the needs of the present without compromising the ability of future generations to meet their own needs is consistent with the concept of sustainable development embraced by NEPA. Control of the environmental effects of coal mining during mining, and concern for continued productivity of the land after mining is finished are central to the concept of sustainability in this context. The Surface Mining Act advances a strong regulatory program in support of sustainable development of coal resources. Thus the concepts of environmental assessment and informed decision making advanced in NEPA were integrated into a regime focused on environmental regulation of one industry, surface coal mining, in the Surface Mining Act. Moreover the Surface Mining Control and Reclamation Act of 1977 is more directly protective of the environment, and more obviously a regulatory statute, than is the National Environmental Policy Act of 1969. Consequently, the permitting process required by the Surface Mining Act is the functional equivalent of the environmental assessment process required by NEPA.[2]

Notes

1 Dollar equivalents in this chapter were calculated at average 2003 exchange rates of 8800 rupiah per dollar.
2 Provision of the functional equivalent of an environmental impact statement through an administrative decision making process was recognized as legitimate in the U.S. Court of Appeals decision for the District of Columbia Circuit in: *Environmental Defense Fund Inc. et al v. Environmental Protection Agency and William D. Ruckelshaus et al*, 489 F.2d 1247 in 1973 (O'Leary, 1993, p. 61).

Chapter 5

Lost Profits, Royalties and Environmental Quality

Feeding the Lembu Suana

> An Indonesian fisherman early one morning saw his neighbor on the coal-black banks of the Mahakam River slowly tossing bits of paper into the water. On closer inspection, he saw the neighbor was carefully dropping 100 rupiah notes one by one onto the muddy water, where they floated away downstream. Thinking his neighbor had gone *orang gila* (crazy man), the fisherman asked what the other was doing. The neighbor replied: "Feeding the Lembu suana. Fishing is poor, my children are sickly. Maybe it will leave more fish for my family."

The Lembu suana is a mythical four-legged beast reputed to live in the Mahakam River on the island of Borneo. With an elephant's trunk and tusks, a water buffalo's ears, horns and tail, the wings and feet of an eagle, the teeth and body of a lion, and scales like a fish, large statues of the Lembu suana guard the entrance to the Mulawarman Museum Complex in Tenggarong, formerly the palace of Aji Sultan Muhammad Parikesit, the last sultan of the Kutai Kingdom of East Kalimantan. Small statues are sometimes sold to tourists as souvenirs. Comparable to the legendary Loch Ness Monster in the Mahakam River: the Lembu suana is mysterious and terrible to imagine, but seldom seen.

Many coal mining operations in East Kalimantan have been feeding the Lembu suana a significant portion of their profits each year since they started production. Visitors to Indonesian coal mining operations have repeatedly observed difficulties with control of fine coal particulates (<2 mm) at coal preparation (prep) plants and loading facilities. Quantities of fine coal lost from these mines are large enough to raise concerns about the efficiency of operations and the impact on both aquatic and downstream human environments. This chapter reviews a series of such observations, illustrating several problems with implementation of Indonesian mining environmental policy during the 1990s. It includes estimates of the value of lost profits and government royalties from failure to capture this product, and refines previous estimates of capital recovery periods for investments in coal washing plants based on new information. The period required for recovery of new investments in fine coal recovery systems at Indonesian coal prep plants appears to be shorter, and the required investments more attractive than previously estimated.

The recovery of fine coal in washing and prep plants at mining operations is apparently a new idea in the rapidly expanding Indonesian mining industry.

Only four of 35 active coal mining operations in Indonesia were known to use fine coal recovery systems in 1999. Those using such technology were all owned by multinational corporations having experience with operations in more industrialized countries. Lack of such equipment results in significant economic inefficiencies and negative environmental effects.

Inefficiencies of Indonesian coal mining operations and the financial and environmental consequences of fine particulate coal lost from washing operations at coal prep plants were identified in a seminal article in *Indonesian Mining Journal* in 1998. Lost coal fines were identified as a significant source of water quality degradation affecting human health. Based on the collective professional judgment of several individuals with vast experience in the American mining industry who visited several Indonesian coal mining operations, losses were estimated in the range of 10–20 percent of annual production during the period 1990–1996 (Hamilton, 1998a). Short capital recovery periods were described as "policy drivers" indicating a need for new mining environmental policy in a second article in *Indonesian Mining Journal* in 2000. Based on coal production data for the period 1990–1998 and cost comparisons to a benchmark coal prep plant in the U.S., capital recovery periods were estimated for new investments in fine coal circuits at 1–2 years for some larger mines, 3–10 years at smaller mines, and were projected to produce increased profits every year thereafter (Hamilton, 2000b).

Alternative regulatory policies requiring national performance standards for sediment and erosion control, regular maintenance of settlement ponds, and a "clear water" discharge standard to reduce surface water contamination were evaluated in a third article in the European mining business journal *Minerals and Energy* (Hamilton, 2001). Estimates of capital recovery periods for installation of fine coal recovery systems at existing plants were revised downwards to 4–5 months in the Australian journal *Natural Resources Management* in 2004. Revised estimates were based on annual coal production data for the period 1990–1999, interviews with mine personnel that confirmed Indonesian mining operations were losing at least 20 percent of annual production, laboratory analysis of coal refuse samples, and site visits to determine what additional equipment would actually be needed (Hamilton, 2004).

Significance of Coal Reserves for National Development

In Indonesia, all natural resources are considered the property of the nation. No resource can legally be extracted or developed without authorization and payment of royalties, a share of production, or concession fees to national and provincial governments. Indonesian mining operations significantly increased their contribution to the national economy and payments to government as production of coal increased by a factor of more than ten from 10.6 million metric tonnes in 1990 to nearly 114.3 million metric tonnes in 2003 (Ministry of Energy and Mineral Resources, 2004, p. 13; Ministry of Mines and Energy, 1997).[1] Coal is expected to eventually displace oil as the principal energy source for domestic use as oil production declines in this OPEC country (Kuntjoro, 1993).

For Indonesia, coal exports provide an increasingly important source of foreign exchange. Indonesia emerged as the fifth largest steaming coal exporter in the world in 1993, shipping the majority of coal imported by Asian countries, principally Japan, Taiwan and South Korea (U.S. Department of Commerce, 1994). Twenty-two mining operations producing more than 200 000 tonnes per year accounted for over 99 percent of the countries coal production in 2003 (Ministry of Energy and Mineral Resources, 2004, p. 13). The proportion of coal exported ranged from zero to 100 percent of production at five mines, averaging about 75 percent from twenty-one mines with exports in 2003.[2] This indicates sophisticated overseas marketing capabilities on the part of many producers.

Lost Profits, Lost Royalties and Environmental Damage

Expectations by investors about the rate of recovery of coal reserves determine the financial feasibility of new coal mining operations. With higher rates of recovery, greater profits are realized by mine operators, and greater royalty revenues are paid to the government. Both results advance development goals and strengthen economic performance, important national goals in a newly industrializing country like Indonesia. When the rate of recovery is lower than it could be because losses of coal product are high after it is removed from the ground, operations are less efficient, productivity and profits are lower, and royalty revenues received by the government are less than they should be. These results reduce contributions to the economy, retard national development, and produce greater environmental disruption than more efficient operations.

In April 1992, members of a delegation from the U.S. Office of Surface Mining Reclamation and Enforcement visited five large mines on the islands of Sumatera and Kalimantan, finding: "Environmental controls for sediment ... were constructed in available space and were probably removing settleable solids but were not removing colloidal clay or coal fines" (Hamilton, Whitehouse and Tipton, 1992, p. 6). At one mine on the banks of the Mahakam River in East Kalimantan it was observed that:

> Prep plant and loadout are built on filled wetland. Small stream with runoff from prep plant, and ditch carrying water from equipment both discharge black water loaded with fine particulate coal directly into river. Downstream river banks are as black as volcanic sand from coal fines.
>
> (Hamilton, Whitehouse and Tipton, 1992, p. 12)

Rivers are highways for commerce in Indonesia, and many operations transport coal from the mine pit by truck to nearby facilities that load coal on river barges, which take it downstream and transfer it to larger ships for delivery to foreign and domestic markets. A portion of the fine coal generated at a prep plant at one mine was captured from a sump pond, shipped to Denmark and Germany, and made into briquettes used for residential heating. Domestic sales of coal briquettes for cooking in Indonesia during 2002 totaled 23 336 tonnes, up 9.4 percent over 2001 (Ministry of Energy and Mineral Resources, 2003, p. 32).

Fine coal particulates of less than 2 mm contain the same heat value as larger pieces of coal and are highly combustible. In underground mines, fine coal particulates suspended in air are so flammable they present a recognized threat of explosion, which must be controlled to guarantee the safety of miners. Coal received by large industrial facilities such as the Suralaya Power Plant in South Sumatera must be ground to a fine powder of particulates considerably smaller than 2 mm before it is injected into boilers and burned to generate electricity. Coal fines are routinely sold as part of the product shipped from mines for industrial use in the United States and other countries. Thus coal fines are widely regarded as product with resale value. Consequently one may wonder why fine coal particulates are sometimes discarded as waste in Indonesia.

In June 1995 a five-person interdisciplinary team visited two mines on the Mahakam River. Members of this group included two geologists, two mining engineers and an environmental policy specialist. Three of these individuals were experienced mine inspectors, and three had combined mining experience exceeding 45 years with U.S. firms such as Anaconda Company, New Jersey Zinc, and Buffalo Coal Company. The team examined two of the same mines visited in April 1992. They indicated:

> At the washing plant at the first mine, there was money all over the ground, laying several centimeters deep, and in some places more than a meter deep in the ditches, in full sediment ponds, and being dumped in the Mahakam River in the form of fine coal particulates. Discharges from the washing plant ran directly into the river and were uncontrolled, producing black beaches on the shoreline of the river for hundreds of meters downstream. Like all coal, this coal almost certainly contains heavy metals and other harmful substances that accumulate in tissues of fish consumed by human populations in the immediate vicinity.
>
> (U.S. DOI/OSM, 1995a, p. 1)

Little had changed in three years since the previous visit, and substantial amounts of coal continued to escape from mining operations.

Environmental Health Effects

Lost coal fines constitute a serious environmental problem which can threaten public health. Most coal contains trace amounts of heavy metals such as lead, mercury, cadmium, chromium, cobalt, nickel and other substances such as selenium and radionuclides which are harmful to human health. Some of these materials are taken up by fish and accumulate in fatty tissues which may later be eaten by local people, where they accumulate in human body tissues. Lead for example, slowly accumulates and causes damage to the brain and other nerve tissue in humans, producing behavioral disorders, mental retardation,

Figure 5.1 **Best practices prep plant with fine coal recovery circuit, zero water discharge and settlements ponds, raw coal in background, Buffalo Coal Company, West Virginia, USA**
Source: Michael S. Hamilton

convulsions, and possibly death. Most severely affected are children, who may suffer developmental disabilities (Greenberg 1987, pp. 81–82, 130). Recovering this product for processing and sale would help clean up one of the most serious environmental problems caused by Indonesian mining operations.

With storm water surges, discharges of coal fines to Indonesian rivers move downstream to river deltas and the ocean, where they adversely affect commercial fisheries. Although record keeping practices at most Indonesian hospitals make epidemiological research difficult or impossible, and there is virtually no published data on the incidence of local health problems, reducing future discharges of fine coal into the rivers of Indonesia should reduce adverse impacts on public health in communities along downstream portions of these rivers, and may reduce the medical costs of care for affected persons. The observable impact of mining operations on land and surface water in the immediate vicinity will also be reduced, perhaps improving often tense relations between mine managers and local communities.

Extraction Costs Already Incurred

By the time this coal arrives at processing and loading plants, the capital, labor and other operating costs to mine it have already been expended. Because operating costs for extracting and processing coal lost at washing and loading facilities are already paid before it is lost, productivity, profits, and efficiency of the mining operation are reduced by such losses. No royalties are paid on lost production, despite the fact a government resource has been extracted from the ground, and its value to society is lost. In legal theory royalties are payable on all coal extracted from the ground, not merely on coal sold or shipped. However records are kept and royalties paid only on coal sold and shipped. No attempt has been made by the government to collect royalties on coal lost after extraction but prior to shipping. If this lost product was captured and sold, profits and royalties would increase, with no increase in the cost of extracting it. The only increase in costs to mining operations would be those necessary to capture and segregate the product from coal refuse at prep plants prior to shipping.

Conditions at the Mahakam river mines did not improve noticeably after 1992, leading visitors in 1995 to observe: "The market and royalty value of coal fines lost from some Indonesian coal cleaning and loading facilities are substantial, possibly sufficient to pay for increased expenditures for environmental controls and improved operating efficiencies" (Hamilton, Wiryanto and Whitehouse, 1996, p. 73). Indonesian mines display vastly different levels of sophistication in management and operations, but even some apparently well-run operations lack engineered fine coal recovery systems at crushing and washing plants. Sediment and erosion controls are rudimentary. Some operations display little concern for controlling either onsite or offsite environmental impacts of storm water runoff from coal storage, prep plants, or loading facilities. Perimeter ditches often discharge directly into rivers and streams rather than settling ponds. Where settling ponds are evident, they are poorly maintained, infrequently cleaned out, often full of solids and generally ineffective.

Estimating the Value of Lost Coal

Using annual production figures reported by the Ministry of Energy and Mineral Resources, the value of lost profits from production of lost coal fines was calculated for five mines located along the Mahakam River during the years 1990–2003, using the average sale price of Indonesian coal for that period of US$30/ton. These estimates approximate the value of lost profits to the mine operator because coal was lost after the costs of extracting and processing it were expended. Mahakam River mining operations are well-known to international coal brokers for difficulties in controlling the quality of coal they ship. Often these mines have been assessed significant financial penalties (specified in terms of purchase agreements) for failing to ship coal that complies with contract specifications for heat content, sulfur and ash. Some brokers have refused to accept future shipments from some Mahakam River mining operations for repeated violations of contract terms.

Consequently the actual price per tonne received after penalties may have been considerably less than US$30/tonne for some shipments, perhaps averaging as low as US$23/tonne for Mahakam River coal in the period

**Figure 5.2   Blackwater discharge of fine coal from prep plant and loadout on
bank of Mahakam River, East Kalimantan**
Source: Michael S. Hamilton

examined. The price of US$30/tonne should therefore be viewed as the price
that may be earned by an efficient, well-run mining operation. Because
royalties paid to government in Indonesia are based on the price actually paid
per tonne of coal sales, the government has received less revenue than it should
have for sales on which penalties were assessed in this period.

Previous estimates that Indonesian coal mining operations on the Mahakam
River may be losing 10 to 20 percent of production (Hamilton, 2001) to rivers
and streams were confirmed in interviews with mine managers of three mines in
February 2001, all of whom stated their losses were at least 20 percent. During
interviews in July 2001, one of the same mine managers reported production
losses of 27 percent. The Australian manager of a fourth mine with similar
equipment stated losses were only about 5 percent, but that all particles smaller
than 2 mm were disposed of as waste. It is reasonable to surmise that the fourth
mine is discarding as waste 20 percent or more of its production by making no
effort to capture particles smaller than 2 mm. Thus lost or wasted production
often exceeds 20 percent at Indonesian mining operations.

Previous studies of the time required to recover capital invested in fine coal
recovery systems at Indonesian mining operations compared the value of lost
product to the cost of building a coal prep plant similar to one built by the
Buffalo Coal Company in the U.S. in 1983 (Hamilton, 2000b, 2001). After
crushing and washing, fine coal down to 150 mesh (~ 0.10 mm)^3 is removed at
225 metric tonnes per hour. Wright described this plant in detail, noting

"minus 150 mesh coal is discarded, as moisture removal would prove too costly without thermal drying" (1984, p. 79). The Buffalo Coal prep plant has for many years upgraded coal to 13 100 Btu and less than 1 percent sulfur using raw coal of about 12 100 Btu and 4 percent sulfur from nine pits. Recovered fine coal is dewatered to about 30 percent moisture and blended with drier run-of-mine coal screened to 1 $\frac{1}{2}$ inches, producing a product with total moisture of about 6 percent, well within contract specifications. As sulfur-bearing pyrites are liberated, water in the refuse sump often has an extremely acid pH of 2, but the plant is a closed loop facility (Duckett, 1989), so there is no water discharge. Acid-generating pyrites are returned to the coal pit with other refuse and buried, isolated from ground water. Consequently, this plant demonstrates what is feasible using commercially available technology, and is considered a "best practice" benchmark for both water quality and economic efficiency.

The US$7.5 million cost of the Buffalo Coal prep plant included equipment for handling, crushing, sizing and washing coal that is already in place in many Indonesian prep plants. Because this equipment is already in place, installation of a fine coal recovery system at such plants will be less expensive and the time necessary to recover capital invested will be less than previously estimated. To calculate the period in which improved recovery of fine coal would pay for an investment in additional equipment to capture it, the estimated value of lost profits was compared with the cost of installing a fine coal recovery circuit in existing prep plants at each mine. Samples of about 4 kilograms each of solids were acquired from waste material discarded at prep plants in three coal mining operations on the Mahakam River in February and June 2001.

All samples were shipped to the United States and analyzed at Standard Laboratories Inc., an accredited laboratory in the coal country of West Virginia. Samples were analyzed for percentage of coal content and quality (moisture, sulfur, ash, volatile organic compounds, fixed carbon and heat content in Btu/lb) as received and after drying, using standard industry methods. Screen analyses were done to determine particulate size of coal in the waste material, and washability tests were run at various specific gravities to determine what portion of the coal might be recoverable (Hamilton, 2004). Results of these tests assist designers of prep plants to determine what types of equipment are necessary to recover fine coal.

To determine what equipment was already in place and identify what additional equipment would be needed, several Mahakam River mining operations were visited in June 2001 and April 2002 by an individual who previously designed, constructed and operated prep plants with fine coal recovery systems in the United States, including the Buffalo Coal plant. This study was intended to find out if Indonesian mines made investments in fine coal recovery circuits at existing prep plants, how long might it take for increased profits from recovered coal to pay for the investment? That is, what would be the likely capital recovery period for such investments?

Lost sales and royalties were estimated for all coal mines in Indonesia lacking fine coal recovery systems, providing estimates of total royalties unpaid to government during a thirteen year period, 1990–2003. The value of lost profits at five mines located on the Mahakam River was compared to the capital investment necessary to recover coal fines at each mine. The value of lost royalties due the government was calculated at the average usual rate during the period of 13.5 percent of production. Lost royalties represent revenues not available for national development or social welfare expenditures, important policy considerations for an industrializing country with the fourth largest population in the world.

Market and Royalty Value of Lost Coal Fines

Extrapolating an average of 20 percent losses of fine coal to all coal mines in Indonesia which do not use fine coal recovery systems in their prep plants, total lost production is estimated at over 513 million tonnes in the period 1990–2003, as illustrated in Table 5.1. The value of lost coal sales in this period is estimated at nearly US$3.08 billion. Unpaid royalties due the government over these fourteen years are estimated to be worth nearly US$415.6 million. These figures are within the range of earlier estimates, but are both more current and more realistic because they include additional years of recent production, and exclude losses estimated previously for four mining operations subsequently found to have operating fine coal recovery systems.

These figures represent significant inefficiencies in some Indonesian coal mining operations, and permanent losses to the government and the national economy of Indonesia which cannot be recovered. Additional future losses might be avoided through investments in fine coal recovery systems at existing mining operations. Would such investments be financially viable at Indonesian mines? Closer examination of a few mines sheds some light on this question.

Capital Recovery Periods for Fine Coal Circuits

Laboratory analysis in March 2001 of coal samples from three Mahakam River mines indicated it would be feasible to recover more than 80 percent of the total weight of each sample. That is, over 80 percent of the waste examined was marketable coal. For example, analysis indicated prep plant waste from the PT Kitadin mine would yield a product which is 83.17 percent of the total weight of the sample, with 5.59 percent ash, 1.07 percent sulfur, and 11,702 Btu/lb heat-content (Whitehouse, 2001; Standard Laboratories Inc., 2001). This low sulfur, high heat-content coal would probably be blended with lower quality coal in the U.S. to produce a product meeting contract specifications of a purchaser (Duckett, 2001a). Other Mahakam River mines produce coal with similar quality and characteristics. Based on these lab tests, equipment was priced for a generic fine coal recovery system with zero water discharge that might be installed at any of several Mahakam River mines,[4] with minor modifications to accommodate existing equipment configurations. The cost of purchasing and installing equipment to process 500 000 tonnes per year on a single shift was estimated at about US$1.2 to US$1.8 million in 2001 (Duckett, 2001b; 2002).

The Mahakam River, like many rivers in Indonesia, flows in a bed comprised mostly of unconsolidated soils and sands, and carries a high sediment load.

Table 5.1   Estimated Value of 20% Lost Production: Sales and Royalties (1990–2003), All Indonesian Mines (thousands)

	Production	Production Lacking Fine Coal Recovery*	20%Tonnes	20%Sales (US$)	20%Royalty (US$)
1990	10 616	8 664	1 733	51 990	7 019
1991	13 932	9 604	1 921	57 630	7 780
1992	23 120	12 765	2 553	76 590	10 340
1993	27 844	14 813	2 963	88 890	12 000
1994	32 593	17 354	3 471	104 130	14 058
1995	41 316	24 731	4 946	148 380	20 031
1996	50 346	30 071	6 014	180 420	24 357
1997	54 822	34 621	6 924	207 720	28 042
1998	61 931	39 942	7 988	239 640	32 351
1999	73 777	49 075	9 815	294 450	39 751
2000	77 040	53 850	10 770	323 100	43 618
2001	92 540	64 577	12 915	387 450	52 306
2002	103 372	71 867	14 373	431 190	58 211
2003	114 278	81 134	16 227	486 810	65 719
Totals	777 527	513 068	102 613	3 078 390	415 582

* Excludes annual production from PT Arutmin, PT BHP Kendilo Coal, PT Kaltim Prima Coal, and PT Gunung Bayan Pratama, which have fine coal recovery systems.

Source: Calculated from annual production data in: Ministry of Mines and Energy, 1997; Ministry of Energy and Mineral Resources, 2004. Totals may not add due to rounding.

Generally brown in appearance, it is similar to stretches of the Colorado River in the western United States. This water is not well-suited for washing coal because it contains clays and slate slimes that mine operators wish to wash out of their coal. It tends to plug up spray nozzles in coal washing equipment while adding sediment to settlement ponds. Consequently, reusing washing water after cleaning solids from it provides a higher quality process water and more efficient operations than continually taking new water from the river. Equipment for a water cleaning and reuse cycle are therefore included in the generic plant estimates.

At 20 percent lost production on 500 000 tonnes, losses amount to 100 000 tonnes per year. Installation of the fine coal recovery system would recover at least 80 percent of these losses, or 80 000 tonnes per year of product. If sold at US $30 per tonne this would generate US$2 400 000 per year in new revenue. After recovering 100 percent of capital invested in less than one year, and paying additional royalties to the government of US$324 000, this would leave the mine operator with net revenues of US$600 000 to US$876 200, depending on actual cost of equipment installed.[5] Each subsequent year would produce an additional US$2 400 000 of revenue. This equipment may be run in double shifts with no additional capital expense, cleaning up to 1 million tonnes per year, producing additional revenues.

Because the product from a fine coal circuit has been washed to reduce impurities, direct benefits from installing the recovery system described here include:

- improved overall coal quality (and possibly sale price) due to reduced ash content and increased heat content of the final product;
- increased coal recovery;
- increased royalties paid to government; and
- increased productivity in the form of greater company profits (Duckett, 2001b).

Indirect benefits from reuse of cleaner water in the prep plant include:

- improved operating efficiency of washing equipment (due to removal of clays and slimes);
- improved appearance of the coal product;
- reduction of consumptive water use by about 80 percent;
- zero discharge of water; and
- reduced costs for cleaning settling ponds and ditches (Duckett, 2001b).

Table 5.2 illustrates the value of lost production and royalties at two shipping prices for five mines located on the Mahakam River. Production figures are for 2003, the most recent year for which such information has been published, except PT Fajar Bumi Sakti for which 1996 production was used because it is more representative of actual production capacity. Alternative shipping prices were used to assess sensitivity of estimated capital recovery periods to differences in sale price of coal. One of these mines, PT Multi Harapan Utama, is owned by foreign investors; all but PT Fajar Bumi Sakti exported more than 71 percent, and PT Bukit Baiduri exported 100 percent of annual production in 2003.[6]

The period required after expenditures before recouping them at five Mahakam River mines are shown in Table 5.3, based on comparing the value of recoverable production after payment of increased royalties to the capital cost of installing fine coal circuits in existing prep plants. Capital recovery periods are estimated at 1–2 months for some larger mines processing 500 000 to 1 000 000 tonnes per year, and 9–10 months at smaller mines processing 250 000 tonnes per year, reflecting variations in capacity of the recovery system, coal sale price, annual production, and percentage of production washed per year at each mine. These capital recovery periods are considerably shorter than previously estimated for the same mines (Hamilton, 2004), due to lower capital costs and increased coal production at some mines in 2003.

Initial estimates were based on installation of a generic fine coal recovery circuit capable of processing 500 000 tonnes per year in single shifts at a capital cost of US$1.2 million to US$1.8 million at four mines (Hamilton, 2004). But not all Indonesian mining operations wash 500 000 tonnes of coal per year. PT

Table 5.2   Estimated Value of Lost Production: Recoverable Sales and Royalties, Mahakam River Mines

	PT Bukit Baiduri	PT Fajar Bumi Sakti*	PT Kitadin**	PT Multi Harapan Utama**	PT Tanito Harum
2003 Production (tonnes)	2 416 917	626 958	2 291 249	1 620 380	2 178 576
%Washed	40	40	0	100	31
Washed or Discarded (tonnes)	966 767	250 783	458 250	1 620 380	675 359
Lost Production (20%)	193 353	50 157	458 250	324 076	135 072
Recoverable (tonnes)	154 682	40 126	366 600	259 261	108 058
Value of Recoverable Production					
(US$30/tonne)	4 640 460	1 203 780	10 998 000	7 777 830	3 241 740
(US$27/tonne)	4 176 414	1 083 402	9 898 200	7 000 047	2 917 566
Value of Recoverable Royalties					
(US$30/tonne)	626 462	162 510	1 484 730	1 050 007	437 635
(US$27/tonne)	563 816	146 259	1 336 257	945 006	393 871

* PT Fajar Bumi Sakti is year of highest production, 1996.
** Lost Production and Recoverable based on estimates of coal discarded in waste.

Source: Calculated from annual production data in: Ministry of Energy and Mineral Resources, 2004.

Kaltim Prima Coal washes about 20 percent of their production, and other mines such as PT Anugerah Bara Kaltim say they wash none. Interviews with mine managers at five Mahakam River mines indicate PT Multi Harapan Utama washes 100 percent, PT Bukit Baiduri and PT Fajar Bumi Sakti wash about 40 percent, PT Tanito Harum washes about 31 percent, and PT Kitadin washes none while discarding about 20 percent of annual production from the top and bottom of each coal seam. Consequently it was assumed PT Bukit Baiduri would operate this equipment in double shifts approximately 93 percent of the time to process about 996 800 tonnes per year, and PT Tanito Harum would operate in double shifts about 35 percent of the time to process about 675 400 tonnes per year. Estimates for the fifth mine, PT Multi Harapan Utama, assumed installation of fine coal circuits capable of processing 1 million tonnes per year in single shifts, costing US$2.4 million to US$3.6 million, running double shifts about 62 percent of the time to process 1 620 000 tonnes per year. Because this equipment would result in substantial excess capacity to process coal above the amounts needed at one mine, calculations

Table 5.3 Estimated Capital Recovery Periods for Fine Coal Circuits at Five Mahakam River Mines, 2003

| | Capital Recovery Periods (months) | | | | | |
| | $0.8 m plant* | | $1.2 m plant | | $1.8 m plant | |
Mine	US$30 tonne	US$27 tonne	US$30 tonne	US$27 tonne	US$30 tonne	US$27 tonne
PT Bukit Baiduri**	–	–	3.6	4.0	5.4	6.0
PT Fajar Bumi Sakti	9.2	10.2	13.8	15.4	20.7	23.0
PT Kitadin	1.0	1.1	1.5	1.7	2.3	2.5
PT Multi Harapan Utama***	–	–	4.3	4.8	6.4	7.1
PT Tanito Harum****	–	–	5.1	5.7	7.7	8.6

* Fine coal circuit capable of processing 250 000 tonnes per year in single shift.
** Assumes PT Bukit Baiduri operates generic 500,000 tonne fine coal circuit in double shifts about 93% of the time to process 966 800 tonnes per year.
*** Assumes PT Multi Harapan Utama operates 1 000 000 tonne fine coal circuit in double shifts about 62% of the time to process 1 625 000 tonnes per year, at capital cost of US$2.4–US$3.6 million.
**** Assumes PT Tanito Harum operates 500 000 tonne circuit in single shifts about 35% of the time to process 675 400 tonnes per year.

Source: Calculated from values in Table 5.2.

for PT Fajar Bumi Sakti were also made based on installation of a smaller fine coal circuit capable of processing 250 000 tonnes per year in single shifts, costing about US$800 000.

Fluctuations in the price of coal apparently do not produce significant change in capital recovery periods. As illustrated in Table 5.3, a difference of US$3/tonne in coal sale price produced an insignificant difference of 9 percent, or a maximum of about two months increase in length of capital recovery period for the generic 500 000 tonne per year fine coal recovery circuit. Investments in the larger US$1.8 million fine coal system may require estimated capital recovery periods 50 percent longer than those for the US$1.2 million system at some mines, yet the largest increase remains an insignificant 6.9 months at PT Fajar Bumi Sakti, the mine recovering the smallest amount of product per year. If they do not foresee greatly expanded production in future years, PT Fajar Bumi Sakti and PT Kitadin might install the less costly fine coal circuits, thereby further reducing their estimated capital recovery periods by 50 percent, to 9.2–10.2 months.

In the United States, a capital recovery period of ten years is generally considered favorable for large-scale industrial equipment; less than two years is extraordinarily attractive. Capital recovery periods of 10.2 months or less for the five mines examined here would be phenomenal. Because the useful life of a fine coal recovery system is usually 10–20 years, the captured value of previously lost coal would provide significantly increased profits of 20 percent or more in each year after the initial investment was recovered. Such rapid returns to capital should have great appeal to investors, and would improve environmental conditions dramatically at each mine.

Underinvestment in Profitable Activities?

Why have these rates of return failed to attract investments at some Indonesian mines? A few of the larger mines operated by foreign firms have invested in fine coal systems at their prep plants. These include PT Kaltim Prima Coal, until 2002 owned by British Petroleum (BP, now BP-Amoco) and Rio Tinto (formerly CRA Limited) of Australia; PT Arutmin Indonesia and PT BHP Kendilo Coal Indonesia, both owned by Broken Hill Proprietary Co. Ltd (BHP), one of the largest Australian mining firms (Carter, 1995a; 1995b), and PT Gunung Bayan Pratama, based in Singapore. There was little evidence of concern for fine coal at other coal mining operations in Indonesia prior to 2002. The technologies necessary to recover fine coal have been commercially available for many years and may be sized to accommodate any scale of coal production. They are commonly in use in Europe, the U.S. and Australia. Consequently, it is puzzling to contemplate why fine coal recovery technology has not been widely used in Indonesia to increase profits of mining operations.

Need for Water Quality Standards

A situation similar to what is now found in Indonesia was evident many years ago in the United States. At one time, coal was mined by dredge in the Big Sandy River of Kentucky following many years of inefficient mining operations which discharged fine coal particulates into it. Some areas of the Mahakam River might now be mined by small-scale dredge, provided they could find markets for the product. Sand dredges are now common on the Mahakam River, and coal is usually worth more than sand. However neither coal nor sand may be dredged legally in Indonesia without government authorization, and anyone applying for one would likely face claims that the coal was already under permit to the discharging mine.

A lack of enforcement of appropriate performance standards for water discharges from mining operations to rivers and streams may have discouraged use of fine coal recovery technology. Black water discharges of coal fines exceed the Indonesian standard for total suspended solids of 400 mg/litre for coal mining operations (Republic of Indonesia, 2001; Ministry of Environment, 1995), but have produced neither cessation orders nor other significant enforcement actions. Until 2003 there was no specific discharge standard for coal prep plants, and the new standard of 200 mg/litre for total suspended solids from such facilities (Ministry of Environment, 2003) greatly exceeds a comparable U.S. effluent limitation of 70 mg/litre (*U.S. Code of Federal Regulations*, 2000, Title 40, §434). This would constitute adequate cause for an order to cease operation of a coal mine in the United States until required equipment was installed and operating. More effective enforcement of a more appropriate standard would provide incentives for investment in fine coal recovery.

Need for Improved Minerals Policy

Some mining companies may have avoided profitable investments in fine coal recovery systems in order to recover capital in the shortest possible time. Mine operators can take the "easy coal" with less investment than would be necessary to recover a higher percentage of reserves. That is, some efficiencies in production may have been deliberately sacrificed to accelerate overall capital recovery. This could be due to reluctance of foreign investors or inability of domestic firms to commit additional capital. However it should be noted that most current coal mining operations were authorized in a period of rapid economic expansion and easy credit in Indonesia, prior to the economic collapse and departure of President Soeharto in 1998. Existing mines were planned and built in an era of plentiful capital.

There are indications some mines in Indonesia are "high-grade" operations which remove the high value or easy coal and leave the rest, displaying little concern for mineral conservation or maximizing royalties paid to government for developing national resources. At one Mahakam River mine[7] visited by the author in 1995, there was a large bulge in the highwall where mine employees said the overlying rock was too hard to mine through, and they would not go to the trouble or expense of blasting. The overburden at that point was only about five meters thick and two exposed seams were probably two meters thick, providing very economic stripping ratios, but mine operators planned to cover this coal with spoil and leave it in the ground. Moreover there was a steady fall of water over the highwall, indicating the operation had mined through a stream instead of diverting it. Portions of the floor of the pit were covered with more than a meter of water, which was being pumped out of the pit at some expense to allow truck and shovel access. Equipment was repeatedly extracted from deep water and mud on the pit floor. This delayed and increased the cost of production, and was not efficient mining practice.

At another Mahakam River mine[8] visited in June 2001, the mine manager stated they wash all their coal for marketing purposes. This makes the product appear black and shiny and enables the mine operator to tell their Japanese buyer it has all been washed clean. This mine discards everything smaller than 2 mm as waste, and is probably discarding 20 percent or more of their product. Merely labeling the fine coal fraction of a resource as waste, rather than recovering it, is not efficient mining practice. Moreover it does not give the government fair value for the portion of the resource discarded.

At a third Indonesian mine in April 2002, visitors were told operators leave about a foot of coal in overburden trucked from the pit at the bottom of each coal seam and discard another four to eight inches on the top to keep from getting clay and mud into the product. Most Indonesian coal operations mine multiple seams in each pit, and this mine has eight minable seams, so they were wasting almost a half-meter of saleable coal per seam. This practice was observed at a fourth mine in March 2002 (Whitehouse, 2002). Personnel at the fourth mine estimated they could recover another 600 000 tonnes per year from adding a cleaning plant and producing the discarded coal.[9] In April 2002 this mine requested bids for construction of a washing plant so they could recover the floor and top coal. Using 2003 production figures, at US$27 per ton this would increase sales by about US$16.2 million per year, and increase royalties paid to government by about US$2.19 million per year from the portion of coal previously disposed of as waste. Other Indonesian mines that do not wash coal probably discard the top and bottom coal to keep the product they ship relatively clean.

It seems evident underinvestment has occurred because current Indonesian minerals policy does not require mining operations to be efficient, and there are no sanctions for failing to make investments necessary to recover a higher percentage of reserves. There is no requirement in Indonesian mining policy for "maximum economic recovery" of a mineral resource, unlike other countries which require that: "Surface mining activities shall be conducted so as to maximize the utilization and conservation of the coal, while utilizing the best appropriate technology currently available to maintain environmental integrity ..." (*U.S. Code of Federal Regulations*, 2000. Title 30, Part 816.59). Such policies are intended to maximize production of resources and avoid wasting portions of the resource while minimizing environmental disruption, by requiring that all economically recoverable coal in a particular location be mined at one time, even if some seams are of marginal value at current market prices. High-grading or taking only the easy coal, discarding marketable fine coal, floor and top coal as waste are inefficient mining practices effectively prohibited by policies requiring maximum economic recovery.

For comparison, relevant provisions in the standard Indonesian Contract of Work authorizing mining operations state:

> The Company shall, in accordance with prevailing environmental protection and natural resource preservation laws and regulations of Indonesia from time to time in effect, conduct its operations under this Agreement so as to minimize and cope with harm to the environment and utilize modern mining industry practices to protect natural resources against unnecessary damage, to minimize pollution and harmful emissions into the environment, dispose of waste in a manner consistent with good waste disposal practices, and in general provide for the health and safety of its employees and the local community. The Company shall not take any acts which may unnecessarily and unreasonably block or limit the further development of the resources of the area in which it operates.
>
> (Republic of Indonesia, 1998, p. 53)

These provisions do not effectively prohibit inefficient mining practices such as high-grading, or discarding marketable product as waste: They do not require mining operations to produce as much coal as is economically recoverable. Moreover such general statements do not provide an adequate basis for enforcement action because they do not specify a performance standard that indicates how laws and regulations are to be applied at particular mining operations.

Indonesian mine inspectors do not estimate the volume of coal extracted from the ground based on changes in elevation of the mine pit, unlike volumetric calculations which are routinely made by U.S. inspectors to verify production. Indonesian production figures and mineral royalty payments to government are based on shipping or sales invoices, and no provision is made to independently verify the accuracy of such data. This is not prudent accounting practice. Officials of regional and local governments have questioned the accuracy of production data published by the Ministry of Energy and Mineral Resources for many years, claiming it is understated. Absent a requirement for maximum economic recovery of the resource, and routine use of standard procedures to verify production data, there are insufficient incentives for coal mine operators to make additional investments to increase productivity. A more appropriate policy would be for the Indonesian government to collect royalties based on all mineral reserves extracted from the ground, not merely on figures in shipping or sales invoices, and verify production through volumetric calculations. This would make the mine operator more accountable for wasting a resource owned by the people and managed by the government, and it would provide incentives for more efficient mining practices. There appears to be no barrier in the 1945 Constitution to establishing such a policy.

Need for Erosion and Sediment Controls

An important first step in recovering fine coal is to capture it at coal processing facilities and prevent it from escaping into streams and rivers. Usually this can be done while preparing coal for shipment. Little effort to accomplish this is currently evident at most mining operations in Indonesia. From June 1995 to August 1998, Indonesian and American personnel participating in the BLT – OSM Mining Environmental Project visited every large coal mine and many small mines in Indonesia, most of them on multiple occasions. During that period, they did not see a single effective settlement pond at any coal mine (Hamilton, 1998b, p. 36).

Except for mining operations that have installed fine coal recovery systems as noted above, fine coals are not valued or systematically captured by most Indonesian mine operators. Often they are allowed to flow from coal stockpiles, crushing and washing equipment into settlement ponds, which are generally poorly designed, undersized, and infrequently maintained. Sometimes coal that could be recovered without washing is treated as waste material and disposed of in spoil dumps where it cannot be easily recovered in the future. Settlement ponds are often full of coal product which could be blended with run-of-mine coal and shipped to purchasers if it was separated from coal waste, colloidal clays, and slate slimes. Full ponds allow movement of water bearing particulate matter to flow out of them without dropping fine coal and associated impurities, carrying heavy load of solids into rivers and streams.

A lack of adequate performance standards for erosion and sediment control at coal mines and prep plants has allowed mine operators to claim settlement ponds are adequate to control water discharges when they are not adequate. These facilities may be inspected only once or twice per year. For comparison, all coal mines are inspected once per month in the United States. Moreover mine operators in Indonesia receive advance notice of inspections, make travel reservations, pay travel expenses and provide housing for mine inspectors. Consequently, inspectors often find prep plants are shut down and settlement ponds are being mucked out when they arrive, but it is unclear whether ponds have been cleaned since the previous inspection. When inspections occur only once or twice per year, mine operators know they need not make the effort or expenditure to maintain settlement ponds in working order in the interim.

Effective maintenance of settlement ponds might require them to be cleaned out 4–12 times per year, depending on their size relative to water volume and production capacity. When observers report six-months growth on shrubbery in the center of a settlement pond during a visit, it seems unlikely the pond is receiving adequate maintenance. There is ample evidence on the ground and in the rivers demonstrating that mine operators in Indonesia will not effectively control storm water runoff unless required to meet specific performance standards.

To be effective, settlement ponds must allow water carrying suspended solids to move across an area with a volume and rate of flow slow enough to allow suspended particulates to fall to the bottom of a reservoir by force of gravity. This may be aided under controlled circumstances by the addition of chemical floculents which assist aggregation of smaller particles into larger masses. Effective settlement ponds generally require a flow rate of less than 4 cubic feet per second (cfs) and a pond depth of greater than 48 cm. Water carrying suspended materials at a higher rate of flow over shallower depths will generally not allow materials to settle out before leaving the impoundment. Fast water does not have time to drop its load of particulate impurities. This is important because it means that structures which are not maintained with sufficient depth and area to allow slow rates of flow are not effective as settlement ponds. Reservoirs which are full of sediments function more like streams than settlement ponds, because the amount of suspended solids in water flowing out of them is not greatly reduced from the amount in water flowing into them.

A properly designed, sized and maintained system of settlement ponds usually includes four to six ponds through which water flows slowly in sequence. The most effective design will usually allow water to flow into and out of settlement ponds at the water surface. Subsurface inflows are less effective because they tend to stir up sediments, and subsurface outflows may flush some sediments into the next pond. Best practices in the United States indicate inflows and outflows should be designed so that any three ponds can remain in operation while a fourth one is being cleaned out or repaired. Impoundments must be large enough in area and depth to handle expected volumes of storm water and sediment with a slow flow, and must be regularly maintained and cleaned out to be effective. This requires removal and appropriate disposal of sediments with sufficient frequency to allow slow flows and settlement to occur continuously.

Water entering the first pond may be black or brown, heavily laden with suspended materials like coal dust, dirt and clay, but should exit only after most solids have settled out. Water entering subsequent ponds should be visibly cleaner or less turbid than that entering previous ponds. Water entering the last pond in the sequence (the finishing pond) should be virtually clear, not cloudy, and water leaving the last pond should be actually clear. With proper sizing and design, it is not difficult to meet these requirements. Inspecting them is a rather straightforward matter of onsite observation of outflows from the finishing pond, and sampling for control of pH and total suspended particulates for comparison to applicable performance standards.

Effective enforcement of existing performance standards for water discharges from mining operations would encourage more effective control of storm and process water at prep plants. Enforcement of a "clear water" performance standard for all water discharges, including those from settlement ponds and prep plant runoff, would require effective design, operation and maintenance of facilities for control of storm water and process water used at coal prep plants. In the United States, many state regulatory agencies require properly designed settlement ponds to be cleaned out when they are about one-third full of sediment. Inspectors file notice work must begin within 30 days or a violation will be cited, which will require payment of a penalty. If the work is timely begun, no penalty is assessed. Some states allow sediment to fill 50 percent of holding capacity before an inspector notifies the mine operator it must be cleaned out. There is no comparable requirement or performance standard for maintenance of settlement ponds applicable to Indonesian mining operations.

Effective enforcement of performance standards for erosion and sediment control would encourage fine coal recovery at prep plants and more diligent maintenance of settlement ponds. These standards would require regular mucking out of settlement ponds so they could actually function to settle solids out of runoff and process water from prep plants. This would encourage fine coal recovery at the prep plant so settlement ponds would not fill up as rapidly, or require cleaning so frequently.

Construction of the fine coal recovery and water reuse system described in this chapter would virtually eliminate the need for settlement pond construction and maintenance, because washing plant water would be reused and there would be zero water discharge from the prep plant. Even if the fine coal circuit is constructed without a refuse dewatering circuit, use of a clarifying thickener cleaning the recirculating water would reduce discharge water to settlement ponds by about 1/3 to $\frac{1}{2}$, reducing solids discharged to settlement ponds by 15 percent to 40 percent, so ponds would take twice as long or more to fill up, depending on recovery rate. Material entering settlement ponds would essentially be mud, which after drying by evaporation in the tropical sun can be returned to the coal pit and mixed with backfill or covered during pit reclamation (Duckett, 2001b). Mined-out pits provide excellent opportunities for burial of settlement pond sediments during reclamation. Costs for transporting refuse back to the pit, and moving only clean coal product (e.g., minus rock, clay and slimes) to the load out facility would be minimized by locating washing plants near the mine pit, instead of the usual Indonesian practice of locating them near a load out facility beside a river on filled wetlands.

Technical Assistance for Recovering Fine Coal

The theory of international environmental relations suggests that for policy transfer to occur between countries, "sustained diplomatic efforts must be initiated or explicit threats communicated" (DeSombre, 2000, Ch. 1). Yet technology diffusion (Rogers, 1995), policy transfer, and reformulation of environmental policy by one government may be stimulated and assisted by another through technical training activities, advice on regulatory program development, and assistance in solving environmental problems. In this instance neither sustained diplomatic efforts of persuasion nor explicit threats may be evident, but policy change may be facilitated by a more subtle but deliberate exchange of ideas. Little has been published about the spread of new ideas between countries through science and technology cooperation, technical assistance arrangements, or other small instruments of international relations. Yet foreign policy objectives have often been pursued through means other than direct diplomacy, utilizing personnel and expertise of technical agencies of government beyond those charged with the conduct of diplomacy. Such objectives may include improved understanding and maintenance of friendly relations between the United States and a country having the largest Muslim population in the world, Indonesia.

Development of performance standards and technical guidelines for recovery of fine coal at Indonesian prep plants and training in implementation of new standards were part of a technical assistance project funded by the U.S. Department of State in 2001 (U.S.DOI/OSM, 2000). A half-day workshop for regional and local government officials concerning estimates of lost profits and royalties due to lost production was held near the Mahakam River in Tenggarong, East Kalimantan in February 2001. Visits to Mahakam River mines by personnel skilled in design and operation of coal prep plants with fine coal recovery systems took place in June 2001 and April 2002 (Whitehouse, 2002).

Following the second visit, PT Bukit Baiduri hired an experienced Australian expatriate to manage their port, prep plant, and keep coal quality within contract specifications. After four weeks there were visible improvements in water management from the prep plant and coal stockpile areas, the settlement ponds were cleaned out and functioning, and there was work underway upgrading the coal cleaning plant. PT Kitadin also cleaned out their settlement ponds, and water discharges to the Mahakam River from these two operations were noticeably cleaner (Whitehouse, 2002b). Apparently these mine operators accepted the new concept of fine coal recovery and were moving to adopt measures to implement it.

In March 2002, a study tour brought personnel from the Ministry of Energy and Mineral Resources Training Center in Bandung, regional government Dinas Pertambangan and BAPEDALDA to the United States to examine fine coal recovery circuits at several prep plants. These individuals will have a role encouraging Indonesian mine operators to invest in similar equipment, so they were introduced to American personnel capable of providing design services. They visited several plants with fine coal recovery systems in Maryland and West Virginia, guided by individuals who designed, built and run this equipment as well as some plants built in Indonesia. Subsequently, two American prep plant designers visited mining operations on the Mahakam River in April 2002 and presented seminars for coal company representatives and government officials in Tenggarong, East Kalimantan, and Jakarta on design and operation of fine coal recovery systems.

Another study tour brought a second group of staff from the Ministry Training Center, the Ministry of Environment, regional and local government planning agencies, Dinas Pertambangan and BAPEDALDA to the U.S. to visit prep plants in June 2003. A second round of discussions was held with senior officials of Dinas Pertambangan, BAPEDALDA and mining firms in East Kalimantan during July 2003.

Fine Coal and Performance Standards

Fine coal recovery systems hold great promise for increasing productivity and profits of existing and planned coal mining operations in Indonesia. Construction of prep plants utilizing fine coal circuits would provide significant opportunities for development of new environmental businesses, as discussed in Chapter 8. Increased royalties paid to the government on recovered fine coal would assist in securing environmentally sound sustainable development by reducing environmental degradation of surface waters from Indonesian mining operations, while providing capital for investments in needed infrastructure and non-extractive industries.

It would be a significant improvement in Indonesian mining environmental policy if royalties were required to be paid to the government on all quantities of coal actually removed from the earth, not just on coal sales. Steps are needed to ensure fine coal particulates are captured and royalties are actually paid on them at existing and future mining operations. Mine inspectors can easily be trained in techniques to estimate the volume of coal removed from the ground and verify production at the pit, instead of merely at the shipping invoice. This would introduce prudent accounting practices to ensure fair value is actually received for mineral resources extracted from the earth. It would provide an incentive for Indonesian mine operators to improve both the efficiency of mine operations and environmental quality at the mine site and surrounding environs.

It is the general policy of the Republic of Indonesia to require royalties are paid to the government on all coal removed from the earth. This policy is not effectively implemented. Royalties are only collected on coal shipped from the mines. Steps are needed to ensure fine coal is captured and royalties are actually paid on all coal removed from the earth at existing and future mining operations. Such steps will increase royalty revenue and assist the government in reducing visible degradation of river environments near mining operations. These steps will also advance national and regional goals of economic development by ensuring coal extraction and profits from existing investments provide maximum value to local economies.

Through bilateral technology cooperation and technical assistance, deliberate efforts to transfer knowledge about fine coal recovery were initiated, and a few "innovators" and "early adopters" (Rogers, 1995, pp. 263–264) of these innovations are now evident in the Indonesian coal mining industry. Establishment of new performance standards for soil and erosion control, effective enforcement of water discharge standards for coal prep plants, development of a requirement for maximum economic recovery of the resource and procedures to verify production data would all provide significant incentives for more efficient mining practices at Indonesian coal mines. Efforts to assist the Indonesian Ministry of Energy and Mineral Resources in reformulating mining environmental policy consistent with these observations are discussed in the next chapter.

Notes

1 Unless otherwise noted, all production figures for Indonesian coal are given in metric tonnes.
2 Calculated from export/production data by mine in Ministry of Energy and Mineral Resources, 2004, pp. 12, 18.
3 A 150 mesh screen has 150 open spaces per inch, or about 59 open spaces per centimeter, allowing particles of <0.10 mm to pass.
4 In the fine coal cleaning circuit, this equipment includes: one glandless slurry pump and sump; three 10″ (254 cm) heavy-medium (e.g., magnetite) cyclones; one set of spirals; two dewatering sieve bends; and one screen bowl or modified centrifuge. In the integral water clarification circuit: one 40′ to 50′ (12 m to 15 m) thickener; one 6′ (~2m) underflow pump and one disc filter (optional), or discharge of sludge to pond for settling. Integrated into the water clarification circuit are the necessary magnetite storage bin; mixing sump; gravity controls; sump level controls; magnetic recovery separators with dilute media sump and pump (Duckett, 2002).
5 Using a more conservative price of $21/tonne for increased sales produces 100 percent capital recovery on the US$1.2 million system in one year with profit of US$198 200, and payment of additional royalties of US$226 800 (Duckett 2001c). On the US$1.8 million system, break-even capital recovery would occur in about 15 months at $21/tonne.
6 Calculated from export/production data by mine in Ministry of Energy and Mineral Resources, 2004, pp. 12, 18.
7 PT Tanito Harum.
8 PT Multi Harapan Utama.
9 PT Indominco Mandiri has the same corporate management as PT Kitadin, and is located on the Sengatta River near Kutai National Park north of the Mahakam River.

Chapter 6

Developing Mining Environmental Policy in Indonesia

Development of more effective mining environmental policy was the major goal of an innovative project in which technical assistance was provided by the U.S. Office of Surface Mining Reclamation and Enforcement (OSM) to the Ministry of Mines and Energy (now Ministry of Energy and Mineral Resources) and funded by The World Bank beginning in April 1995 (U.S.DOI/OSM, 1995b). A new Bureau of Environment and Technology (Biro Lingkungan dan Teknologi-BLT) was established in the General Secretariat of the Ministry to counterpart this project. The initial BLT-OSM Project was designed to develop institutional capacity for reformulating and implementing mining environmental policy in Indonesia. This involved actions directed towards several objectives, three of which were: (1) application of general environmental policy to mining operations, (2) improvement of procedures for review of environmental assessments, management plans, and inspections, and (3) strengthening the relationship between enforcement and permitting processes.

Initial Assessment

After relocating the duty station of a management-grade OSM Project Director in Jakarta, the first activity in the project involved performing an assessment in Indonesia to identify priority needs for improvement of both policy and procedural aspects of environmental impact analysis and inspection programs in the Ministry of Mines and Energy. An Assessment Team comprised of four OSM employees with substantial mining and technical experience, the Project Director, and the author visited mines in East Kalimantan and Sumatera in June 1995 to get on-the-ground impressions of what Indonesian mining and enforcement practices were like. This team included two geologists, two mining engineers and an environmental policy specialist. Three of these individuals were experienced mine inspectors, and three had combined mining experience exceeding 45 years with U.S. firms such as the Anaconda Company, New Jersey Zinc Company, and Buffalo Coal Company. Five were management and professional staff with long experience in OSM and one was a consultant from a university setting. The Assessment Team was accompanied by the BLT Director, a former Director of DTPU and mine inspector, and the Chief of the BLT Division of Technology Analysis and Construction Services, a former Section Head for Mining Activities in the Directorate of Mines who was also a

163

mine inspector. Members of this team would later participate on one or more teams providing technical assistance in the areas of Policy, Program Management, Technical Training, and Technology Cooperation.

The Assessment Team interviewed Ministry training personnel at research and training facilities in Bandung; regulatory personnel and mine inspectors in Jakarta and some regional offices of the national government; and mine managers and technical staff during visits to several operating coal mines. The Assessment Team found the technical challenges of Indonesian mining operations are substantial, as described in Chapter 2. During this visit and after their return to the United States the Assessment Team evaluated information and prepared recommendations for technical assistance concerning performance standards for mining and reclamation of land and information requirements for environmental assessments of proposed mining activities. They evaluated procedures for calculating reclamation performance bonds, guidelines for mine inspections, and criteria for identifying lands which are not suitable for mining. The Assessment Team recommended development of a system to assess penalties for violations of AMDAL documents and environmental standards, fees for processing mining approval applications, and bond forfeitures for sites left unreclaimed after mining ceased. These recommendations set the initial agenda for technical assistance provided by four joint-teams in the areas of Policy, Program Management, Technical Training, and Technology Cooperation during the three-year duration of the project. Findings and recommendations for improvement of regulatory program management are discussed in Chapter 7. Technical training needs and related technology transfer and assistance for development of more sophisticated capability for analysis of data are described in Chapter 8. Recommendations and technical assistance in the area of mining environmental policy reformulation are discussed below.

Policy Development

OSM and BLT established a joint Policy Team charged with examining Indonesian coal mining environmental policy to determine whether adequate legal authority and technical guidance were provided for implementation of an effective regulatory program. The Policy Team was charged with exploring alternatives for improvement of programs, administrative procedures, and technical guidance which relate general environmental policy and regulations specifically to mining activities. Concerns about enforcement capability were among those identified by the Assessment Team. Work providing a foundation for reformulation of mining environmental policy was initiated shortly after the project began. To facilitate mutual understanding, effective communication and appreciation of the cultural and legal traditions affecting the role of enforcement in mining environmental policy implementation in Indonesia, three Indonesian members of the Policy Team visited the U.S. in July 1995 to receive training with U.S. personnel in OSM courses. Indonesian members of the Policy Team who participated in these meetings included individuals who served in key positions affecting policy development and legal advice concerning review of ANDAL and the conduct of inspection and enforcement in the Ministry of Mines and Energy.[1]

These individuals participated in regularly scheduled OSM training courses on: (1) Enforcement Procedures (3 days); (2) Evidence Preparation and Testimony (3 days); and (3) Administrative and Legal Issues in Security Bonding (3 days). Because all natural resources are owned by the government in Indonesia, these courses (discussed in Chapter 8) were followed by a one-day training session on federal coal mine permitting in the United States, which is most similar to the Indonesian procedure for granting Mining Authorizations. An additional one-day session was conducted demonstrating use of a computer-based Technical Information Processing System (TIPS) for permitting and evidence preparation (discussed in Chapter 8). Completion of these courses provided the basis for a series of seminar discussions between Indonesian Policy Team members and their OSM counterparts concerning similarities and differences in U.S. and Indonesian legal systems, mining environmental regulatory policies, mine permitting procedures, and alternatives for policy development.

The Policy Team explored procedures and information requirements for permit applications and environmental assessment reports, reclamation performance bonds, mine inspections, and enforcement penalties for violations of requirements. Thus early Policy Team training and discussions provided an essential orientation and introduction to the program of work on the project for key personnel on both sides. Interviews conducted by the author in June 1996 indicated OSM team members' understanding of Indonesia's mining environmental policies and practices were greatly assisted by discussion papers presented in these seminars by Indonesian team members, suggesting this approach was an effective way to initiate policy discussions for the project. These discussions identified issues and topics for consideration during a review of Indonesian mining environmental regulatory programs, which was presented in Jakarta by OSM members of the Policy Team in October 1995. Development of new policies and procedures were subsequently undertaken within the Ministry of Mines and Energy, as discussed below.

Regulatory Analysis

Following the meetings described above, a comprehensive regulatory analysis was developed through examination of existing laws, regulations, and decrees, extensive interviews with knowledgeable individuals within the Ministry from both central and regional offices, and review of previously published literature on environmental regulation in the Republic of Indonesia (Miller, 1996a). This analysis provided an overview of the regulatory structure and institutional arrangements where additional policy development might increase program efficiency and effectiveness. Subsequently, it was used as a reference for policy development throughout the project. U.S. members of the Policy Team visited Indonesia in March 1996 for discussions on the role and function of regulatory performance standards and methods for their development; approaches and technologies for conducting inspection and enforcement actions; and methods for integrating mine permit decisions with environmental protection and management requirements.

Reclamation Performance Bond Policy

One of the first issues to attract the attention of Indonesian Policy Team members concerned the U.S. requirement that coal mine operators post a reclamation performance bond before mining commences. The purpose of requiring the bond is to provide a financial incentive for mine operators to reclaim disturbed lands for other productive postmining uses before the mine operation ceases, and to ensure that funds will be available for the government to reclaim the land if the mine operator is unable or unwilling to do so. In the U.S., often lands mined before enactment of the Surface Mining Control and Reclamation Act of 1977 were abandoned in a disturbed and unproductive condition after mining ceased, causing long-term environmental problems and threats to public safety and health. Indonesian Policy Team members had observed similar problems in their own country, and the use of financial incentives to obtain compliance with reclamation requirements apparently had substantial appeal.

In determining an appropriate amount for a reclamation bond, the costs of reclaiming mined land are estimated based on the depth and area of material to be disturbed during mining, characteristics of topsoil and water available for revegetation, and the planned postmining land use for the mine site. The amount of the bond must be sufficient for the government to contract with a third party to reclaim the mine site in the event the mine operator fails to meet its reclamation responsibility and the bond is forfeited. Thus reclamation is assured if a mine operator becomes financially insolvent before reclamation is completed. Because contemporaneous reclamation during mining is required, mine operators post a bond sufficient to reclaim the maximum amount of acreage estimated to be disturbed at any one time during mining. Portions of the bond are refunded to the operator in phases as the land is actually reclaimed successfully.

The OSM course on Administrative and Legal Issues in Security Bonding provided Indonesian members of the Policy Team with information on how to estimate the costs of reclamation and evaluate bonding instruments and liability insurance. The course covered business organizations (corporations, partnerships, joint ventures, actions requiring permit transfer such as mergers, acquisitions or dissolutions, and the authority to do business) and financial instruments. It explained the pros and cons of using various instruments, their evaluation in particular circumstances, financial status of banks and sureties, how they work during bankruptcy and insolvency, surety bonds, letters of credit, certificates of deposit, bond forfeiture, alternative bonding systems, public liability insurance, real property collateral, securities, and self-bonding. A half-day introduction to bond calculation involving estimation of the costs of reclaiming mined land was also conducted. Recognition of this basic technique was apparently not commonplace among policy makers or inspectors in Indonesia, and new concepts to insure sustainable uses of mined

Figure 6.1 Fast growing softwood reclamation revegetation on eroding ungraded spoil with invasive ground cover, after about 18 months, East Kalimantan
Source: Michael S. Hamilton

land were introduced to Indonesian members of the Policy Team for the first time during these sessions.

As a result of this training and subsequent discussions over a period of several months, the Ministry decided to adopt a performance bonding program for mining operations. On 17 July 1995, a Decree of the Minister of Mines and Energy instituting a Reclamation Guarantee Program came into force (Ministry of Mines and Energy, 1996c). The Ministerial Decree is general in nature, and work on a more specific decree from the Director General of Mines to implement it continued during the next twelve months. The OSM Project Director participated in these discussions, provided comments on drafts of the Director General's decree on more than six occasions, and published an article about reclamation bonds in the Ministry's *Buletin Informasi Lingkungan* (*Environmental Data Bulletin*) which assisted in discussion of these issues (Whitehouse, 1996d). A method for calculating a reclamation performance bond was described, along with data needs and useful sources of information concerning estimation of construction costs, equipment productivity and performance guidebooks (e.g., see Dodge Building Cost Services, 1987; Engelsman, 1985; Waier and Balboni, 2003).

On 1 August 1996 the Director General of Mines mandated that all coal mining companies post a reclamation guarantee before beginning production, to serve as a deposit to insure timely and proper reclamation of mined areas (Ministry of Mines and Energy, 1996c). Decree 336/1996 implementing this new requirement for reclamation guarantees (e.g. bonding) brought to fruition many months of policy discussions between OSM personnel and the Ministry. This was a significant step toward the goal of sound environmental management through mine reclamation.

The Decree requires that companies plan for mine closure and reclamation before beginning mining operations, and provide funds for the Ministry to implement the plan if the company fails to do so. This was a departure from previous practice, which allowed mine operators to prepare mine closure plans in the final few years of operation. Under Decree 336/1996 when planned reclamation is completed by a company, the guarantee will be refunded along with any accumulated interest. The reclamation guarantee must be posted before authorization to begin the exploitation phase will be issued. The guarantee may be in the form of a time deposit, an accounting reserve, or a third party guarantee (e.g., surety bond or irrevocable letter of credit from an approved bank).

The amount of the reclamation guarantee is based on reclamation costs specified in the mining operation's Annual Plan of Environmental Management (RKL) covering a 5 year period, which under Decree 1211/1995 was required to contain an approved reclamation plan (Ministry of Mines and Energy, 1995b). Costs covered by the Indonesian reclamation guarantee include: rent for heavy equipment; refilling the mine pit; land surface management such as regrading, spreading fertilized soil and revegetation; erosion control and water management including costs for preventing and handling acid mine water. Cost calculations must be based on the assumption reclamation will be carried out by a third party, so indirect costs for heavy equipment mobilization and demobilization, costs for administration, planning and design, and contractor profit with prevailing taxes will be included. Significant portions of these costs can be avoided if a mine operator performs reclamation work in a timely manner, so there is an incentive for contemporaneous and effective reclamation.

The requirement for posting a reclamation guarantee applied retroactively to all existing mining operations. Existing mines were given 180 days to calculate and post their guarantees. Unfortunately, the short implementation deadline and lack of training in reclamation cost estimation created some confusion within the industry about what they should do, and in the Ministry over how they were to review industry submissions. Confusion caused delay and at the end of 180 days, there were still many companies which had not submitted proposals for reclamation guarantees to the Ministry. Authorization and procedures for deposit and release of reclamation bonds were included in the 2000 Ministry of Energy and Mineral Resources decree formally decentralizing approval of Mining Authorizations, CoW, and AMDAL documents to regional governments (Ministry of Energy and Mineral Resources, 2000, Article 5).

Reclamation guarantees reduce risk to the government and society. By requiring companies to post a reclamation guarantee which will cover costs implementing an approved mine reclamation plan, the government limits its financial risk to reclaim the land if the company fails. Mine operators have a financial incentive to minimize the area of disturbed land so as to minimize the amount to be reclaimed at any time, and therefore minimize the amount of the bond. Companies have a further incentive to reclaim mined land because they will receive a complete refund of funds after reclamation is completed and approved. Moreover mining firms have an incentive to explore innovative techniques to reduce the actual costs of reclamation below estimated costs.

Amounts of reclamation guarantees in the U.S. range from a statutory minimum of US$10 000 to whatever amount is necessary to accomplish an approved reclamation plan. The government holds a guarantee until reclamation is complete, which is defined as between 2–10 years after planting, depending on annual rainfall and success of revegetation efforts. Releases of guarantees in the U.S. are staged according to the following formula: 60 per cent after rough grading of the last parcel mined, 20 per cent after topsoil has been spread and seeds have been planted, and the final 20 per cent after a waiting period to determine overall success of reclamation. In cases of noncompliance or company bankruptcy, the government should have sufficient money to contract with a third party to complete any remaining reclamation according to the approved plan.

Indonesia patterned its reclamation guarantee decree after the reclamation bond concept in the U.S., including provision for phased release of the bond. Many questions were raised by the industry concerning how the Ministry would calculate the reclamation bond required, how they would evaluate reclamation plans, and the process for returning the money after reclamation work is done. The decree did not specify a process for appeal of these decisions and offered few standards to measure performance or results, so there was quite a bit of uncertainty. Several important provisions missing from the Indonesian decree and regulatory program led to confusion in the industry and the Ministry. These include requirements for an approved postmining land use plan; an approved mining plan providing materials handling plans, excavation sequences, schedules, and adequate maps; an approved reclamation plan describing postmining topography, revegetation and locations; and standards for measuring reclamation success (Whitehouse, 1999, p. 66). The missing components are considered prerequisites for effective calculation and implementation of reclamation guarantees, and determine when mining operations recover their bond funds. Without these components, it is doubtful the Indonesian reclamation guarantee can be implemented successfully.

In the U.S. regulatory system a postmining land use for each mining area is stipulated and approved before any permit to mine is issued. The postmining land use determines the contours and placement of material for reclamation, which in turn determines the costs of moving and handling these materials, backfilling and grading, and revegetation. Mining plans in the U.S. require detailed excavation schedules, pit and waste dump designs so material handling and potential decommissioning costs of the reclamation plan can be calculated for each stage of mining. Concern for minimizing reclamation costs encourages more efficient mining practices and one-time placement of overburden so it does not require movement to a second location, with significant savings in fuel, labor and equipment maintenance. The reclamation plan includes not only a revegetation plan but also any water quality amendments or additional work that must be done to achieve the postmining land use. Standards for measuring reclamation success determine the timing of final bond release to the mining operation.

Most existing mines operating in Indonesia were not required to identify a specific postmining land use prior to beginning operations, and do not have a closure plan on file in the Ministry or local government offices. Most mines operate without specific requirements to return the land to some agreed condition or use after mining ceases, and with only rudimentary mining and reclamation plans. Consequently, overburden, mine waste and settlement ponds have often been located on surfaces where there are minable coal reserves, and therefore required double-handling before reaching a final resting place, at great additional expense to mining operations. Planning documents and maps are not required to contain sufficient detail to allow a reasonable calculation of reclamation costs. Difficulties experienced during attempts by mine operators to calculate required reclamation guarantees in the absence of an identified postmining land use, mining plan schedule, and adequately detailed maps highlighted this shortcoming. One must know to what configuration the land will be reclaimed if one is to effectively estimate the costs for reclaiming it.

A policy bias in favor of large publicly traded companies is apparent in the form of reclamation guarantee required, and in provisions governing its release. Only publicly listed firms with an equity position of US$25 million that are registered in a stock exchange inside or outside Indonesia may post a reclamation guarantee in the form of an accounting reserve (Ministry of Mines and Energy, 1996c, Article 6). After a request for release of all or a portion of the reclamation guarantee is received by the Director General of Mines, it must be granted within 45 days or it is considered approved (Ministry of Mines and Energy, 1996c, Article 13.3). Any inspection of the reclaimed site must be carried out within 15 days of the request for release, an exceedingly tight time frame. There is no procedure for a firm to claim a reclamation guarantee posted in the form of a time deposit held "in a state-owned bank in the name of the Director General" or third party guarantee if released by such automatic approval, but a guarantee posted in the form of an accounting reserve would be under the sole control of the mining firm, and require no additional paper work for release.

A requirement for contemporaneous reclamation is implicit in the form of authorization for the Director General of Mines to temporarily close a mining operation and assign a third party to conduct reclamation activities using part or all of the reclamation guarantee if the mine operator neglects to implement the reclamation plan (Ministry of Mines and Energy, 1996c, Article 17). However no procedure is specified and it is unclear whether the reclamation guarantee must be replenished before operations may resume.

Decree 336/1996 contains a provision barring mining firms and their majority stockholders from securing future mining authorizations if their mining operations have been terminated due to negligence or failure to carry out their reclamation obligations (Ministry of Mines and Energy, 1996c, Article 17.3). This provision is similar to one in the Surface Mining Reclamation and Control Act of 1977 which bars issuance of mining permits to firms which have previously demonstrated a pattern of willful violations of regulatory requirements (30 *U.S. Code* 1260(c)). Implementation of this provision in the U.S., where there are over 8000 inspectable coal mine units, required creation of an Applicant Violator System providing a centrally-maintained data base of information about previous mining applications, permits, and violations sufficient to determine whether an applicant for a new permit owns or controls operations cited for previous violations of regulatory obligations. In Indonesia, where less than 1000 Mining Authorizations and Contracts of Work have been issued, there is no comparable system for checking ownership and control of mine operators. In a decentralized regulatory system with increasing numbers of mine operators, there is no mechanism for implementing this provision of the Decree, other than manually checking paper files which may not be located within the jurisdiction of a regional government.

Finally, one of the most important pieces missing in the Indonesian regulatory program is a standard for measuring reclamation success, on which a process may be based for returning performance guarantees after reclamation is completed. Moreover the decree does not specify a process for appeal of decisions concerning financial adequacy or refusal to return reclamation guarantees by the Ministry. Without established definitions and measurements for success, mining companies and Ministry staff will be unable to measure compliance objectively, or determine when final bond release is appropriate. Reclamation success standards do not need to be complicated. They can be as simple as specifying the number and spacing of living trees on a reclaimed area at the end of some period after planting suitable to local or regional climate and soil conditions. In the U.S., such periods range from 5–10 years (*U.S. Code of Federal Regulations*, 2002, Title 30, Part 816.116), and are largely determined by degree of aridity, because more arid areas require longer periods for successful establishment of vegetation capable of self-regeneration and plant succession. Standards for reclamation success should be part of the authorization to mine and could be easily inspected by both company and ministry staff during mining and before final bond release.

Calculation of reclamation guarantees   Responding to initial confusion over calculation of reclamation guarantees, in late August and early September 1996 two U.S. members of the Policy Team visited Indonesia to assist the Ministry in developing a strategy for effectively implementing the decree on reclamation guarantees. Policy Team member David Lane, one of OSM's principle instructors in the area of reclamation cost calculation, conducted three separate seminars on Cost Estimation for Reclamation Guarantees for 13 senior members of the Ministry's Central AMDAL Commission, 20 mine inspectors, and 19 representatives of the mining industry representing 16 of the

20 largest mining operations. These seminars provided open forums for discussion, allowed personnel from the Ministry to better understand industry concerns, and created opportunities to familiarize interested parties with the operation of the new requirement for reclamation guarantees. Additional training was provided in February 1998 in a seminar sponsored by BLT in Jakarta that was attended by about 100 persons from various units of the Ministry of Mines and Energy (including Kanwil regional offices) and the mining industry. In March 1998 another seminar attended by about 40 mining industry representatives was presented at the Manpower Development Center for Mines (now Education and Training Center for Mineral and Coal Technology) in Bandung. Thus methods for estimating reclamation bond amounts were widely disseminated in the government and coal mining industry in Indonesia during this period.

Interest in reclamation guarantee policy may reflect an Indonesian aversion to direct conflict while searching for solutions to recognized problems, and a corresponding preference for use of financial incentives over enforcement of command and control regulatory requirements. Use of financial instruments to guarantee land reclamation softens some of the more confrontational aspects of regulation by encouraging an expectation that funds will be refunded if desired behavior is demonstrated. If a reclamation bond is forfeit, the result is analogous to payment of compensation for harm done, consistent with traditional *adat* law concerning the uses of land and associated natural resources, discussed in Chapter 1. Moreover payment of compensation to the government for restoration of damages done to natural resources is consistent with requirements in the Act on Basic Provisions for Management of the Living Environment of 1982 (Republic of Indonesia, 1982, Article 20; 1997, Article 34). Presumably the funds would be dedicated to restoring the damage, but this is not explicit in existing decrees or regulations.

However creation of a major new program requiring mining operations to provide time deposits, accounting reserves, or a third party guarantee under the control of Ministry or local government officials provides new opportunities for extraction of informal "user fees"[2] from mine operators in the absence of a sound basis for estimating the costs of reclamation, and explicit standards for determining reclamation success. Disagreements about financial adequacy of the reclamation bond amount to reclaim a mine site may delay issuance of mining authorizations, and ambiguity concerning the timing of final bond release may be resolved with user fees if discretion is left to officials to determine these issues. Specification of a postmining land use; an approved mining plan with materials handling practices, excavation sequences and schedules; an approved reclamation plan with maps describing postmining topography and revegetation; and standards for measuring reclamation success would constrain the use of official discretion and reduce opportunities for extraction of user fees. Establishment of a fixed reclamation cost per hectare would simplify the bond calculation process, and if set high enough to reclaim mined land under worst-case conditions, this might induce mine operators to provide sufficiently detailed maps and plans from which more accurate reclamation costs could be calculated on a case-by-case basis.

In any case it is evident the idea of a reclamation performance guarantee was successfully transferred from the United States, a new policy was established, and extensive efforts were conducted in support of further diffusion of this innovation within Indonesia to government decision makers, enforcement staff, and mining industry personnel. Reclamation bond requirements and procedures of Decree 336/1996 were specifically incorporated into Decree 1453/2000 devolving these responsibilities to regional governments in 2000 (Ministry of Energy and Mineral Resources, 2000). It remains to be seen if this requirement will be implemented in a manner that effectively achieves reclamation of mined lands.

Guidelines for Preparation of AMDAL

A second area in which project activities produced early results concerned the adequacy of information requirements for environmental assessments of proposed mining activities. New draft guidelines for preparation of AMDAL were prepared by Ministry personnel which utilized concepts and techniques discussed during Policy Team meetings in Denver, Colorado and Jakarta during 1995. The OSM Project Director and Policy Team were asked to review and comment on several drafts of this guidance, and final Guidelines for Environmental Impact Analysis for Mining and Energy Activities were published by the Ministry of Mines and Energy on 9 August 1996 in Decree 1256/1996 (Ministry of Mines and Energy, 1996a). Development of these guidelines was an important step towards improving policy and procedures for review of AMDAL environmental assessments, and strengthening their relevance and utility as a basis for mine inspections and enforcement of environmental requirements, two important objectives of the project discussed below.

The Policy Team considered several significant changes concerning procedures for preparation and review of environmental assessments, and focused efforts on refining the substance and process of ANDAL review in seminars in Indonesia in the fall of 1996. These efforts were especially significant because the Ministry of Mines and Energy's AMDAL Central Commission had no internal directives to guide review of environmental assessments, leaving near total discretion to approve operations regardless of environmental impacts. Discussions continued during the next twelve months concerning the need for environmental assessments and permits for mineral and coal exploration that involve substantial excavation to prove reserves.

Penalties for Violations of Standards

Comments were provided on a draft policy proposing a schedule of penalties for violations of environmental standards by mining operations during 1995. Mine closure was the only sanction available to mine inspectors if a mine operator failed to comply with instructions for correction of inspected deficiencies in a timely manner. Because inspectors are understandably reluctant to take such severe actions where government revenues for development are involved, mine closure by itself is an ineffective enforcement tool where non-life-threatening violations are found. No Indonesian coal mining operation has ever been ordered closed for environmental violations (Miller, 1999, p. 325). A schedule of intermediate sanctions, tailored to the severity and frequency of the violation and cognizant of the mine operators previous record of compliance, would improve the effectiveness and consistency of enforcement at many mines. The Minister of Mines and Energy agreed a policy should be developed in this area, but the collapse of the Asian economy in 1997–98 and the struggle for succession to the Presidency after the ouster of President Soeharto in 1998 stalled mining environmental policy making in Indonesia for the better part of two years. No decree was published before the project ended in 1998.

Sediment and Erosion Control Guidelines

Lack of effective erosion and sediment control is one of the most serious environmental problems caused by mining in Indonesia, and efforts were directed towards this issue from the beginning of the project in early 1995. When the project began, the Ministry of Mines and Energy had no regulations or recommended practices for erosion and sediment control at mining operations. Technical guidance published by the Ministry focused on mineral extraction, soils were viewed as somewhat of a nuisance to be moved out of the way, and interviews suggested the Ministry organizational culture displayed little interest in soils or plants necessary for revegetation of mine sites. This lack of interest was also reflected in the low priority given enforcement of reclamation requirements at Indonesian coal mine operations.

To explore possible remedies for this deficiency, technical training of Indonesian personnel in erosion and sediment control was conducted in the U.S., with field exercises and tutorials provided to improve understanding of the problems, and feasible prevention techniques. When the project brought a specially designed training course on erosion and sediment control to Indonesia in October 1996, the OSM instructor spent a week consulting with Ministry personnel, answering questions, offering suggestions and assisting in drafting guidelines. Handbooks on erosion and sediment control from the United States provided the basis for these discussions. In November 1996 the OSM Project Director was asked to visit the PT Adaro coal mine in South Kalimantan where the mine operator reported one of its sediment ponds had breached and dumped its contents into a branch of the Balangan River, damaging water quality in a nearby village. The Project Director was asked to help evaluate sediment control practices being developed by mine managers following the failure of this mud pond.

This was a large, open-cast surface coal mine producing approximately 8 million tonnes per year. Surface water in drainage areas above the mine pit had not been diverted and was allowed to flow into the active working area. Much of the active coal removal area was covered with water, to depths greater than one meter in some places. Two large floating pumps were observed transferring a water/mud slurry created by movement of trucks and mining equipment to a

Figure 6.2   Prep plant settlement pond being cleaned out, showing shrub with six months' growth under arm of backhoe, East Kalimantan
Source: Michael S. Hamilton

first stage settling pond located outside the pit in a small stream. The slurry pumped from the mine contained a high percentage of fine coal and colloidal clays. A slightly thickened slurry was pumped to an adjacent pond approximately 70 by 20 meters in size. It was this pond that breached.

PT Adaro's production manager was one of the industry students who attended a project training course on Erosion and Sediment Control a week before this visit. During the visit, his staff was already actively applying many of the techniques learned in the training course. A second stage settling pond was nearly doubled in size to increase retention time, and construction was underway on three shallow third-stage settling ponds. Negotiations were in progress to acquire more land to construct additional ponds. These were substantial improvements, which indicated the BLT-OSM project was beginning to produce change in mining environmental practices on the ground in Indonesia.

Further evidence of policy change was apparent when recommendations by U.S. project personnel were adopted on 26 December 1996, and the Director General of Mines issued Technical Guidelines for Erosion and Sediment Control for Mining Activity in Decree 693/1996 (Ministry of Mines and Energy, 1996d). These guidelines included a formal listing of each recommended erosion or sediment control practice, structure and standard with its purpose, technical specifications, recommended application and methods for installation and maintenance. The guidelines do not have the force of regulation but express the Ministry's recognition and concern for the problem. Despite the fact the guidelines are not themselves binding, the Director General articulated his expectation that they would provide guidance for mine operators in controlling erosion during mining activities, serve as a reference for mine inspectors in conducting inspection and enforcement actions in the field, and that recommended erosion control efforts would be included in every future plan for a mining operation (Ministry of Mines and Energy, 1996d, Article 2, 3). Presumably this would include reclamation plans in the required Environmental Monitoring Plan (RPL), Environmental Management Plan (RKL), and annual updates to them submitted by all mining companies. When these management plans are approved, they become binding and enforceable.

Decree 693/1996 may be characterized as a preliminary step in an educational process intended to inform Indonesian mining operations and government officials about the effects of erosion from mining operations, and what must be done to control them. Thus the Decree resembles a brief tutorial in erosion and sediment control. It is not as specific as the recommendations which preceded it. Following a pattern discernible in many Indonesian Decrees providing technical guidelines, it is general in nature, describing factors influencing erosion (e.g., rainfall, slope conditions, and soil and rock properties), and the undesirable impacts of uncontrolled erosion: loss of soil, degradation of soil productivity and surface water quality, decrease of vegetative quality and water retention, increase of landslides, transportation of toxic contaminants downstream, deposition of silt resulting in decreased transportation capacity of rivers, loss of fertility in deposition areas, and erosion of stream banks and bottoms.

The decree describes planning activities that might be integrated into mine planning to accomplish effective soil and erosion control during mining operations, but is not well-related to other relevant decrees. It does not specify when these activities must be completed, at the feasibility study stage for a Mining Authorization or Contract of Work, or how they relate to preparation of an ANDAL, RKL, or RPL. Decree 693/1996 does contain several general principles for effective erosion and sediment control which do not appear in other regulations or technical guidance concerning Indonesian mining operations, as illustrated in Table 6.1.

These prescriptions were briefly elaborated in the Decree as principles, and constitute policy recognition of the importance of soil erosion and sediment control at Indonesian mining operations, but they were not stated as required actions mining operations must take. Consequently, the technical guidelines constitute only a preliminary step in the direction of effective sediment and erosion control at Indonesian mining operations, because they do not prescribe required actions with sufficient specificity to be useful to mine inspectors in enforcement actions. Subsequently, further efforts to encourage the diffusion of these ideas into the Indonesian mining industry and related offices of government were conducted, including development of a four volume training manual around these guidelines (Maynard 1998a, 1998b, 1998c) for a course Indonesian instructors were trained to deliver to mine inspectors and industry personnel, described in Chapter 8.

Table 6.1 Indonesian Principles for Effective Erosion and Sediment Control

---

- Fit the development to the existing terrain and soil.
- Develop an erosion and sediment control plan before land disturbing activities begin.
- Retain existing vegetation on the construction site wherever possible.
- Minimize the extent and duration that bare soil is exposed to erosion by wind and water.
- Keep sediment on the construction site as much as possible.
- Divert off-site runoff away from disturbed areas if possible.
- Minimize the length and steepness of slopes.
- Stabilize disturbed areas as soon as possible.
- Keep velocity of runoff leaving a site low.
- Inspect and maintain erosion control measures regularly.

Source: Ministry of Mines and Energy, 1996d, Appendix.

For comparison, specific soil erosion and sediment control practices are required in several national performance standards for surface coal mining operations in the United States. The use of surface water diversions, sediment control measures, siltation structures, and impoundments are specifically authorized and regulated to minimize disturbance of the hydrologic balance within mining operations and adjacent areas, and to prevent material damage outside mining areas. Diversions, sediment control measures and siltation structures must utilize the "best technology currently available" to prevent, to the extent possible, additional contributions of suspended solids or sediment to streamflow or runoff outside the mining area (*U.S. Code of Federal Regulations*, 2002, Title 30, Parts 816.43-816.46). Determining what constitutes "best technology currently available" is a matter for decision by an informed and trained enforcement officer, subject to administrative review by superiors, and may change over time as technology improves. Surface water flow from undisturbed or reclaimed areas may be diverted away from areas disturbed by mining by means of temporary diversions designed to pass safely the peak runoff of a 10-year, 6-hour precipitation event, or permanent diversions adequate to pass safely a 100-year, 6-hour precipitation event, as certified by a qualified registered professional engineer.

In the United States, approved sediment control methods include but are not limited to the items listed in Table 6.2. Some of these practices are quite similar to those in Decree 693/1996, but in the U.S. they are stated as requirements, not general principles or guidelines. Moreover in the U.S. all surface drainage from the disturbed area must be passed through a siltation structure before leaving the mining area. These structures must be constructed before surface mining begins and certified by a qualified registered professional engineer, must be maintained, and cannot be removed until at least two years after the disturbed area has been stabilized and revegetated. One of the most common siltation structures is one or a series of sedimentation ponds which must be designed, constructed and maintained to provide adequate sediment storage

Table 6.2   Approved Sediment Control Methods in the United States

- Disturbing the smallest practicable area at any one time during mining operations through progressive backfilling, grading, and prompt revegetation.
- Stabilizing backfill material to promote a reduction in the rate and volume of runoff.
- Retaining sediment within disturbed areas.
- Diverting runoff away from disturbed areas.
- Diverting runoff using protected channels or pipes through disturbed areas so as not to cause additional erosion.
- Using straw dikes, riprap, check dams, mulches, vegetative sediment filters, dugout ponds, and other measures that reduce overland flow velocity, runoff volume, or trap sediments.

Source: U.S. Code of Federal Regulations, 2002, Title 30, 816.45(b).

volume, contain a 1-year, 24-hour precipitation event, and provide adequate detention time to allow any discharge to meet applicable water quality standards (*U.S. Code of Federal Regulations*, 2002, Title 30, Part 816.46). All discharges of water from areas disturbed by surface mining activities must comply with applicable water quality laws, regulations and effluent limitations.

Disposal of excess spoil and mine waste must also be accomplished in a manner that minimizes the adverse effects of leachate, surface water runoff and erosion on surface and ground waters, including mandatory slope protection in the form of revegetation, and stabilization of all reclaimed surface areas. Rills and gullies which form in reclaimed areas which either disrupt the approved postmining land use or contribute to violations of water quality standards must be filled, regraded or otherwise stabilized, topsoil replaced, and the area reseeded or replanted (*U.S. Code of Federal Regulations*, 2002, Title 30, Parts 816.71; 816.81; and 816.95). Compliance with these requirements may be determined through observation during mine inspections by qualified personnel.

Coal Sector Policy Review

The Joint Policy Team in August and October 1996 provided comments on a Draft National Coal Policy developed for the Ministry by a consultant funded by Kaltim Prima Coal. Little was found in the draft policy statement which could be considered binding on the industry or the Ministry. There were no specific requirements for industry to meet, only general directions and goals. The draft policy document contained only general targets relating to increased coal production and suggested several other areas for consideration. Recommendations were that more specific targets should be developed which support accomplishment of the broad goals and targets of the coal policy.

In response to a subsequent request from the Ministry of Mines and Energy, the OSM Project Director in 1997 provided comments on a Draft Final Report for a "Coal Sector Policy Study for Indonesia" prepared by a consultant for the Ministry. He commented that the report was based on assumptions of economic growth and demand for electric power and coal production that might be overly optimistic because there were indications the economy was weakening. This was shortly before the sharp economic downturn of 1997–98. He suggested production and demand estimates should be revised downward, and that the effect of these downward trends might be salutary in that the Ministry might cope more effectively with the significant increase in workload certain to accompany even modest growth in the coal mining industry (U.S.DOI/OSM, 1998a).

Only brief treatment was given to the role of environmental management and strengthening of regulatory capacity in this coal policy study. Additional assessment of the capacity of the Ministry to conduct meaningful environmental analysis that would result in more useful input into the AMDAL process would have been useful. The report recommended changes in the regulatory scheme focusing primarily on the need for better control of and standards for water quality, and the need for some flexibility in the authority of inspectors to assess violations. The OSM Project Director suggested there was a need to move to more performance based standards in mine regulation and a need to address, through the regulatory scheme, all environmental aspects of coal mine operations (U.S.DOI/OSM, 1998a).

The report discussed resource quality and availability, but did not include a discussion of the concept of maximum economic recovery or the importance of proper mine operation to ensure that mining is conducted in such a way that the resource is fully extracted and conserved. For example, only slight mention was made of the diminution in coal quality and quantity by existing mining practices, which appeared to take only the highest quality and easiest coal to mine and failed to recover fine coal particulates from washing plants. The report established general time schedules for reorganization of coal regulatory functions in the Ministry, but did not establish schedules for related activities, particularly as they concerned necessary revisions in the Coal Contract of Work and the regulatory program. It was suggested it might be useful to have a separate document with action steps and target dates. Moreover the OSM Project Director suggested it might be useful to have a temporary organizational structure located at a high level in the Ministry with authority and responsibility to track, evaluate, and report upon the pace of adoption of recommendations, to ensure their successful implementation (U.S.DOI/OSM, 1998a). Absent constant high level management attention, it did not seem likely the recommendations would be fully implemented, and subsequently they were not implemented.

Understanding the process of policy innovation requires a time perspective of a decade or more to determine whether new ideas and technologies have been adopted and successfully implemented (Sabatier, 1988, p. 131). That there were some significant policy changes stimulated by the OSM project within a three year period is therefore cause for cautious optimism, and provides some evidence of success in planned diffusion of innovations. Matters such as the ones described above often require a substantial period of time for discussion within the Ministry before sufficient agreement is evident for action to be taken. Actions and recommendations of the Policy Team with regard to the policy objectives described at the beginning of this chapter are related to and often overlapped with activities of the Program Management Team. Consequently, the efforts of the two teams complement and reinforce each other in attainment of these objectives. Efforts of the Policy Team are supported by recommendations of the Program Management Team discussed in Chapter 7 below.

Notes

1   They included: Kadar Wiryanto, Chief, Division of Technology Analysis and Construction Services, Bureau of Environment and Technology, who also served as Expert Program Manager for this project; Mahyudin Lubis, Chief, Sub-Directorate of Environment, Directorate of Mines, which had responsibility for mine inspections and assisted in technical review of ANDAL prepared for proposed mining operations; and Satry Nugraha, Attorney, Legal Bureau of the Ministry of Mines and Energy, which provides legal advice to the Ministry.
2   Although the author was unable to find a credible estimate for the payment of user fees for illegal mining activities (which are widespread), the annual value of user fees paid for illegal fishing have been estimated at US$2–4 billion; for illegal logging at US$4–5 billion, and the total for all activities at US$23 billion per year in Indonesia (Lachica, 2004). However unless these funds actually leave the country, or are hoarded in unproductive savings, the assertion that such payments are a drain on the national economy is probably inaccurate, as they likely circulate in local economies purchasing goods and services.

Chapter 7
Improvement of Program Organization

Improvement of administrative organization, procedures and management of the regulatory program of the Ministry of Mines and Energy was one of the principal objectives of technical assistance provided by OSM beginning in 1995. An Assessment Team of six Americans with over 45 years mining experience traveled to Indonesia for two weeks in June 1995, visiting the headquarters of the Ministry of Mines and Energy in Jakarta, the Ministry Manpower Development Center in Bandung, two large mines in East Kalimantan, and the Ministry Kanwil regional office in Samarinda. Assessment Team members interviewed mine operators, Ministry personnel, and discussed their assessment with Indonesian counterparts. They identified a priority need for development of procedures which would formalize and direct communication between Jakarta mine inspection personnel and inspectors in Ministry Kanwil offices concerning mine inspections and input into requirements of the Terms of Reference (KA) and technical reviews of ANDAL. In identifying these needs, organizational and procedural aspects of strengthening the relationship between enforcement and permitting processes were emphasized to improve the enforcement strategy of DTPU.

A joint Program Management Team consisting of five OSM professional staff and five Indonesian personnel was established later in 1995. This team was charged with assessing the effectiveness and efficiency of existing organizational arrangements through a review of lines of responsibility, coordination and communication between program managers, technical and inspection staff and Kanwil offices of the Ministry. U.S. members of the Program Management Team visited Indonesia in April 1996 to present a seminar about principles of regulatory program management, and to review existing organizational arrangements of the Ministry. The Program Management Team interviewed numerous individuals involved in mining environmental programs of the Ministry. Their findings and recommendations, presented near the end of the first year of effort on the project, generally reinforced and extended discussions held earlier in the year within the Joint Policy Team.

Program Organization

The Program Management Team review in 1995 found that inspection and enforcement responsibilities for mining were dispersed within the Ministry. Mine inspectors located in Kanwil offices of the Ministry reported to their superiors who in turn reported to the Secretary General of the Ministry of Mines and Energy. However the Secretary General ordinarily has charge of staff and auxiliary administrative functions of the Ministry, but not line functions such as mine inspections. Mine inspectors located in the central office in Jakarta reported through a different chain of command to the Director General of Mines, who had line responsibility for mine inspections and enforcement of regulations. This made coordination of effort and sharing of information between mine inspectors in the central and Kanwil offices of the Ministry slow and cumbersome. Although all mine inspectors in the central and Kanwil offices had similar credentials, training, and were appointed by the Director General of Mines, inspectors in the central office were said to receive preference in assignments and status. This preference constrained training, professional development and career advancement opportunities for Kanwil inspectors and limited the value and travel cost savings the Ministry could gain by using them for routine inspections, follow up inspections and for increasing the frequency of mine inspections.

The Program Management Team also found the mine inspection process was not well-related to the AMDAL environmental assessment process. Pre-mining inspections were rarely conducted during the environmental assessment process because of cost and time constraints. Mine inspectors did not report their findings back to the Ministry's AMDAL Central Commission in a routine or timely manner, and violations of ANDAL requirements, conditions or agreements were not regularly enforced. Mine plans in ANDAL, RPL and RKL were not of sufficient detail to provide an adequate basis for mine inspections.

The ANDAL was not adequate as a decision making tool for determination by the Ministry of whether mining should or should not take place as proposed. ANDALs contained too little data, and the time allowed to review them was much too short for a meaningful technical evaluation to occur. Maps provided in ANDAL were of too large a scale and were not always tied to established benchmarks. Ministry guidance for preparation of ANDAL, RPL and RKL did not require much development of site-specific information regarding affected resources as part of the environmental impact assessment process. Environmental mitigation requirements or standards were not specifically described in the ANDAL or RKL. Lack of site specific information in both mine plan and environmental documents, compounded by very general environmental standards, made effective inspection and enforcement difficult.

The required feasibility study was completed before the environmental assessment, and therefore did not include environmental mitigation or reclamation costs necessary for the AMDAL Commission to make a sound financial decision about the feasibility of proposed mining operations. ANDAL could be amended by annual environmental reports submitted by mine operators to the Director General, without review by an AMDAL Commission. Thus mining and environmental management plans were changed annually without AMDAL Commission review. As described in Chapters 2 and 4, postmining land use and mine closure plans were not required before mining operations began. Land reclamation was not done concurrently with mining operations, but delayed until the end of mining when there might be insufficient revenues being generated to pay for it.

The Program Management Team found that responsibility for mine inspection and enforcement were within the same Directorate that was responsible for mineral development, attracting foreign investment, and collection of royalty revenues, presenting conflicts between commodity production and environmental protection missions. Inspections were too infrequent because of travel distances to the mines, personnel and budget shortages. Personnel and technical capability were in very short supply. Consequently, enforcement actions were rare and inconsistently applied to mining operations. There was no workable penalty system for violations of environmental standards.

There was little communication between mine inspectors in the Kanwil offices and Jakarta. A lack of coordination was apparent between the Kanwils, the AMDAL Central Commission, and inspectors in the Directorate of Mines. The Kanwils had many superiors and a confusing reporting system. Yet, there was evident an increasing public awareness of environmental problems and increased public concern about the environmental impacts of mining operations.

Reorganization?

In its report the joint Program Management Team recommended:

- Regulatory responsibility should be separated from functions concerning mineral development, production of royalty revenues and technical assistance within the Ministry. This would provide two units focussing on distinct functions at the same level of organization who could bring their concerns to a superior decision maker if necessary.
- Either a new regulatory organization should be created at the Directorate level or the Ministry AMDAL Commission should be provided with a permanent staff reporting to the Chair of the Commission. This staff would perform technical reviews of all AMDAL documents, recommend changes or conditions to projects proposed, manage agreements contained in approved ANDALs and recommend document amendments or changes to the Commission during the life of each project. This staff would manage the day-to-day requirements of environmental programs within the Ministry.
- Responsibilities for mine inspection and enforcement should be decentralized to Kanwil offices and all mine inspection reports should be routed through the new regulatory Directorate/staff to the Chairman of the AMDAL Commission.
- Standard Terms of Reference or a standard application form should be developed specifying the minimum information required for submission of AMDAL documents. AMDAL applications which do not contain the required information should be rejected by the AMDAL Commission after an initial review of the adequacy of information provided. Rejected applications should be accepted for technical review only if they contain all of the minimum information required.

- Mining Authorizations should be changed from permission to construct (explore, exploit, preparation plants, etc.) to permission to occupy a parcel of land for the preceding purposes for specified minerals at some specific royalty rate. The approved ANDAL should become the permission to begin construction.
- Economic feasibility studies should continue to be part of the basis for the Mining Authorization, but technical details of the mining plan and postmining land use should be required as part of the ANDAL for evaluation of environmental impacts.
- Exploration may require some kind of environmental permit depending on the degree of disturbance exploration activities would cause. A threshold level of disturbance (e.g., some number of hectares of disturbed land) should be identified which would trigger a requirement for such a permit.
- A penalty system should be finalized with increasing sanctions for increasingly serious violations of environmental standards, to improve compliance with ANDAL (U.S.DOI/OSM, 1996b).

Subsequently discussions focussed on development of the policies and decrees described in Chapter 6, and on modes of interaction and relationships between national and regional offices of the Ministry. As part of ongoing discussions, during June 1997 Indonesian personnel from the Bureau of Legal Affairs and BLT met with American personnel from the OSM Regional Office in Pittsburgh, Pennsylvania. The focus of these meetings was discussion of intergovernmental relations concerning U.S. procedures for amendment of state regulatory programs previously approved by the national government, and national government field office responsibilities. These individuals also visited Harrisburg, Pennsylvania to meet with staff and managers in the State of Pennsylvania regulatory authority and the OSM Field Office to discuss national government oversight of state regulatory programs for surface mining. Finally the group traveled to Washington, D.C. for meetings at OSM headquarters with managers and staff on national government oversight of approved state regulatory programs, interagency cooperation and divisions of authority and responsibility. After returning to Jakarta, discussions continued about possible actions for improvement of program organization and management, focussing on decentralization of mine inspection authority and revisions to the AMDAL process.

Program Reorganization Aborted

After considerable discussion of recommendations within the Ministry of Mines and Energy over a period of more than two years, it appeared ready to consider management reorganization and reform of mining and inspection policy. Recommendations were presented to top management of the Ministry in March 1998 during a Quarterly Business Colloquium at the Manpower Development Center for Mines in Bandung (Whitehouse, 1999). Recommendations concerned reorganization of the Ministry to establish a stronger organizational unit for mining environmental inspections, and decentralization of some inspection responsibilities to Kanwil offices of the Ministry. Detailed recommendations were provided concerning reform of the oversight and inspection function, proper structuring of the relationship between minerals development and environmental inspection and enforcement functions, and establishment of inspectable performance standards.

In July 1998 the Minister of Energy and Mineral Resources announced he had discussed this advice with his staff and Directors of the various Directorates in the Ministry, had accepted nearly all of the recommendations, and had charged the BLT Director with the task of convening meetings of appropriate Ministry personnel and developing a plan to implement the recommendations. This willingness to reorganize the Ministry represented a substantial accomplishment directly attributable to the BLT-OSM Mining Environmental Project. Thereafter a proposal was developed to extend the OSM technical assistance project for an additional two years to accomplish the task of creating a higher-level Directorate of Environmental Management which would include a revised and enhanced AMDAL process and environmental inspection functions then housed in separate, lower-level units of the Ministry (U.S.DOI/OSM, 1998a).

If these recommendations had been implemented, they would have required substantial reform of mining environmental policy, procedures and organization within the Ministry of Mines and Energy. However the proposed reorganization was overtaken by larger events and the project extension was not approved. Collapse of the Indonesian economy in 1997–98 and the replacement of President Soeharto by Vice President Habibie produced an unsettled period during which Indonesia attempted to conserve its financial resources, to the extent that the Ministry was instructed to minimize expenditure of foreign loan funds. A new Minister of Mines and Energy was named by President Habibie, and ultimately over US$200 000 of US$3.2 million borrowed from The World Bank for the initial project remained unexpended when it ended in 1998. Enactment of the decentralization laws in 1999 extended this period of uncertainty into the present.

Improving Organization

Successful agencies focus on a fairly limited set of core functions or activities identified by management as essential for mission accomplishment. Agency leadership often recognizes there are many functions which could be performed, and that taking on functions which are not essential tends to detract from the core mission and limits success. Multiple roles within the same agency can also create conflict over what aspect of agency mission is truly a priority, and what activities are subordinate. The current mission and organization structure of national and regional mining regulatory authorities in Indonesia tends to submerge the regulatory mission that is central to effective environmental protection. For example, in the Ministry of Energy and Mineral Resources, key aspects of the regulatory function are scattered throughout the ministry and have the potential to conflict with other ministry activities.

The Director General for Mining is not only responsible for mine safety and environmental inspection, but also for mineral research and development, mining business development (both coal and non-coal Contracts of Work between the government and foreign partners), Mining Authorizations (KP), mineral conservation, inspector training and certification, and training of Mine Technical Managers (Kepala Teknik). Thus inspection and enforcement activity is located within the same Directorate that has responsibility for mineral development, foreign investment, and royalty revenue where activities extend across jurisdictional lines between kabupaten and kotamadya. Regional governments have responsibility for these functions where mining activities occur within a single kabupaten or kotamadya. The mission of inspection and enforcement activity is oriented towards achieving regulatory compliance and may conflict with development and revenue generation functions in the same organization. In addition as mineral development accelerates, it will be important to have a visible inspection and enforcement function which can serve as a focal point for ensuring regulatory and environmental compliance (U.S.DOI/OSM, 1996b).

Three functions which are important to effective regulation, but not necessarily central to the actual regulatory activity include: analysis of economic viability, decision making regarding leasing, and royalty collection activity. Activities such as technical assistance and research which are not directly related to applications for mining authorizations may need to be located close to the regulatory function and activity, but need not be located in the same unit. Finally, such activities as postmining employment programs and workplace safety functions can be located in other units of the larger organization.

Subnational governments in Indonesia have replicated the internal organizational structure of the Ministry of Energy and Mineral Resources in regional mining regulatory authorities called Dinas Pertambangan at provincial, kabupaten and kotamadya levels, but have not done so consistently. Some, like Sawahlunto Kotamadya in West Sumatera have a Dinas Pertambangan for mine inspection but no separate environmental protection agency (BAPEDALDA-Badan Pengendalian Dampak Lingkungan Daerah) to handle other environmental matters. Others in mining areas like West Kutai Kabupaten in East Kalimantan and Balikpapan Kotamadya in South Kalimantan have BAPEDALDA but no Dinas Pertambangan. BAPEDALDA are not authorized to perform mine inspections and their jurisdiction is limited to off-site impacts outside mining concessions, which are often substantial, especially concerning water pollution (Whitehouse, 2003).

Consequently, one result of Indonesian decentralization and regional autonomy is a chaotic assemblage of regulatory jurisdictions and organizational forms replacing the previous national exercise of authority. Related results are a rapid increase in illegal mining, especially of coal and gold, a virtual halt in foreign investment by responsible mining firms, and a series of unfulfilled investment promises from shaky foreign and domestic companies. There was no discernable improvement in control of the environmental effects of mining operations in Indonesia since the regional autonomy laws were enacted in 1999, and some evidence of less control being exercised.

The Ministry of Energy and Mineral Resources and Dinas Pertambangan should clearly identify the core functions of a mining environmental regulatory program, and assign other functions to other units of their respective governments. Consideration should be given to which functions might be brought together to form a visible, cohesive regulatory presence and what functions might be more appropriately located elsewhere (U.S.DOI/OSM, 1996b). At a minimum, the regulatory responsibility should be separated from mineral development, royalty revenue collection and industry assistance functions of the Ministry. This would provide organizational subunits with distinct functions who could bring their respective cases to a superior decision maker if necessary to resolve mission conflicts.

Decentralization of Mining Regulation

Regulatory program responsibilities have not been reorganized so as to provide an effective on-the-ground presence of government authority in mining areas. Greater efficiencies could be realized if responsibilities which are best carried out by national authority were more clearly identified. At the same time, authority and responsibility for other activities could be located in regional government offices, allowing for faster action on problems, more efficient use of enforcement resources and more effective oversight of mining activities. Decentralizing inspection and enforcement functions will require a substantial commitment to additional training for field inspectors, if it is to be effective. Needed training is described in Chapter 8.

Functions which are most appropriately located in national offices of the Ministry of Energy and Mineral Resources include development of national policy and performance standards for mining operations; planning; oversight of regional government implementation of national standards; intergovernmental coordination such as development of Memoranda of Understanding; and administrative and legal appeals. Functions which could be most efficiently carried out by regional government offices include issuance of mining authorizations and review of environmental assessments; field verification of permit and environmental assessment data; reclamation performance bonding and bond release; mine inspection and enforcement; monitoring of mining and reclamation; and enforcement actions against illegal mining. All inspection reports should be routed through Dinas Pertambangan staff, BAPEDALDA, and to the Chairperson of the appropriate AMDAL Commission to coordinate activities and data acquisition.

Need for Coordination

Although the national government continues to have policy making, standard setting and oversight responsibilities, and continues to inspect mines operating under CoW issued prior to 1999, communication between regional government inspectors and the Jakarta office of DTPU is infrequent. Lack of effective communication about mine inspections and enforcement of national policies

**Figure 7.1   Acid mine drainage at toe of partially reclaimed overburden dump in
mine pit**
Source: Alfred E. Whitehouse

can create problems because of the current organization of the mine inspection
cycle, which divides inspection and enforcement responsibilities between
national and regional government inspectors. Communications would be
improved by more regular contacts between inspectors in regional and national
government offices who often view each other as competitors for jurisdiction
over existing mining operations. Regular quarterly meetings of personnel
should be considered, particularly where policy changes or regulatory
requirements are being implemented which would affect their job performance.

There is a need to improve cooperation and communication with other
agencies within the national and regional governments, particularly in those
cases where mineral development is likely to affect resource values managed by
other Ministries, such as forestry and areas set aside for protection of
biodiversity and endangered species. Development of Memoranda of Understanding (MOU) and mutually agreed upon standard operating procedures
with other Ministries, regional and provincial governments affected by mineral
development activity would help clarify roles and responsibilities (U.S.DOI/
OSM, 1996b). For example, establishing a procedure for clearance of Mining
Authorizations and CoW for mining operations in forested lands by the
regional forestry authority after they are considered by the regional AMDAL
Commission but before they are approved by the regional mining regulatory
authority would help reduce jurisdictional tensions and delays.

Another example concerns water quality standards established by regional governments, which must be at least as stringent as national standards determined by the Ministry of Environment. If regional governments must supervise enough mining activity to justify maintenance of both Dinas Pertambangan and BAPEDALDA, it would be most efficient and effective to authorize mine inspectors from Dinas Pertambangan to enforce these standards on mining operations. BAPEDALDA need not give up its enforcement authority, but could supplement it with additional inspections by Dinas Pertambangan mine inspectors with specialized mining expertise. An MOU between OSM and the U.S. Environmental Protection Agency authorizing mine inspectors of both national and state regulatory agencies to enforce water quality standards has been implemented successfully for many years in the United States.

Setting Priorities

Demands for regulatory change and development, training, improvement of analytical capability, increased inspection and enforcement duties, and increasing AMDAL review workload have generated significant pressures on national mining regulatory authorities in recent years. These conditions seem unlikely to lessen in the foreseeable future, although some of the workload has now been decentralized to regional governments. Clear priorities should be established by management of national and regional government organizations among their many competing objectives; personnel and budget resources must be reallocated accordingly (U.S.DOI/OSM, 1996b).

Improving Operating Procedures

It is not evident how operating procedures within the Ministry of Energy and Mineral Resources and regional mining regulatory authorities are documented. Because of the rapid increase of mining development activity in Indonesia and continuous changes in procedures and policy being developed, it is imperative that written procedures be revised to reflect new operating requirements (U.S.DOI/OSM, 1996b). It is important that new procedures and policy be widely understood, not only within the agency, but in the regulated community as well. This is especially significant where recent regulations have either changed technical requirements or revised jurisdictional responsibilities or standards for compliance.

A review of existing internal guidance should be undertaken by management of each national and regional government agency to identify areas where additional guidance is needed. Areas should be identified where policy or operational change is anticipated and it should be determined whether additional operational guidance will be required as a consequence of those changes. Significant changes in requirements and procedures should be explained carefully to the mining industry, and where requirements have a substantial technical component, the Ministry should consider providing training opportunities to appropriate industry representatives. Meetings and conferences of the Indonesian Mining Association provide opportunities for continuing education of personnel in these areas, but should be supplemented by regular seminars or curriculum at regional universities in provinces having significant mining activity. In the U.S., considerable training is provided each year by OSM in over 50 courses offered at decentralized locations throughout the states having significant coal mining activity (U.S.DOI/OSM, 2001).

Regulatory Workload

Overall agency workload is considerable in both the national and regional governments. Staffing levels, budget and travel constraints all act to prohibit an effective level of inspection of existing mining activities. As planned new mines come on line, pressure on enforcement activity can be expected to increase. New mines are likely to be larger, more complex and disturb more area in shorter periods of time, thus heightening the need for technical knowledge and more effective enforcement (U.S.DOI/OSM, 1996b). The current practice of an annual inspection coupled with a semi-annual inspection and follow-up cannot realistically be expected to keep up with the pace of activity and land disturbance at large, fast-moving mining operations.

As discussed in Chapter 2, due to limited personnel, budget and substantial travel distances from Jakarta to mine sites, detailed pre-mining inspections have rarely been conducted and regular inspections have been too infrequent to effectively manage environmental effects of coal mining operations. Mine plans and maps available to mine inspectors lack sufficient detail to facilitate effective inspection of mines. Mine Book entries and requirements for remediation of violations imposed by mine inspectors have not been reinspected to judge compliance within a reasonable period of time. When a company ignores or fails to comply with the inspector's instructions in the Mine Book, the same requirements are usually re-entered during the next annual inspection without penalty to the operator. Because environmental requirements are insufficiently detailed to be useful during inspections, they have been mostly *pro forma*. Inspectors have nothing to measure so they try their best to evaluate a mine's performance against their own internal standards. Each new inspector brings his or her own different internal standards, making consistency impossible (Whitehouse, 1999, p. 66).

Following decentralization to regional governments of mine inspection responsibilities, the frequency of inspections can be increased at least to require quarterly inspections, with followup inspections by the same inspectors on a monthly basis if there are significant violations. Regional government mine inspectors need considerably more technical training than they have received to date, including "continuing education" refresher courses to keep them up to date with new mining and reclamation practices and technology, as discussed in Chapter 8.

Self-evaluation Reporting Systems

The most important inspections of mining operations are those performed by mining companies. The best violations are those that never occur. The second best violations are those that are identified immediately and corrected by the mining company before significant adverse effects are suffered off the mine site. Accordingly, the regulatory program should require and encourage mine operators to inspect their own operations, and to monitor their own compliance with requirements of their mining and reclamation plans.

At a minimum, self-inspection and self-monitoring results should be entered in the Mine Book, and be sent to mine inspectors upon request. Information concerning serious violations should be immediately reported to the nearest mine inspection office. The program's fine and penalty provisions should include reductions or waivers of fines or penalties for self-reported violations, depending upon the seriousness of the violation and the speed with which the mine operator abates the violation. However a pattern of violations may require penalties from a regulatory authority. Failure to conduct regular self-inspection and monitoring, or failure to report violations, should be cause for significant penalties or fines. The intent is to encourage compliance with self-inspection and monitoring requirements, and provide information needed to evaluate an operator's performance under its mining authorization.

Operators should be encouraged to perform periodic evaluations of their own performance using information gained from self-inspection and monitoring. In these evaluations, mine operators should view information about violations as symptoms and look for underlying causes. Properly performed, self-evaluations would induce mine operators to reassess the validity of assumptions in their mining and reclamation plans, and to evaluate whether they are fully implementing approved plans. At a minimum, the self-evaluation results should be entered in the Mine Book, along with any corrective actions that were taken in response to the self-evaluation. If the need for an amendment to the mining and reclamation plan is indicated, the self-assessment analysis should be submitted in support of the proposed amendment. Failure of the Mine Book to show adequate, periodic evaluation of self-inspection and monitoring data should be subject to stern fines, as should any false entries in the Mine Book.

The Indonesian regulatory program contains an excellent tool for industry education and training. The requirement that the government certify a technical mine manager (kepala teknik) for each mine provides good opportunities to educate the industry concerning the need for regulation, performance standards, the regulatory process, and available techniques for achieving performance standards. By requiring periodic continuing education to maintain their certifications, the government has the ability to train kepala tekniks concerning self-inspections, monitoring and self-evaluations.

Improving AMDAL Programs

Broadly stated, AMDAL requirements need to be revised so the AMDAL process serves as the basis for making management decisions about mining feasibility and desirability and provides a platform for enforcing environmental protection and reclamation requirements. This can be accomplished under existing legislation by revising BAPEDAL guidelines. Making the AMDAL process a better decision tool can be accomplished through development of standardized Terms of Reference (KA-ANDAL) which should serve as an application form describing minimum information required for submission of AMDAL documents. Separate standardized KA-ANDAL can be developed for mining projects, electric power generating facilities, and oil and gas facilities, tailored to their distinctive features and probable environmental impacts.

Standardized Terms of Reference

A standardized format for coal mines should require description of detailed environmental information, more specific plans of proposed mineral development activity and operations than is currently required, and the declaration and approval of a postmining land use. This would improve the scope of the ANDAL, RKL and RPL submitted, improve the quantity and quality of the data, and reduce the review time required because of standardization. Mining companies would benefit from advance knowledge of what is expected so data can be gathered during the entire exploration, feasibility and mine development process (Whitehouse, 1999, p. 68). Consideration should be given to reviewing the form for a standardized KA-ANDAL with individuals from industry who have responsibility for preparing them and for preparing ANDAL. In addition, the review process should be restructured to insure that mining activities, including premine construction and exploration, do not begin prior to final approval of AMDAL documents.

Within the context of BAPEDAL guidelines for preparation of AMDAL documents, more detailed guidance should be prepared by BAPEDAL or the Ministry of Energy and Mineral Resources concerning the scoping process and contents of the KA-ANDAL for mining projects, focusing on development of relevant baseline data concerning environmental conditions at proposed mine sites. Baseline data on physical characteristics, including climate, air quality, soils and geology, hydrology, topography, protected areas and biological resources currently required in contents of the ANDAL (BAPEDAL, 2000b, Attachment II.B.5.3a and II.B.5.3b) should be specifically described in the KA-ANDAL as the required product of scoping studies (BAPEDAL, 2000b, Attachment I.A.8, I.B.2.2 and I.B.3.1) which must later be evaluated in the ANDAL.

Biodiversity assessment Study methods described in the KA-ANDAL (BAPEDAL, 2000b, Attachment I.B.3.1) should specifically require a rapid assessment of biodiversity based on field observation by recognized experts. Studies of tropical forest fragments reveal that ecosystem viability is substantially negatively impacted by fragmentation and reduction of habitat (Bierregaard, et al., 1992). The continued existence of large areas of intact habitat is central to long term ecological viability of tropical ecosystem functions because it allows maintenance of keystone predators and viable fruit dispersal systems (Terborgh, 1992). Eradication of keystone predators may have a ripple effect on other species, producing dramatic change in an entire ecosystem. Early identification of "hot spots" with high biodiversity is therefore important if these values are to be preserved intact (Conservation International, Inc., 1995, p. 3). Allowing mining to proceed adjacent to such areas is therefore preferable to discovering belatedly that it has been authorized in the midst of them.

Mining in Indonesia almost always occurs in rainforest where there may be as yet undiscovered "hot spots" or pockets of biological or genetic resources that may be worth more for medical, scientific or ecotourism uses than for the value of the mineral resource. All acres are not equally rich, but by surveying areas slated for development, hotspots of biodiversity may be found that should be protected if possible. By doing this, the majority of an area's critical species may be preserved. Because destruction of these resources could be avoided by judicious mining nearby (perhaps using underground rather than surface mining techniques), with little sacrifice of mineral resource values, special attention should be directed to assessment of biological resources at potential mine sites.

Excellent compendiums of previous ecological research have been prepared for Sumatera (Stone, 1995, Whitten, 2000) and Kalimantan (MacKinnon, 1997), but more site-specific field research is needed for adequate environmental assessments. In recent years, rapid assessment methods have been

Figure 7.2   Extremely endangered orangutans reintroduced in North Sumatera
Source: Carol J. Boggis

developed to provide an inventory of biological resources in poorly known areas that are potentially important biodiversity conservation sites. These methods involve expert systems using sampling methods designed to be fast, reliable, simple and cheap (Coddington, et al., 1991). Typically they involve teams of leading tropical field biologists and host country scientists who generate first-cut, on-the-ground assessments of the biological value of different sites (Conservation International, Inc., 1995, p. 1).

A rapid assessment of biodiversity should be sufficient to answer the following three questions: Is there unusual or exceptional diversity of species (e.g., number of different species) present? Is there any unusual concentration of endemic species (e.g., a single species found nowhere else) present? Are there any known rare or protected species present (Baas, 1996)? The objective of this requirement would be to improve the amount and quality of biological data provided in the AMDAL, and establish a more credible baseline for evaluating and monitoring environmental change both prior to authorizing mining operations and during inspection of ongoing operations. A rapid assessment of biodiversity should include:

- Description of drill core data soil types as they relate to vegetation diversity and evaluation for indicators for specific habitats like mangroves, limestone vegetation, kerangas, ultrabasic soils.
- Evaluation and description of the pre-mining flora and fauna in the project area, with identification of primary and secondary vegetation types. Any biological associations within the project area should be described and biodiversity quantified. Wherever a natural community of tropical vegetation is present, a description of its location, composition, structure and degree of maturity should be given. A biogeoclimatic zone map of the natural vegetation covering the site should be required in the ANDAL.
- Evaluation and description of the relative abundance of both formally designated and locally perceived economic species present and their potential value to human economy and ecology.
- Documentation of rare or protected species present and their general location. Exact locations should not be reported due to the risk of publication of this information leading to illegal taking of the species.
- Description and location of any communities of unique and undisturbed vegetation or fauna, due to their historical value, beauty of the landscape and recreational values, and specific economic and ecological values, such as original stands of ironwood forest, dipterocarpaceae forest, conservation forest, traditional settlement and so forth.
- Evaluation of existing destructive processes which modify vegetation in the project area, and description of any potential linkage between conditions created by the proposed mining project, associated infra-structure and acceleration of these destructive processes.
- A statement of the Ministry of Forestry designation of the project area.
- Contractors must be authorized to collect specimens and ship them out of country if necessary to compare them to existing collections (Kessler, 1996).

The KA-ANDAL (BAPEDAL, 2000b, Attachment I.B.3.1) should therefore specifically require that the ANDAL include a report of the information and evaluation produced by the rapid biodiversity assessment described above.

In the mid-1990s, a rapid biodiversity assessment of 40 000 hectares of secondary forest was completed by staff of the Rijksherbarium/Hortus Botanicus, University of Leiden for PT Kaltim Prima Coal in East Kalimantan. To obtain the necessary data, a team of three experienced Indonesian technicians and one expert Dutch field botanist carried out just over one month of fieldwork. In the process 11 transects of 10m × 300m were established and analyzed; field identification of about 3300 trees was carried out; flowering and fruiting material was collected (500 accessions); useful and cultivated plants were inventoried; forest communities were identified; and a biogeoclimatic zone map was prepared. The most time-consuming part of this assessment was identifying over 200 specimens shipped to The Netherlands for comparison with those in a large collection at the Rijksherbarium/Hortus Botanicus. Total cost of the study was about US$44 000 in 1996 (Baas, 1996).

A rapid biodiversity assessment of primary forest would be more time consuming than for secondary forest because of its much greater species diversity. A rapid biodiversity assessment of a primary forest area of 100 000 hectares would require much more manpower and expertise. Although the number may be increasing, in 1996 there were only a few people in the world ($\sim$20) capable of reliable pre-identification of primary tropical forest species. Herbarium collections of many taxa would have to be submitted to taxonomists specializing in certain plant families. Use of satellite imagery, remote sensing techniques and overflights in small planes would allow experienced field biologists to rapidly reduce a primary forest area of 100 000 hectares to a manageable number of plots for ground inspection (Kessler, 1996; Roos, 1996). For a 100 000 hectare rainforest or coastal plain region the costs (largely labor of Indonesian counterparts and some skilled forest botanists) might amount to about US$241 000.[1] This assessment would cover only plants and trees. Similar, less expensive assessments for birds, mammals and possibly insects can be made if appropriate after the forestry assessment identifies any likely habitat types which might contain biodiversity hot spots. The price and number of people required would be directly proportional to the information content and quality demanded of the assessment (Baas, 1996).

Thus the requirement for a rapid biodiversity assessment would probably be considerably less burdensome than current requirements for biological baseline data in an EIS done in the U.S. Rapid assessment is a cursory technique, when compared to the usual field biological survey, and gives only a crude indication of the most obvious conditions that may be present on-site, but better than any ANDAL yet produced in Indonesia. Rapid assessment is therefore less expensive and less time consuming than a full-scale biological field survey would be. And the estimated cost would constitute a fraction of expenditures for exploration and proving economically recoverable reserves. Furthermore, the absence of previous field studies (and baseline data) for most areas of Indonesia makes it much more critical that this type of study be done, and that the results be added to some biological data base as they accumulate. Every assessment should include Indonesian counterparts in training, so a cadre of incountry expertise will develop. There are already some Indonesian counterparts in the Ministry of Forestry, but they could easily be drawn from regional universities or engineering firms that prepare AMDAL documents. Staff from the Rijksherbarium/Hortus Botanicus have provided 33 day training sessions for "parataxonomists" used in their projects. A well-organized mining firm could perform a rapid biodiversity assessment during exploration for mineral resources.

Several organizations with experience in Indonesia are capable of and willing to perform rapid biodiversity assessments. Through the Flora Malesiana Foundation, the Rijksherbarium/Hortus Botanicus has developed relationships with the Herbarium Bogoriense and the Botanical Gardens in Bogor, Indonesia, which now has a collection of over 1.5 million specimens. The Rijksherbarium/Hortus Botanicus at Leiden University in The Netherlands has about 4 million specimens from Indonesia, and would like to acquire more. The Center for Applied Biodiversity at Conservation International, Inc. has performed rapid biodiversity assessments at over 29 locations in Africa, Madagascar, Central and South America, and the Asia-Pacific region, including some in Indonesia and Papua New Guinea (Conservation International, Inc., 2005). These organizations employ individuals with great expertise who are usually eager for opportunities to acquire new specimens.

Standardized Environmental Plans

A standardized format should also be specified by the government for preparation of the RKL and RPL, and they should be prepared at the same time as the ANDAL and feasibility study. The present system separates these studies in both time and organization. This separation makes it impossible to realistically determine the economic feasibility of a project because environmental management and monitoring components have not been determined and their costs are not factored into the feasibility analysis. Combining these studies would allow the proponent to improve planning and give the AMDAL Commission, Dinas Pertambangan and BAPEDALDA a more complete and useful picture of a project. This combination would send a clear message to proponents that the government believes environmental protection is an integral part of each project, not extra burden or cost (Whitehouse, 1999, p. 68).

A standard format for the RKL and RPL should require information necessary to describe a detailed mining and reclamation plan, and show how the proposed operation will meet established performance standards. In the United States, detailed information about mine design, cut sequence, design of waste rock disposal dumps, water quality and sediment control, overburden chemistry, special handling of potentially acid forming materials, facilities locations, cleaning plant technology and design features is prepared for environmental assessments during mine permitting. For example, the OSM California mining permit application requires the proponent to submit geological data sufficient to determine all potential acid or toxic forming strata down to and immediately below the lowest coal seam to be mined (U.S.DOI/ OSM, 1988, p. 59). For comparison, the AMDAL process does not require overburden sampling and analysis which could indicate the potential for a mine to cause acid mine drainage. Coal mining operations can be designed to minimize acid mine drainage with advance warning.

The alternative of requiring less detailed baseline data was considered and rejected during development of the surface mining regulatory program in the United States. It was determined that failure to require detailed information would mean the magnitude of impacts on the environment could not be known with any degree of certainty. The data provide a baseline or benchmark against which environmental impacts of mining can be measured, and mining plans show how an operator proposes to comply with performance standards and whether those standards can be met. Not providing this data would require less effort and expense by mine operators, which explains why they resist imposition of such requirements. Fewer hours required by specialists to gather necessary data on various environmental parameters would necessarily result in lower costs to operators who must hire the expertise to meet requirements. However it was decided that lack of detailed information would increase the possibility that regulatory authorities might make faulty determinations in deciding whether to approve mining operations. The result would then be adverse environmental, public health and safety impacts because of inadequate environmental protection during mining and reclamation. Some environmental damage could be irreversible, and after-the-fact activities to bring an operation into compliance with performance standards could cause more environmental harm than if operations had been conducted initially in a manner that would ensure environmental protection. Increased costs might be incurred by mine operators in the long run because of the need to meet standards after violations were identified. This could also result in increased costs to regulatory authorities and society to repair damages caused by improper mining methods that might be approved due to inadequate data and poor decisions made on that basis (U.S.DOI/OSM, 1979, BIV-21).

Towards an Adequate ANDAL

An adequate ANDAL must include a complete and accurate description of the environmental resources that may be affected by the proposed mine development, including information concerning climate, vegetation, fish and wildlife, soils and land-use baseline data. Maps, plans and cross-sections showing geologic data at a scale specified by the Dinas Pertambangan are necessary for effective decision making. A detailed reclamation and operation plan must be described in the ANDAL, including: the mining operation, blasting, air pollution control, fish and wildlife, reclamation of the mine site, groundwater, surface water, postmining land use, ponds, impoundments, diversions, public roads, excess spoil and transportation. The ANDAL must show the operation will be in compliance with specific performance standards stated in law and regulations.

Completeness review Staff of the appropriate BAPEDALDA and Dinas Pertambangan should review every ANDAL to determine if it addresses each requirement of the regulatory program necessary to initiate processing and public review. If the ANDAL is not complete, the regional AMDAL Commission should send a letter to the applicant describing any deficiencies. The applicant must be allowed an opportunity to respond to any deficiency letters, modifying the ANDAL to address any shortcoming. The applicant's responses and modifications to the ANDAL must be reviewed by the AMDAL Commission for completeness. This process must be repeated until all deficiencies are satisfactorily resolved. However if the applicant is unable or unwilling to submit needed information, then the ANDAL must be summarily rejected.

After the applicant responds satisfactorily to all deficiencies, an ANDAL submission should be declared complete by the AMDAL Commission. This is a preliminary determination of data adequacy, not a final approval of the project. The applicant must ensure that a copy of the ANDAL is available for public examination near the proposed mine site, usually at a court house or other public building. The applicant must publish a notice in local newspapers that briefly describes the proposal, advises the public where they may review a copy of the ANDAL and where any comments to the AMDAL Commission should be sent. A period of about 60 days should be allowed for the public to review the ANDAL, submit written comments and request an informal conference. If requested, the AMDAL Commission should chair a public meeting. The public must be given an opportunity to submit written and verbal comments at this meeting. The AMDAL Commission should distribute a copy of the ANDAL to internal offices and to other appropriate government agencies (e.g., Ministry of Forestry and others) for review. National and regional agencies should be allowed about 60 days to submit written comments on an ANDAL. Mine inspectors in the Dinas Pertambangan should review the ANDAL application for on-the-ground "inspectability" of the proposed mine operation. The AMDAL process provides the best opportunity for well informed decision making and to achieve desired postmining land uses, effective reclamation and environmental protection. However accomplishing this will require substantial revision of organization, procedures, staffing and operating practices.

AMDAL process organization The current AMDAL process does not allow for adequate technical review of mining applications. Limitations on reviewers technical capability and the lack of a multidisciplinary approach, adequacy of data required and submitted, and short scoping and review periods allowed in the process also impose significant strains on existing organizational structures. These difficulties are accentuated when there is a need to coordinate reviews among national and regional government staffs, or those in two or more regional government jurisdictions. A dramatic increase in AMDAL submissions since the early 1990s, compounded by decentralization of AMDAL reviews to regional governments have not been addressed by management with organizational changes which would make a decentralized

AMDAL review process effective. Because this structure does not provide dedicated budget or personnel resources for review of AMDAL documents, reviews are not necessarily a high priority for technical or program staff. Improvement of the quality of AMDAL document contents will likely impose significant additional analytical burdens on existing staff and organizational structures.

Time limits   Short periods for review of AMDAL documents and the prospect of automatic approval after some number of days seriously limits their substantive review. The 75 day review period starts on the day an ANDAL is received and sets the schedule for an AMDAL Commission decision meeting. This 75 day period includes the time it takes to distribute documents to members of the AMDAL Commission and any review by a technical team, the general public and other government agencies. In effect the actual review period becomes 51 days because technical team and AMDAL Commission member comments must be received and consolidated before the AMDAL Commission meets. AMDAL documents have gotten thicker and more complicated since those prepared in the early 1990s. This makes a thorough review more difficult, if not impossible, within 51 days. Yet if an AMDAL Commission fails to act within the specified period, the documents are automatically considered approved.

The automatic approval requirement should be abolished and adequate time allowed for completion of a complete technical review, with extended periods allowed for large, complex projects involving multiple mine pits. AMDAL Commissions should be authorized by BAPEDAL to set time limits appropriate for individual projects, provided the time limits are consistent with the environmental protection purposes of the Act on Environmental Management of 1997 (Republic of Indonesia, 1997). In setting time limits, AMDAL Commissions should take into consideration distinctive character- istics of each project, including the potential for environmental harm; size of the proposed project, number of persons and agencies affected; degree of public need for the project; availability of relevant information and the amount of time required to obtain it; adequacy of analytic techniques; and the degree of controversy attending the proposal.

Regulations normally should allow a minimum of 90 days for public comment, interagency consultations, and technical review after submission of an ANDAL before an AMDAL Commission meeting is scheduled. AMDAL Commissions should be authorized to extend the 90 day period for increments of 30 days if necessary for review of complex or controversial projects, and should not make a decision until after consultation with other agencies and technical review of all comments are completed, even if the review period must be extended for additional periods of 30 days. The community interest in a complete review outweighs the proponent's interest in speedy approvals of major development projects causing significant environmental impacts.

Staffing   Staffing for the AMDAL function is not appropriate given the rate at which applications are being submitted. The major burden of document review falls on a small administrative staff of the AMDAL Commission, who are expected to deal with mine designs, power plant designs and oil and gas facilities. Whether AMDAL Commissions of regional governments have been staffed with adequate breadth of expertise to handle such diverse projects since decentralization began in 1999 is highly questionable. Some regional governments such as Kotamadya Samarinda BAPEDALDA in East Kalimantan and Kabupaten Muara Enim Dinas Pertambangan and BAPEDALDA in South Sumatera have begun to hire multidisciplinary staff, but this experience is very uneven. In the United States, multidisciplinary teams of technical specialists assigned to review applications for mining permits and environmental assessments typically include mining engineers, hydrologists, soil scientists, ecologists, environmental specialists, wildlife biologists, air quality specialists, geologists, archaeologists and project managers. In Indonesia there is no requirement for prior notification of an AMDAL Commission by a proponent that an ANDAL is about to be delivered, so it often catches staff by surprise or during times of high workload. There may be several AMDAL documents being processed by staff of a single AMDAL Commission at any given time, which further restricts time available for a thorough review of any document.

The increase in AMDAL review responsibilities of regional governments should be acknowledged. Reviews should be made a primary responsibility of some number of permanent, interdisciplinary technical staff in each BAPEDALDA dedicated to the AMDAL process, and not continue as a secondary or "extra duty" assignment. One option for consideration is to provide each AMDAL Commission with permanent staff reporting to the head of the Commission. This staff would be responsible for managing the AMDAL process for a regional government: conduct required AMDAL reviews, recommend changes or conditions to proposed projects; manage agreements contained in approved AMDAL documents; and recommend amendments or changes to the AMDAL Commission during the life of a project to achieve on-the-ground environmental performance promised in AMDAL documents.

An Integrated Organization?

Because both the Mining Authorization and AMDAL processes are regulatory processes, it would be possible to create a new regulatory organization within each regional government that could manage the day-to-day requirements of mining environmental programs. Its mission would be to regulate the environmental effects of mining to ensure citizens and the environment are protected during mining, postmining pollution and safety hazards are prevented, and the approved postmining land use is attained. Its staff would conduct administrative and technical AMDAL reviews for mining operations only; recommend changes or conditions to proposed projects, manage agreements contained in approved AMDAL documents, and recommend amendments or changes to decision makers during the life of the project. Mining Authorizations issued by this group would be the final approvals needed to begin exploration and mining. Different staff units or teams within a larger organization could process authorizations and AMDAL reviews for oil and gas, electric power, and other major facilities with specialists in soil science, ecology, wildlife biology, air and water quality, and cultural resources moving between project teams as needed.

For such an organization to be effective, it should have some clarity and unity of mission. Responsibility for mineral development, recruiting foreign investment, and collections of land rents and royalty revenues should be assigned to other organizations. Collection of land rents and mineral royalties would more appropriately be performed by a department of finance or tax collection for a regional government. BAPEDALDA could then focus their efforts on other industrial polluters and environmental protection issues, coordinating with the regional government mining environmental authority through AMDAL Commissions and by sharing inspection reports. In the United States, local and state governments often contain a Department of Economic and Community Development tasked with recruiting new businesses to the area to increase the property tax base and employment opportunities. Often they sponsor trade delegations to foreign countries and seek overseas partners. They receive direction from local and state leaders (e.g., mayors and governors) and actively seek to increase commerce and trade. Similar units could be given responsibility for recruiting foreign mine operators to specific provinces, kabupaten and kotamadya, separately from a mining regulatory authority.

Consideration should be given to revising the present Mining Authorization system from permission to construct and begin work on general survey, exploration, exploitation, processing and refining, transportation and sales to permission to occupy a parcel of land (e.g., lease) for the preceding purposes for some period of time at some agreed fee or royalty rate. This would reduce the number of separate approvals required for different stages of each development. The approved Mining Authorization should incorporate the ANDAL, RKL and RPL and become the permission to begin operations, as well as the trigger to begin inspections.

Integrating AMDAL Compliance

There is a continuing need for establishment of an administrative penalty system to improve compliance with AMDAL requirements. An integrated mining environmental regulatory organization could use the same penalty system for violations of ANDAL, RKL, and RPL that it uses for violations of Mining Authorizations and Contracts of Work, similar to one discussed in Chapter 3. Any system should include a written warning followed by a series of increasing fines and other penalties for noncompliance with AMDAL provisions. Penalties would be in addition to a requirement for correction of noncomplying activity within a specified period of time. An administrative system can be established providing penalties which are closely tied to the severity of infractions. A graduated scale of penalties would focus mine operator's attention on those aspects of environmental protection which are most important and speed compliance.

Decentralization of mining environmental inspection, enforcement, and AMDAL programs has created a pressing need for additional technical training of personnel employed by regional governments. Adequate technical review of AMDAL documents, inspection of mining operations, and protection of the public health, safety and right to a clean environment will not occur in the absence of substantial efforts to develop the human resources of Indonesian regional governments. Needed training is described in Chapter 8.

Note

1 Adjusted using U.S. inflation rates 1994–2004 totaling 20.64 percent available at InflationData.com/Inflation/Inflation_Rate/InflationCalculator.asp (accessed January 2005).

Chapter 8

Development of Institutional Capacity

Development of Ministry institutional capability to identify and solve potential environmental problems from mining activities before they degrade the human environment is a major objective of this project. Efforts to develop the human and technological resources of the Ministry have produced substantial results on several fronts: (1) enhancing the permanent capability of the Ministry to provide technical training to its employees, (2) developing the expertise of Ministry employees in relevant technical subjects, (3) providing technical consultations in the U.S., field training and site visits to develop the knowledge of Ministry technical personnel related to specific environmental problems of mining in Indonesia, (4) transfer of state-of-the-art technologies for analysis of mining environmental problems and inspection of mining operations, and (5) training in the use of these technologies.

The principal beneficiaries of these efforts were the Directorate General of Mines, especially mine inspectors in the Directorate of Technical Mining (DTPU) in Jakarta and in at least 14 Kanwil regional offices of the Ministry; faculty and students at the Manpower Development Center and the Mineral Technology Research and Development Center in Bandung; senior managers on the AMDAL Central Commission; and professional staff in the Bureau of Legal Affairs and the Bureau of Environment and Technology. In addition, personnel from the Directorate of Coal, Directorate of Environmental Geology, Directorate of Oil and Gas, PLN, BAPEDAL, the Ministry of Forestry, seven offices of Provincial governments, PT Timah, PT Tambang Batubara Bukit Asam (PTBA) and at least 10 other mining companies benefitted from training courses, workshops, seminars and conference presentations sponsored by the BLT-OSM project.

Use of Joint Teams

Because institutional and human resource development were major goals of the technical assistance project, it was deemed essential that Indonesian personnel who would implement any recommended changes acquire a sense of ownership of the recommendations. Because some aspects of mining environmental policy as implemented in the U.S. are probably not appropriate to different ecological, cultural, social, economic and political circumstances in Indonesia, OSM approached the project as a unique problem solving activity. That is,

OSM did not expect it would merely export the Surface Mining Control and Reclamation Act of 1977 to Indonesia.

Fortunately, the laws and principal regulations of Indonesia are published in English and most educated Indonesians speak at least one foreign language, often English. Although Indonesian team members had good language skills, differences in language between U.S. and Indonesian personnel were significant and obscured cultural differences that were sometimes difficult for both sides to appreciate. Anticipating such difficulties, the project was conducted through a series of training activities, technical consultations and problem-solving seminars by teams comprised of both American and Indonesian personnel, who jointly prepared recommendations. Emblematic of the spirit of partnership embraced in the entire project, each team had Indonesian and American co-chairs.

Seminar discussions covered methods and relevant techniques of analysis, and involved advance preparation and presentation of information by both U.S. and Indonesian team members, to ensure all members of the team would remain on an equal footing. A phased seminar format was designed to present general principles in initial meetings, using discussion to identify alternative courses of action and obtain feedback from Indonesian agency personnel, before holding later meetings to discuss and prepare recommendations. For example, postponing development of recommendations until later meetings of the joint Policy Team allowed consultations between Indonesian team members and their respective superiors and organizational units to influence the recommendations, as described in Chapter 6. This approach was expected to maximize transfer of skills and information to Indonesian personnel, while allowing them to provide cultural and organizational filters for team recommendations. Thus the approach was designed to increase the probability work products would achieve an appropriate fit with Indonesian cultural circumstances and political will.

Personnel Training in Technical Subjects

Priority needs for technical training were identified for mine inspectors to improve their ability to carry out environmental inspection responsibilities, and for reviewers of ANDAL to improve their productivity and technical competence. OSM and BLT established a Joint Training Team consisting of eleven U.S. team members and nine Indonesian personnel. Initially, U.S. team members were those providing training for the OSM Branch of Training and Technical Information, but later instructors from state regulatory authorities in the U.S. were also employed. BLT identified experienced instructors to receive additional training in the U.S. and provide instruction to personnel in Indonesia.

Technical training activities were designed to develop the permanent training capability of the Ministry of Mines and Energy as well as the level of subject-matter expertise of Ministry personnel. These activities utilized a multicultural, pluralistic approach (Bauzó, 1987, pp. 253–254) in a three-step "training the trainers" method developed for this project. First, English-speaking Indonesian instructors were trained in technical subjects (e.g., principles of mine inspection, spoil handling) with OSM and other American personnel in regularly scheduled courses presented in the U.S. In the second step, this training was followed immediately by joint development of similar new technical courses by the same Indonesian and OSM instructors. In the third step, these courses were delivered in the Indonesian language by Indonesian instructors in Indonesia, with coaching from their U.S. instructors. Consequently, several small groups developed 17 new technical training courses and delivered them in Indonesia, some several times between 1995–1998.

Instructor Training

For example in April 1995 OSM accommodated three Indonesian members of the Training Team[1] in regularly scheduled OSM courses in the U.S. in Blasting and Inspection, and Principles of Mine Inspection. After completion of these courses, the Indonesian Training Team members completed the OSM Instructor Training course, which is designed to teach persons with technical and programmatic subject-matter expertise how to develop lesson plans, training aids, and instructional materials and to present training courses effectively. Instruction was provided in development of speaking skills and voice, use of instructional and projection media, psychology of adult education, classroom management and learning styles, with emphasis on use of the presentation/discussion format and training conference method of instruction and practical exercises (U.S.DOI/OSM, 2001, p. A-25; Mayo and DuBois, 1987, pp. 64–68). They then designed courses of their own on instructor training and mine inspection for the Indonesian context, with assistance of OSM instructors.

The Principles of Inspection course focused on how to inspect an active mining operation for environmental performance and the importance of documentation which supports observations made during inspections. Inspectors learned how to recognize and record pertinent information during inspections for future use, using methods and technologies applicable to the mining and reclamation process. The course covered interpretation of photos and maps; topsoil removal, storage, protection and replacement; backfilling and grading; revegetation; blasting; mine safety; plant identification; and sampling with a focus on what the inspector should look for, which observations should be recorded, and how the record of findings should be

Three American members of the Training Team traveled to Indonesia in August 1995 and assisted these Indonesian team members in 30 days of training for another 20 instructors, all senior mine inspectors from Jakarta or Kanwil offices of the Ministry. Training included five days for an Indonesian Instructor Training course and another five days in the classroom for a course in Principles of Inspection at the Ministry's Manpower Development Center and two days of field exercises at the PTBA Airlaya Mine in Tanjung Enim, Sumatera. Some Indonesian trainees were initially concerned that this course would duplicate portions of the Ministry's required six-month inspector training program, but those concerns were quickly set aside during the instruction.

documented. Instruction reinforced the importance of making accurate observations and recording them in formal documentation (U.S.DOI/OSM, 2001, p. A-29). Previously these skills were described by Ministry personnel as weak.

Subsequent interviews in June 1996 with persons who received this training both in the U.S. and in Indonesia suggested the new one-week Principles of Inspection course compared quite favorably with the six-month course for mine inspectors conducted at the Manpower Development Center. All persons interviewed who had been required to take the six-month course before attaining their positions as mine inspectors, said they acquired new knowledge and techniques in the Principles of Inspection and Instructor Training courses. All indicated these courses would change the manner in which they conducted inspections and teach future courses for other inspectors (Hamilton, 1996).

At the end of this first round of training, 27 Indonesian employees of the Ministry representing mine inspectors from 12 Kanwil offices and Jakarta, and several professional staff from Jakarta and the Manpower Development Center were able to teach others how to be instructors in their specialties, and three had actually done so successfully with OSM supervision. All 27 were certified by OSM as instructors for an Indonesian Instructor Training course. Subsequently by August 1998 these 27 instructors in turn had provided training in Principles of Inspection to essentially all 90 national government mine inspectors in Indonesia and many additional professional staff.

Training Needs Assessment

During training for mine inspectors in April 1995 a need was identified for an Inspection and Enforcement Handbook which describes in narrative form with checklists and by pictorial example the methods to be used to conduct inspections, sample water quality, and evaluate reclamation success. This manual would be used to assure inspection and enforcement consistency and as a training tool for new recruits as they are hired. Preparation of such a manual was discussed by the Training Team during the first year of the project. Action was initially postponed until decisions could be made by the Ministry concerning proposed changes in the mine inspection process and the AMDAL process which might relate inspections more closely to AMDAL documents. The content of such a manual would be greatly affected by any changes in these processes and resulting requirements. During the first quarter of 1997, a draft Mine Inspector's Handbook modeled after the *Basic Inspection Handbook* used for many years by OSM to train U.S. mine inspectors (U.S.DOI/OSM, 1991) was prepared and circulated for review and comment by Ministry personnel. Comments were incorporated into a second draft, which was circulated for additional Ministry review in the third quarter of 1997. A final version was completed in the second quarter of 1998, and provided to BLT for translation into Bahasa Indonesia.

During training courses held in July 1997 a survey of trainees was conducted concerning their perceptions as to what kinds of training were most needed by their offices. Results from this "needs survey" helped to guide what additional training was provided through technical assistance. Mine inspection personnel and their supervisors were asked: "What future training should OSM provide mine inspectors that would help them become better environmental inspectors?" Their responses indicated a rather surprisingly high degree of consensus among DTPU inspectors, Kanwil inspectors and supervisors that the subjects in which training was most needed were: what to look for at the mine site; acid mine drainage; revegetation; backfilling and grading; erosion and sediment control; spoil handling; postmining land use; and water monitoring and sampling techniques. There was less agreement, but some sentiment that additional training was needed in the following subjects, but these were clearly of lower priority than the subjects above: bond calculation; preparing for an inspection; writing mine book entries or violation notices; interpreting field data; field testing equipment; surface and ground water hydrology; presenting findings to mine operators; recording field data; reading mine maps; presenting findings to supervisors; and air quality. A total of 21 responses were received from 7 DTPU inspectors, 7 Kanwil inspectors and 5 Kanwil supervisors.

Erosion Control and Acid Mine Drainage

Erosion and sediment control are probably the most widespread mining related problems in Indonesia. Soils are highly erodible and the rainfall is both high and intense. Although every mine has a reclaimed area which has been successfully revegetated, it is usually a very small percentage of the total disturbed area. Companies are frustrated by high maintenance costs of sediment control structures but do little to prevent erosion. The Ministry is concerned about the problem but they believe many companies do not know what actions to take. Acid mine drainage is beginning to be recognized as a water quality problem in Indonesia. The Ministry is fortunate that its mining industry is still young and widespread impacts from acid mine drainage have not yet occurred. Acid mine drainage has been observed at both coal and metal mines, but prior to 1996 there was little evidence of attempts to deal with it. Overburden and predictive analyses are not required as part of the AMDAL process, and remedial techniques focus on "end of pipe" methods such as chemical treatment, which are expensive and require continuous effort. Little attention is given to prediction and prevention as an alternative way to meet Indonesian water quality standards and to minimize costs at existing installations. Low grade ore stockpiles and some waste rock piles, coal yards, refuse and tailings disposal sites and processing areas are likely locations of future acid mine drainage problems.

In August 1996 four members of the Indonesian Training Team[2] traveled to the US for 30 days to attend regularly scheduled OSM technical training courses and special tutorials in Acid Mine Drainage Planning and Prevention, Surface and Groundwater Hydrology, Erosion and Sediment Control, Spoil Handling, and Bond (Reclamation Guarantee) Calculation. These individuals had all previously completed the new Instructor Training course in Indonesia. Each of these training courses included a field study component so instructors could demonstrate practical aspects of each subject and trainees could use skills learned during classroom sessions. Tutorials were conducted in the field so overburden and water samples could be taken and examined in the context of specific site conditions.

While receiving this instruction, Indonesian personnel accompanied an OSM mine inspector on a complete inspection of a surface coal mine. They observed the manner in which American inspectors perform an environmental inspection and were provided opportunities for discussion of similarities and differences between U.S. and Indonesian regulatory programs. This training reinforced what Indonesian personnel had previously learned during the Principles of Inspection course they had taken in Indonesia during the previous year, providing a refresher course. During training in the U.S., course outlines were prepared for use in October and November 1996, when three U.S. members of the Training Team traveled to Indonesia to assist their Indonesian counterparts in delivering 30 days of training at PT Bukit Asam in Tanjung Enim, South Sumatera. Bukit Asam is a good field training location because the two coal mines, Air Laya and Maura Tiga Besar, provide a variety of teaching examples to supplement classroom training.

Because acid mine drainage and erosion and sediment control problems are so widespread and the environmental risks high, both government and industry personnel were trained in order to speed up the dissemination and application of new technologies, in hopes that some practices would be implemented more quickly. One of the Ministry's major concerns when considering new environmental practices is the ability of the industry to comply with new environmental standards and regulations. The Ministry is especially concerned that some small domestic companies don't have the technical know-how or financial capacity to comply. This concern reportedly led to uneven implementation of the few standards which did exist, limited enforcement actions, and imposition of few penalties for noncompliance. Direct training of the industry speaks to this concern, and may stimulate development of environmental business capabilities in Indonesia.

Technical Training Courses

Technical training courses developed and delivered in Indonesia during the OSM project are listed in Table 8.1. Brief descriptions of each course are provided below.

Soil erosion and sediment control   An Erosion and Sediment Control course was developed and presented twice at the Bukit Asam Base Camp Training Facility for 18 Ministry personnel and 16 representatives of 12 Indonesian mining operations. A specially developed 40 hour expanded version of an OSM one day (8 hour) non-field course was presented. The course was based on local conditions and applied to Indonesian mining practices. It was team-taught by American and Indonesian instructors from the Ministry who were previously trained in the U.S. Each participant group spent two half-day sessions at the

Table 8.1   Technical Training Courses Developed for OSM Project

Instructor Training
Principles of Inspection – Basic
Principles of Inspection – Advanced
Principles of Mining Environmental Inspection
Blasting and Inspection
Inspection Report Writing
Evidence and Expert Testimony
Surface and Groundwater Hydrology
Erosion and Sediment Control
Soils, Erosion, Sediment Control and Revegetation
Acid Mine Drainage Planning and Prevention
Acid Mine Drainage Prediction, Prevention and Treatment
Spoil Handling
Elements of Reclamation
Principles of Reclamation
Mining Reclamation Technology
Cost Estimation for Reclamation Guarantees

Air Laya and Maura Tiga Besar mines. Discussions included problem identification and practices to reduce erosion rates. Areas of focus included: effects of uncontrolled water on outslopes and terraces; outslope length and gradient; excessive sediment and pond effectiveness; use of indigenous vegetation (observed at a thriving experimental revegetation site) including Vetiver grass, and practices to reduce erosional impacts (U.S.DOI/ OSM,1998d; Maynard, 1998a, 1998b, 1998c).

One session of this course in late 1996 was attended by the Production Manager of the PT Adaro mine in South Kalimantan. PT Adaro is a large surface coal mine then producing approximately at 8 million tonnes per year. A few weeks earlier Adaro had reported to the Ministry that one of its sediment ponds had breached and dumped its contents into a branch of the Balangan River, damaging water quality in a nearby village. Their erosion and sediment control difficulties began with surface water management. Surface waters from drainage areas above the mine pit were not diverted and were allowed to flow into the active working area. Much of the active coal removal area was covered with water, to depths greater than one meter in some places. Mine staff said there was still 3–4 meters of coal below the observed water level. At least two large floating pumps were observed transferring a water/mud slurry created by trucks and mining equipment to a first stage settling pond located outside the pit in a small stream. When mine staff were asked about the size of the drainage area above the settling pond and whether it's size was a factor in the pond's design, no one knew.

The mud/water slurry pumped from the mine contained a high percentage of fine and colloidal clays. The first stage pond retention time was far too short for effective settlement and exacerbated by continuous flow. A flocculant aluminum sulfate (alum) was added to improve settlement but proved ineffective. A small amount of surface water in the pond met discharge requirements and was released to a stream by manually removing narrow boards in the crest of the dam one-by-one. The remaining slightly thickened mud/water slurry was pumped to an adjacent pond approximately 70 by 20 meters in size. It was this pond that breached.

The Production Manager was fortunate to receive appropriate training at that time. A week after completing the training his staff was actively applying many of the techniques he had learned. The second stage settling pond was being nearly doubled in size to increase retention time, and construction was underway on three shallow third stage settling ponds. Negotiations were in progress to acquire more land to construct additional ponds. Diversion of the unaffected surface water around the mining pit to minimize the quantity of water needing treatment was being considered. Cationic flocculants were being investigated, which are more expensive but are much more effective in settling colloidal clay. Changing the third stage treatment from flow-through to batch settlement was under consideration, which would eliminate turbulence caused by continuous flow and improve settling. Design of the first stage pond was also recommended, based on predicted contributions of undiverted surface water (a design storm event) to prevent failure during an abnormal storm event. These techniques had not been considered before the training course was delivered, and their rapid adoption provided an indication the training was effective.

Acid mine drainage Training in the prediction and control of acid mine drainage included three weeks of field and classroom exercises for a total of 40 students from 17 units of the Ministry of Mines and Energy (including twelve Kanwil offices) and 12 of the 18 largest Indonesian mining companies. Like the above course, the Acid Mine Drainage course was team-taught by American and Indonesian instructors from the Ministry who were previously trained in the U.S. Course content was developed for the Indonesian mining setting from elements of several regular OSM training classes including: Acid Forming Materials – Principles and Processes; Acid Forming Materials – Planning and Prevention; Surface and Ground Water Hydrology; and HC-GRAM (hydrogeochemical analysis). These elements were specially modified for application to Indonesian mining circumstances. Class materials included prepared notes, reprints of selected papers and problem exercises. Training aids and instructor notes were also provided for Indonesian instructors so additional students can be trained in these subjects in the future (Roberts, 1998). Training for 18 students from the Ministry included a one week session in Jakarta which covered fundamental concepts of geology, chemistry, hydrology, and agronomy needed to recognize and manage acid forming materials. Instruction illustrated the depositional environment of various rocks and minerals, testing procedures, formation of acid and alkaline waters and soils, and impacts to fish, other aquatic life and plants.

A second one week session provided field training in Tanjung Enim, Sumatera. It included a combination of field exercises, classroom lecture and problem assignments at the mine site. This instruction was designed to build on fundamental principles learned in Jakarta. The main focus was on applying concepts learned in the classroom to actual problem identification, data interpretation, acid mine drainage prediction and remediation. Students from Indonesian mining companies were given a five day condensed course of instruction derived from the two week training described above. Considerable field experience and language skills of the industry students allowed some course acceleration without reduction in quality. The industry training took place in Tanjung Enim following the two week government course. Thus new technical training courses in Acid Mine Drainage, and Sediment and Erosion Control were developed for Indonesian mining circumstances and delivered in Indonesia by teams including instructors from DTPU, BLT and two Kanwil offices of the Ministry who were newly qualified to present these courses in the future. Training on Cost Estimation for Reclamation Guarantees was also provided by the project for 13 senior members of the Ministry's Central AMDAL Commission, 20 mine inspectors, and 19 representatives of the mining industry from 16 mining operations, as discussed in Chapter 6.

Interviews conducted in August 1997 with persons who received this training in the U.S. and Indonesia indicated they were pleased with the practical nature of new courses in Acid Mine Drainage and Sediment and Erosion Control. In addition, three U.S. members of the Training Team assisted three Indonesian instructors[3] in presenting technical courses in Surface and Groundwater Hydrology, Principles of Reclamation, Basic Principles of Inspection, and Advanced Principles of Inspection with field components at several mines along the Mahakam River in East Kalimantan. Interviews conducted by the program evaluator with students in these courses indicated a high level of satisfaction with both the subject matter and the practical utility of the training received (Hamilton, 1997b). Several students indicated enthusiasm for participating in additional training on related subjects.

Surface and groundwater hydrology A one-week introductory course on Surface and Groundwater Hydrology was presented in Bandung to 17 students from various units of the Ministry of Mines and Energy and the mining industry. This course was designed for persons who had not had academic or other training in hydrology, and was tailored to Indonesian circumstances. It provided training in basic concepts such as water-holding capacity of soils; stream-flow measurements; erosion and sediment development and hydrographs; techniques for control of water and sediment; groundwater hydrology and effects of mining on it; acid mine drainage; surface and groundwater monitoring, and data interpretation (U.S.DOI/OSM, 2001, p. A-32).

Principles of reclamation A one-week introductory course on Principles of Reclamation was presented in Bandung to 19 students from various units of the Ministry of Mines and Energy (including DTPU and Kanwil offices) and the mining industry. This course was designed for persons who had no training in the study of soils, plant/soil relationships, drainage control, slope stability or revegetation, and was tailored to Indonesian circumstances.

Principles of inspection A one-week introductory technical course on Basic Principles of Inspection was presented in Bandung to 20 students from various units of the Ministry of Mines and Energy. This was the third time this course was offered in Indonesia. These three courses served two general purposes. First, they introduced a total of 56 additional Indonesian students to ideas and techniques that are fundamental to efforts to protect the environment at mining sites. Second, these courses provided three Indonesian instructors with experience in teaching the material presented in the courses. After this experience, and with teaching materials provided to the Ministry of Mines and Energy by OSM instructors, the three Indonesian instructors were able to present similar training courses in Indonesia without additional assistance.

An advanced course in mine inspection was presented in Samarinda, East Kalimantan and at nearby coal mines to 23 students from various units of the Ministry of Mines and Energy over a two-week period which included two days of classroom review of basic principles of inspection, six days of actual inspections at three mines, and a presentation of mine inspection results by students to the Chief Mine Inspector of DTPU. The advanced inspection class was divided into three groups. Each group prepared and conducted a pre-inspection entry briefing with the *kepala teknik* (the head technical officer at the mine, who usually has responsibility for environmental management) at the first mine the group inspected. The groups then conducted two-day, complete inspections at three mines. At the third mine inspected by each group, the inspecting group presented an exit briefing to the *kepala teknik*.

During mine site training, trainees observed mining operations and environmental conditions. They recorded their observations with sketches in an inspector's notebook. Each group wrote a complete inspection report for each of the three mines inspected. Inspection reports and inspector notebooks were used to prepare Mine Book entries of inspection results, and exit briefings with the *kepala tekniks*. Because this was a training exercise, and as a courtesy to the host mines, no actual Mine Book entries were made, but *kepala tekniks* were active participants and learned a great deal about what to expect in future mine inspections. At exit briefings, trainees presented their observations and recommended solutions to problems they identified. In addition, each group prepared and presented a post-mine inspection briefing to the Chief Mine Inspector. This allowed students to demonstrate many of the techniques and skills they acquired from training classes, including presentation skills (use of audio/visual equipment and aids, organization, voice control, eye contact); technical skills (ability to provide a technical justification for observations and conclusions); and inspection skills (identification of environmental problems and the mine practices that created them).

Training included site visits to mining operations of PT Bukit Baiduri, PT Tanito Harum and PT Kitadin on the Mahakam River in East Kalimantan. The Training Team concluded that on-site work is a particularly useful approach for students to obtain hands-on experience with sampling, inspection, and the use of portable equipment needed to collect samples and conduct tests. Equally useful was the opportunity to translate on-the-ground observations into presentations and recommendations for action to representatives of mine operations and to Ministry officials. Course outlines and lesson plans developed in cooperation with Ministry instructors for Spoil Handling and Disposal (U.S.DOI/OSM,1998d), Acid Mine Drainage Prediction, Prevention and Treatment (Roberts, 1998), and Soil Erosion, Sediment Control, and Revegetation (Maynard, 1998a, 1998b, 1998c) were published and provided to BLT for future use.

During February and March 1998 three OSM technical training instructors traveled to Indonesia to assist four Indonesian members of the Training Team[4] in presenting six technical courses, for a total of four weeks of classroom training and field exercises to 20 Ministry mine inspectors and technical staff from Jakarta and Kanwil regional offices. Classroom training was presented at the Ministry's Manpower Development Center and field exercises were conducted at the PT Bukit Assam (PTBA) Air Laya Coal Mine in Tanjung Enim, South Sumatera. Intermittently joining these classes were 14 students from local government mining offices who attended parts of the training in Bandung. Local governments were then responsible for regulating Class "C" construction minerals, and are now responsible for inspecting coal mines. The Ministry is responsible for training and guidance for them.

Mining reclamation technology A five-day introductory course in Mining Reclamation Technology was presented for persons who had no academic or other training or experience with such technologies. It focussed on soil

Figure 8.1 Indonesian sampling acid mine drainage during training, South Sumatera
Source: Alfred E. Whitehouse

handling and stabilization, topsoil preservation and substitution, seedbed preparation, woody plant establishment, and indicators of vegetation success which might be used to determine when a reclamation bond should be returned to the mine operator.

Mining environmental inspection   A five-day course in Principles of Mining Environmental Inspection was presented, which included a special half-day session on the utility of Mining Performance Standards. This course was a modification of the Basic Principles of Mine Inspection course tailored to Indonesian environmental inspections, with field exercises described below.

Inspection report writing   A one-day course in Inspection Report Writing emphasized using plain language as a way of expressing written ideas clearly. Participants learned how changes in writing style, organization and layout can improve the quality of documents. Topics included gathering information, note taking, use of checklists and references; describing information from direct observation using sight and sound; and qualities of good inspection reports (U.S.DOI/OSM, 2001, p. A-16). The course was developed especially for Indonesian mine inspectors using case studies and classroom exercises.

Evidence and expert testimony   A one-day course in Evidence and Expert Testimony focussed on collection and presentation of mine inspection data in a legally defensible manner using graphic, photographic, sketching and other techniques of documentation. It represented a combination and substantial modification of two OSM courses on Evidence Preparation and Expert Witness testimony (U.S.DOI/OSM, 2001, pp. A-20, A-22), but without reliance on the U.S. legal context. Presentation of evidence in a formal hearing setting was emphasized.

Soils, erosion, sediment control and revegetation   A five-day technical course in Soils, Erosion, Sediment Control and Revegetation was presented for the second time in Bandung for 20 persons from various offices of the Ministry, and separately in Balikpapan, East Kalimantan for 16 employees of seven mining companies. This course focussed on Indonesian mining and ecological circumstances (Maynard, 1998a, 1998b, 1998c).

Acid mine drainage prediction, prevention and treatment/spoil handling   A five-day technical course in Acid Mine Drainage Prediction, Prevention and Treatment/Spoil Handling was presented in Bandung for 20 persons from various offices of the Ministry, and separately in Balikpapan, East Kalimantan with a site visit to the PT Bukit Baiduri Coal Mine for 19 employees of seven mining companies. A similar course was presented previously to different trainees in Indonesia (Roberts, 1998).

Field exercises   Three weeks of classroom training for Ministry personnel described above was supplemented by a one-week field exercise at the PT Bukit Assam Air Laya Coal Mine in Tanjung Enim, South Sumatera, where

Figure 8.2   Training class field exercise examining acid forming materials and pit flooded with acid mine drainage, South Sumatera
Source: Alfred E. Whitehouse

instructors demonstrated how classroom lessons could be applied to a specific mine. During these exercises the class was joined by the environmental staff of PTBA, who were especially interested to learn how to apply principles of acid mine drainage prediction and prevention, and erosion and sediment control to their mine. Field exercises were designed especially for this technical training to integrate material in the above courses with applications in the field in Indonesia.

Several mining companies explored opportunities for follow-up assistance on acid mine drainage prediction and erosion and sediment control. In February 1998 the Ministry invited the Training Team's American acid mine drainage instructor to visit the Kaltim Prima Coal (KPC) mine in Sangatta, East Kalimantan and discuss a difficult acid mine drainage problem. KPC was then the largest coal mine in Indonesia, producing over 20 million tonnes/year. One mining area in this large complex presented a high risk of acid mine drainage. KPC and the Ministry wanted to learn how acid mine drainage prediction and prevention science was applied in the U.S., whether KPC was doing all it needed to do to get a reliable prediction or whether they were doing unnecessary, redundant, or wrong things. After touring the mining area, reviewing laboratory reports, and seeing the waste rock disposal site, the group met with KPC mine management, and discussed its observations and recommendations for improvements. The PT Arutmin mines near Balikpapan,

South Kalimantan were also visited for discussions of overburden sampling and AMD prediction.

Transfer of training courses to Indonesian instructors was a significant step in accomplishing one objective of this project, development of institutional capability and human resources, because the Ministry greatly improved its capability to train its own staff. New technical training courses were developed for Indonesian mining circumstances and delivered in Indonesia by teams including instructors from DTPU and BLT, who are now qualified to present these courses in the future. Thus during three years of effort 1995–1998, OSM technical assistance helped the Ministry develop a cadre of instructors from its senior inspector and technical staff who are now able to develop new courses and teach them to mine inspectors, technical staff and mine operators. Moreover Indonesian members of the Training Team who received training in the U.S. now have greater knowledge and field experience relating directly to the functions they perform in the Ministry. This constitutes significant evidence of technology diffusion through a bilateral technical assistance project between the United States and Indonesia. Interviews with the Director of the Manpower Development Center in July 1998 indicated there was continued interest in adding technical training courses similar to those developed for the project to the curriculum of the Center (Hamilton, 1998b). Some faculty of the Manpower Development Center participated in training courses presented by the project, and indicated interest in developing new courses which might be marketed to the mining industry.

Technical Consultations and Site Visits

Formal course instruction and technical consultations in the U.S. involving Indonesian participants were supplemented by site visits and demonstrations designed to increase capabilities of Ministry personnel to identify and solve environmental problems associated with mining activities. For example, during the initial Policy Team discussions in July 1995 an afternoon was devoted to visiting the U.S. Environmental Protection Agency (EPA) Regional Library in Denver, Colorado. Indonesian Policy Team members received assistance in performing research on a variety of topics and were able to obtain a number of reference publications on air and water quality programs, waste management, and environmental law. Perhaps more importantly, they learned how to access the EPA Library through the Internet from Indonesia. This new capability enhanced access to informational resources useful in setting standards and solving environmental problems. Furthermore, Indonesian participants on all OSM project teams now know how to reach their professional counterparts in the U.S. via electronic mail, and have received invitations to do so whenever they need assistance in finding information or solving problems. This invitation and capability extended beyond the term of the technical assistance project, drawing Indonesian participants into an international network of professional peers.

Also during the initial Policy Team discussions in July 1995 OSM conducted a field trip focused on an environmental problem of great concern in Indonesia:

acid mine drainage from constructed gravity drainage tunnels. Participants visited three different U.S. mining areas for gold, silver, and coal to compare conditions. The purpose of the field trip was to show Indonesian participants the costly, long term adverse water quality impacts associated with gravity drainage tunnels. For example, the Argo Tunnel (gold) they visited near Denver has been polluting the receiving stream there for over 100 years. This visit vividly illustrated the long-term liabilities of gravity drainage tunnels for AMD.

Site visits with experts increase the knowledge and expertise of individuals, providing special opportunities for development of human resources and enhancement of institutional capacity to solve mining problems in Indonesia. There was substantial interest in experimental wetland treatment techniques which could have applicability in Indonesia. Three Indonesians including a senior mine inspector from a Kanwil office of the Ministry and a BLT staff geologist visited Knoxville, Tennessee to participate in training workshops at the Annual Meeting of the American Society for Surface Mining and Reclamation in 1995. They attended workshops on the Chemistry of Pyrite, GIS in Mining and Reclamation, Sediment and Erosion Control, and Planning and Design Concepts for Coal Mine Closure. These individuals also visited a Tennessee Valley Authority mine to see a demonstration of acid mine drainage treatment with constructed wetlands, and toured abandoned mine land restoration projects in the Bear Creek Watershed and Big South Fork National River and Recreation Area. Each of these activities helped to develop the human resources of the Ministry of Mines and Energy and increase its institutional capacity to solve similar environmental mining problems.

Because rain forest topsoils are often thin and difficult to preserve during mining, innovative techniques for preparation and use of alternative growing mediums are of some interest in Indonesia. During technical consultations in June 1996 Indonesians toured the Black Mesa-Kayenta Mine and the BHP Navajo Mine in New Mexico and Colorado to view successful demonstrations of topsoil substitution techniques, the Black Mesa coal slurry pipeline, examples of successful mine reclamation and implementation of mine permit requirements.

In October 1997 two Indonesian biologists on the Technology Team visited the United States for 30 days to receive technical training in the use of biological resource information in environmental assessments and permitting processes. This training emphasized hydrologic treatment regimes, mine reclamation, and reclamation bond release inspections, including vegetation sampling and indicators of reclamation success. Classroom time was also used to provide an overview of environmental assessment procedures and focused on cumulative environmental impact assessments which would have applicability to circumstances in Indonesia. A number of field trips were arranged as part of the training, including visits to Buffalo Coal Company's biological water treatment for acid mine drainage, the MAPCO Mettiki Mine trout rearing operation in western Maryland, a wildlife research park on reclaimed mine land in Ohio, and both wetland and anoxic limestone drain treatment of acid mine drainage in Colorado and Tennessee. The two trainees received instruction from a total of 20 American instructors and mining industry representatives, participated in an actual reclamation bond release and post-mining land use inspection at the B and N Coal Company mine in Ohio, and conducted field exercises and site visits at a total of six coal and hardrock mining operations.

A different group of Indonesians in November 1997 visited the Buffalo Coal Company mines, MAPCO's Mettiki Mine, and the Laurel Run Stream Neutralization operation in western Maryland to see successful demonstrations of acid mine drainage treatment, mining and reclamation practices. The Mettiki Mine is an underground longwall operation producing over 2 million tons per year. Their closed circuit, heavy media preparation plant includes a thermal dryer and was designed to process a single coal seam. Unlike coal processing plants in Indonesia, there was no water discharge from this plant. The Mettiki Mine operates near the axis of a syncline which requires pumping many millions of gallons of water daily to keep the mine from flooding. Gravity discharges of mine water are no longer allowed in the United States because of the long-term threat of acid mine drainage, and water pumped from the Mettiki Mine is high in iron and has a low pH. Following chemical treatment and aeration, the mine discharge provides the entire source of water for a State of Maryland Fish and Game Commission trout hatchery on the mine property before being discharged into the receiving stream. This was a most vivid demonstration of successful treatment of mine drainage and a mining company's commitment to clean water. One small slip on the part of the company and the trout in the hatchery would be killed.

The Buffalo Coal A-34 Preparation Plant is also a closed circuit, heavy media plant but was designed to accommodate a wide range of coal qualities in nearly 30 coal seams encountered in the company's many surface mines. Like the plant at the Mettiki Mine, there is no water discharge. Buffalo Coal employs a fine coal recovery circuit in the plant to enhance coal recovery, and uses a belt press filter to dry fine refuse before combining it with coarse refuse on the waste pile. The Lonaconing Preparation Plant is a low-tech, closed circuit hydro-cyclone facility with deister tables. This unique fine coal recovery system uses a vacuum filter belt to separate fine coal from water slurry in the plant. As discussed in Chapter 5, there are very few fine coal recovery circuits in coal preparation plants in Indonesia, where estimates of coal lost typically exceed 20 percent of coal processed.

Unmet Needs for Technical Training

Transfer of knowledge and skills about technical aspects of planning, operating, and regulating mining operations was spread broadly and deeply in the Indonesian mining sector by the OSM project. Persons receiving this training included all of the mine inspectors employed by the national government. Participants included not only persons working in the central government in Jakarta, but also in its regional offices scattered among the provinces, individuals employed by some regional governments in the most important mining districts, and employees of major mining operations in the

Indonesian mining industry. During the OSM project over 460 persons from 17 units of the Ministry and 16 of the 18 largest mining operations received technical training in at least one course. The Ministry was assisted in developing a group of instructors from its senior inspector and technical staff who prepared new courses to fit Indonesian circumstances, and taught them to mine inspectors, other technical staff and mine operators. These individuals are now capable of developing and delivering additional technical courses. The permanent training capability of the Ministry (especially its Education and Training Centers) grew in both breadth and depth, and must now be considered among the most sophisticated in the Asia region.

Interviews conducted in July 1998 with persons who received technical training from the OSM project reiterated the need for additional training for mine inspectors and mine operators in reclamation, erosion control, acid mine drainage, and water quality monitoring. Persons interviewed who participate in review of AMDAL documents suggested they need additional training in groundwater hydrology, surface water hydrology, biology, forestry, water and soil chemistry, roughly in that order of priority. However there were no scheduled assignments or plans for Indonesian instructors in DTPU and BLT who were trained during this project to provide additional training to others after the project ended. An encouraging development as this book went to press concerned indications the Education and Training Center for Mineral and Coal Technology would offer a revised course in Mine Inspection 5–6 times during 2005 for regional government personnel in various locations using DTPU inspector/instructors trained by the OSM project (Whitehouse, 2005).

Continued efforts to foster a more focussed development of capabilities in the Education and Training Center for Mineral and Coal Technology would ensure the permanent institutionalization of this sophisticated technical expertise. Further development of certification requirements and related technical training courses in mine inspection, preparation and review of environmental assessments would help ensure the longevity of this training capability. Periodic refresher courses to keep faculty abreast of new developments in best mining practices and mine monitoring technology are needed to ensure the long-term viability of training programs.

An ongoing need is evident in Indonesia to establish a process to develop and deliver technical training in a systematic manner. There is also a continuing need to establish a coherent model for employee development which sets out a core of technical training courses to be taken by inspectors, ANDAL reviewers and mine operators. A review of technical training curriculum offered by the Education and Training Center for Mineral and Coal Technology would be a useful starting point in such an effort, and was included in a proposal to extend the OSM technical assistance project (U.S.DOI/OSM, 1998a). Moreover there is a need to establish an on-the-job-training program which would include sharing information and skills within an office, and mentoring of younger, less experienced staff by senior staff and managers in Jakarta, Kanwil and regional government offices. Additional training is needed for ANDAL technical reviewers, mine operators and their consultants who prepare such documents. A system for evaluating the effectiveness of training should be established, and some follow-up to employee training should be designed to measure the results of expenditures on training efforts (e.g., see Buckley and Caple, 2000, ch. 9).

Most of the technical training described in this chapter was provided to employees of the national government. As noted in Chapters 2 and 4, decentralization of mining environmental inspection, enforcement, and AMDAL programs has created a pressing need for additional technical training of personnel employed by regional governments. Since 2002, mine inspections have been the responsibility of kabupaten and kotamadya. The expertise of their mine inspectors is extremely uneven, and many of them have not received the education and training required for certification by the Ministry of Energy and Mineral Resources. Results of such inspections are certain to vary considerably from inspector to inspector, and therefor from mine to mine. Adequate technical review of AMDAL documents, inspection of mining operations, and protection of the public health, safety and the right to a clean environment will not occur in the absence of substantial efforts to further develop the human resources of Indonesian regional governments.

As suggested in Chapter 7, there is a pressing need for development of mining environmental compliance and reclamation curricula at regional universities in provinces having significant mining activity. Mining environmental compliance and reclamation curricula are not currently evident in universities in coal mining kabupaten in Indonesia. Most training in these subjects is centralized in the Bandung Institute of Technology or the Education and Training Center for Mineral and Coal Technology of the Ministry of Energy and Mineral Resources, both located in Bandung. There is considerable faculty expertise in these institutions in mining engineering and mining practices, but little, if any expertise in soils, revegetation, or reclamation of drastically disturbed mined lands. While this is also an accurate characterization of regional universities, some of them do have agricultural and horticultural business curricula which would support development of new expertise in these areas. In the United States, curricula supporting mining environmental compliance and reclamation training have long been available in universities and local two-year community colleges in many mining states. Significant improvements in the quality of mine inspections and ANDAL reviews might result from development of mine reclamation curricula at regional universities based on the subjects of training courses discussed in this Chapter.

Data Base Analysis and Technology Transfer

The Assessment Team for the OSM technical assistance project identified a priority need for evaluation of the adequacy of data used for regulatory decision making by the Ministry. A joint Technology Team consisting of five U.S. members and five Indonesian personnel was formed to perform this evaluation and to examine programs providing guidance to mine managers and personnel concerning regulatory requirements, and the technology and practices necessary for compliance with mining environmental regulations. This team reviewed the adequacy of the available data base, data base management and analytical capabilities, computer equipment, software and technical reference library for preparation and review of environmental assessments, regulatory decision making, and control of the environmental impacts of mining operations. Alternatives for improvement of the acquisition, management and analysis of data in mining and environmental decision making were discussed and some were implemented, as discussed below. The OSM project provided new equipment needed by the Ministry of Mines and Energy to inspect mines in the field, identify and solve potential environmental problems associated with mining operations, and analyze environmental assessments.

Overall, the existing computing situation in the Ministry of Mines and Energy in 1995 was encouraging. A majority of the individuals interviewed had clear, realistic expectations and/or perceptions of how computing technology may be applicable to their jobs. For a few applications, there were staff more expert than any in OSM. Unfortunately, each unit seemed to be operating independently and was unaware of the successes and failures of the other groups. There had not been a coordinated policy within the Ministry on data acquisition and management or equipment specifications. This lack of policy allowed several units to spend money to acquire the same data rather than share data, and to purchase new equipment that was inadequate to meet the expanding needs of the Ministry.

Hardware and Software

There appear to have been three major periods of computer modernization in the Ministry of Energy and Mineral Resources: (1) the early 1980s for the Mineral Technology Research and Development Centre (MTRDC) mainframe shop; (2) the late 1980s in MTRDC, the Ministry of Technology and Research, Pertamina (the national oil company), and the Bureau of Planning; and (3) the mid 1990s for most other sites. The most significant of these periods was the late 1980s. Apparently many Indonesian agencies and industries spent a great deal of money on Geographic Information Systems (GIS) software and hardware in this period. This also occurred in the United States during the same period. In the U.S., interest in GIS fell abruptly as people realized that it required a lot of training, design, and data entry before any useful analyses became possible. There were indications that Indonesian installations may have suffered similar experiences (U.S.DOI/OSM, 1995d). These three periods of computer modernization appear to have been the results of forward thinking. However facilities established then in MTRDC later fell into disuse when key personnel were absent for prolonged periods due to overseas education, promotions or other reasons. A structural policy or process for continuous upgrading of equipment and facilities that would not falter or shift as personnel leave or move up the organization was not in place.

A compilation of major remote sensing activities in Indonesia for the period 1988–1993 (Soesilo, 1993) contained extensive lists of GIS installations. Unless these installations were subsequently updated, many had software and hardware which were obsolete or incompatible with market leaders or current industry standards in 1995. Equipment examined at MTRDC and the Bureau of Planning (BOP) supported this observation. Local area networks (LANs) were examined at MTRDC and BOP. Hardware at both sites was manufactured by DEC and was obsolete technology. Similar hardware used in Kentucky and Pennsylvania in the United States in 1995 could only be maintained by independent contractors at an annual cost three times the original purchase price. The advantages offered by Local Area Networks (LANs) of shared printers, documents, and data; office email and scheduling; routine backups; and reduced software cost were not evident in the Ministry in 1995. Experience in Jakarta with network connections, Internet, and e-mail revealed that all dial-up connections were excessively vulnerable to service disruption and ran at very low speeds to minimize data corruption. They could not be used for technical/scientific applications given the unreliable state of telephone service at that time.

Limited database access   The Assessment Team was told in 1995 that large amounts of information and expertise existed in regional offices, other Directorates and Ministries, universities, and the private sector. An extensive, fully-maintained database for forest land inventory and management was said to exist in the Ministry of Forestry, but the Ministry of Mines and Energy had no access to it. A database meeting all of BLT requirements had been developed for DTPU in the Ministry but BLT had no knowledge of the database or access to it (U.S.DOI/OSM, 1995d).

Lack of workstations and training facilities   Personal computers and software available in the Ministry were current versions but were not on every technical person's desk. There were no engineering work stations with analytical software available to technical staff reviewing environmental assessments (ANDALS) and little if any computer analysis of data provided by mining companies was available to predict or verify the proponent's anticipated impacts of a proposed mine on the pre-mining environment. Lower level engineering software was available in Jakarta and training was commercially available for those packages (e.g., AutoCAD, Statgraphics, Reflex, etc.). A computer training facility was installed in the Manpower Development Center. but machines there appeared to be designed for basic computer skill training and used multiple dumb terminals without hard disks tied to an instructor's machine. The entire system had low capacity and would not be suitable for technical training without substantial upgrading to accommodate the hard disk and RAM requirements of technical software (U.S.DOI/OSM, 1995d).

Technical Information Processing System

Ministry data base management and analytical capabilities were substantially improved by installation of a new, state-of-the-art Technical Information Processing System (TIPS), and subsequent training in its use. TIPS consists of

Table 8.2   Capabilities of the Technical Information Processing System

---

(a) Geologic surface modeling for hydrologic impact assessment; distribution of groundwater contaminants; subsidence control planning; earth moving volumetrics to estimate reclamation costs and resource production values; simulation and analysis of post-mining topography; and three-dimensional distribution of toxic and acid-forming materials.

(b) Stratigraphic analysis to quantify continuity of potentially affected aquifer systems, likelihood for subsidence, and extent of toxic or acid-forming materials.

(c) Ground water modeling to predict consequences of surface and underground mining on the hydrologic balance of adjacent and mined lands.

(d) Surface water modeling of changes induced by mining on small watersheds and effectiveness of sediment/erosion control structures in meeting effluent standards.

(e) Slope-stability analysis for design of coal waste and excess spoil fills, backfills, sediment ponds, natural slopes, post-mining slopes, and embankments.

(f) Blasting and noise impact analysis to determine the amount of blasting powder that can safely be used in proximity to structures, impact of equipment and maximum air overpressure related to evaluating blast vibration impacts.

(g) Statistical analysis of baseline conditions and changes induced by mining operations to evaluate water sampling adequacy and water quality reports during mining, and success of reclamation and revegetation prior to bond release.

(h) Geographic information systems analysis and mapping at regional and local levels, and automated drafting of engineering drawings, illustrations, schematics and annotated maps.

---

several off-the-shelf hardware components and software packages that use standard drill log information, coal or mineral and overburden quality data, groundwater quality data, coal-crop information, cross-section data or other digitized information in an integrated system to conduct independent technical analyses and modeling of mine permit application data. This system was then in use by OSM and all 26 state regulatory authorities in the United States, and comparable systems were widely used in mining industries world wide, including several mining firms in Indonesia. The TIPS installation was based on a Silicon Graphics Indigo Hi Impact (UNIX) Workstation with digitizer, plotter and printer which could be networked to additional personal computers. It utilized approximately 20 different software packages which were integrated to share data for several types of analysis, summarized in Table 8.2.

During initial Policy Team discussions in July 1995 a one-day training session demonstrated the utility of the Technical Information Processing System for review of environmental assessments and evidence preparation. The Policy Team also visited the Colowyo Mine, producing about 4.5 million tons/year, to see first-hand how environmental protection practices are implemented on a large surface coal mine in the United States. The company demonstrated computer software it uses to plan the mine and optimize production. A follow-up visit was made to the MINCOM Ltd. office in Denver to see a more complete demonstration of computer software used at the Colowyo Mine to improve production and manage environmental controls. Thus early in the project Indonesian members of the Policy Team developed an appreciation for the use of such technology in controlling mining environmental impacts.

Installation of a TIPS workstation was initially scheduled for October 1995 but was postponed to March 1996 to allow acquisition of the more powerful Silicon Graphics Indigo Hi Impact Workstation which was just coming into the market (in place of the Silicon Graphics IRIS Indigo). This delay also allowed an Indonesian member of the Technology Team to receive training and participate in configuring the workstation in the U.S. before it was shipped to Jakarta.[5] He visited the U.S. in January 1996 for training in Unix Workstation System Administration at the Silicon Graphics, Inc. Training Center in Mountain View, California. After completing the training, he configured and tested the Silicon Graphics Workstation with U.S. members of the Technology Team at OSM offices in Denver. Silicon Graphics had established an office in Jakarta to provide support services, technical assistance and maintenance for over 155 of their UNIX workstations in Indonesia, at least six of which were in use in major mining companies. Computers with comparable capabilities were not then in use by inspectors or AMDAL technical teams in the Ministry's Directorate of Technical Mining.

Three U.S. members of the Technology Team visited Indonesia in March 1996 to install and provide basic instruction in use of the Technical Information Processing System. Installation went somewhat slower than expected due to insufficient electrical circuits and a lack of ground/earth connections. Some electrical cable needed to be procured locally and the TIPS work space was reorganized. Such difficulties are not uncommon in technology transfer efforts where new equipment is state-of-the-art but infrastructure is somewhat primitive. However by the end of the first week the system was operational with an interruptible power source and its own ground wire routed down the outside of the building. In addition to the new Unix workstation, six freestanding personal computers were configured with Windows 95, Microsoft Office, HC-GRAM, StatGraphics, and StratiFact analytical programs.

Nineteen days of TIPS training for personnel in the Directorate of Mines and the Bureau of Environment and Technology began as soon as the workstation was installed and tested. The original training plan anticipated delivering standard OSM training courses in UNIX System Administration, EarthVision, StatGraphics and the Microsoft Access Data Base Management System. After determining the experience level and expertise of Ministry personnel related to both quantification of environmental effects and use of computers, it was necessary to significantly modify the course material and provide courses at a more basic level than expected. Instruction was planned in mining applications of the 14 courses listed in Table 8.3. However only the first three courses in Table 8.3 and some instruction in the use of Global Positioning Systems (GPS) to resolve boundary disputes were actually completed (Hamilton, 1998b).

Table 8.3   TIPS Training Courses Planned for OSM Project

Workstation Basics
Data Import and Export
MS ACCESS Data Base Management System
Statgraphics Statistical Analysis
Slope Stability Analysis (SB-Slope)
Stratifact Geologic Correlation
Surface Water Flow Simulation:
      RUSLE: Revised Universal Soil Loss Equation
      STORM Sediment Control Design
      TR-55 Hydrographic Watershed Modeling
Hydrochemical Graphical Analysis Modeling (HC-GRAM)
Global Positioning Systems (GPS) Technology Applications
Groundwater Modeling (MODFLOW)
Surface Deformation Prediction System (SDPS)
EarthVision (3-dimensional geologic modeling)

Difficulties Encountered

Obstacles that slowed the TIPS training also affect the overall quality of ANDAL review. These difficulties are associated with a lack of field experience, basic scientific and data management skills, and institutional confusion over legal and regulatory requirements on the part of persons being trained. Basic mining concepts were weak and the lack of field experience in BLT staff caused significant delays in computer training. In several cases, trainees were experienced computer programmers but had difficulty understanding basic concepts of digital representation of spatial dimensions necessary for the use of digital data to construct map displays. These concepts are fundamental to understanding Geographic Information Systems (GIS) and three-dimensional geologic modeling. Difficulties in imparting these concepts were attributed to language barriers between instructors and trainees, which might be resolved by securing basic computer mapping course instruction from vendors in Indonesia before attempting more advanced TIPS training. Some local vendor training was acquired during the third year, but more is needed to give a greater number of Ministry technical personnel the foundation they need to prepare them for training in proper use of TIPS. Still, there was a shortage of persons capable of receiving the planned training.

A second significant factor in the TIPS training was that technical staff did not have a clear understanding of regulations under which they reviewed ANDAL. For example as part of the digital geospatial mapping demonstration conducted during training on the Nonny Mining Concession, BLT staff learned for the first time that all Mining Authorizations (KP) and ANDAL must have a surveyed point tied to an actual bench mark of known longitude and latitude. This basic requirement, necessary to locate a project in the real world and relate it to locations of other projects, was not routinely checked during previous ANDAL reviews. BLT staff were not aware it was a requirement (U.S.DOI/OSM, 1996a). A third impediment to effective TIPS training was the lack of data to run computer simulations that project environmental impacts for effective ANDAL review. ANDALs reviewed as part of class projects contained little environmental data on which to evaluate proponents' conclusions. The ANDALs addressed environmental baseline data weakly. Proponents often spent their time and money analyzing low percentage impacts from unlikely sources such as the potential for various organic pollutants unrelated to mining operations, while failing to analyze the pyrite content of overburden in areas of known acid mine drainage production. This lack of environmental data in the ANDAL greatly limited the use of automated techniques for quantifying impacts and reduced the overall effectiveness of the AMDAL process.

Due to this initial experience, plans for delivering TIPS training were substantially revised for the remainder of the project, adopting a three-step approach similar to the one described above for technical training courses. English-speaking Indonesian instructors were trained with OSM and other American personnel in courses presented in the U.S. This training included tutorials and was followed by joint development of new TIPS courses by Indonesian and OSM instructors. Thereafter, OSM instructors traveled to Indonesia where the new courses and tutorials were delivered with Indonesian instructors.

Three Indonesian members of the Technology Team visited the United States in August 1996 to receive TIPS instruction. This included mine site field instruction and tutorials in use of TIPS during permitting and analyzing acid forming materials, erosion and sediment control, and hydrology. A third round of TIPS training was initiated with Indonesian instructors visiting the U.S. for 30 days in July/August 1996. Three American instructors visited Indonesia in March 1997 to assist them in providing TIPS instruction in Unix System Administration (3 days), Data Import/Export (1 day), and tutorials on the use of Global Positioning Systems (3 days). Seminar demonstrations and tutorials were conducted with Indonesian personnel who review environmental assessments in the use of geospatial data, computer mapping, computer impact assessment, and basic coal geology, using three Indonesian projects as examples. These concerned the Ombilin Mine resource assessment in West Sumatera, the Nonny Mining Concession mapping effort, and the Pulau Sebuku Island Coal Mine ANDAL. For example, using GPS readings obtained in the field, a boundary dispute between the Nonny Mining Concession and an oil and gas lease was successfully resolved, demonstrating the analytical power and utility of this technology.

A fourth round of TIPS instruction in 1998 involved one Indonesian member of the Technology Team traveling to Denver, Colorado during April for three weeks of individualized training in EarthVision, use of Global Positioning Systems (GPS), and Technical Information Processing System administration and maintenance. One American member of the Technology Team traveled to Jakarta in August 1998 to consult with individuals responsible for information processing, and provide follow-up training and maintenance of the BLT Technical Information Processing System work-station.

Interviews conducted in August 1997 and July 1998 with persons receiving this training indicated greater satisfaction with the revised mode of delivery for TIPS training in Indonesia. Use of the Technical Information Processing System briefly increased the capacity of the Ministry to evaluate environmental assessments and monitoring plans, reclamation plans, and monitoring of mining operations. This new technological capability was as sophisticated as any in use by mining operations in Indonesia, but further development of Ministry human resources will be required to utilize it fully. Concerns remained about whether the appropriate persons actually received training, had continuing access to the TIPS equipment, and whether they would be able to receive sufficient training in its use to fully utilize its more sophisticated capabilities. However because the Ministry allowed software licenses to lapse at the end of the OSM technical assistance project in 1998, the US$200 000 TIPS workstation located in Ministry offices in Jakarta has been unusable and unused since that time.

A proposal was prepared for extension of the OSM project which included provision of equipment and training to substantially upgrade permanent capabilities of the Manpower Development Center (now Education and Training Center for Mineral and Coal Technology) to train Ministry personnel in use of the Technical Information Processing System (U.S.DOI/OSM, 1998a). Such training is consistent with the mission of the Education and Training Center for Mineral and Coal Technology, and would provide a permanent home and ongoing training capability which would be easier to sustain over time than seems likely with a single machine in an administrative staff office of the Ministry. However this proposal was not approved due to onset of the Asian financial crisis of 1998. Development of a permanent institutional capability to utilize the Technical Information Processing System would be advanced significantly by implementation of such a proposal.

Interorganizational Competition Over Technology

A potentially serious difficulty became apparent in interviews in June 1996 which indicated there was some misunderstanding or disagreement within the Ministry of Mines and Energy about the role and mission of BLT, some of which may reflect its structure, and some of which may reflect concerns over program jurisdiction within the Ministry. This misunderstanding became apparent during a demonstration of the TIPS workstation shortly after its installation in March 1996 but had not been resolved in June 1996. If this matter had remained unresolved, it might have adversely affected training and utilization of the Technical Information Processing System. BLT had three divisions: (1) Construction Services and Technology Analysis, (2) Environ-mental Management and Spatial Zoning, and (3) the Secretariat for the AMDAL Central Commission of the Ministry, a coordinating function involving members from other Ministries (e.g., Forestry, Environment, Agriculture). However the heads of the Divisions of Construction Services and Spatial Planning actually directed or performed most of the work of the AMDAL Secretariat in addition to the work of their own divisions. This created some difficulty within BLT, because personnel there had both "structural" and "functional" responsibilities which sometimes conflicted. BLT Division heads of Construction Services and Spatial Zoning had a "structural" responsibility to the Ministry to be sure the work of the AMDAL Secretariat got done, but had no "functional" authority to direct the work of staff in the AMDAL Secretariat.

Moreover the AMDAL Central Commission was charged by statute with responsibility for reviewing and approving all ANDAL involving development of resources within the jurisdiction of the Ministry and its Directorates General (e.g., oil and gas, electricity, geology, and minerals like coal and gold). Also by statute, technical review of ANDAL by the Commission may be "assisted" by technical teams comprised of individuals with temporary or collateral assignments from the relevant Directorates General (e.g., for coal ANDAL, technical team members came mostly from the Directorate General of Mines, with a few from other Ministries as needed for review of projects affecting forests, agriculture, etc.). However the actual approval decision belongs to the AMDAL Central Commission headed by the Secretary General of the Ministry, who must sign off on all ANDAL approvals for the Ministry.

There is a general division of responsibility and jurisdiction between the Secretary General and the Directorates General which is variously described as a distinction between administrative and technical functions, or policy making and policy implementing functions. The Secretary General has responsibility for administrative functions such as budgeting, personnel, inter-Ministry coordination, and sometimes policy making. These are similar to what is commonly referred to as staff or auxiliary functions provided internally by one unit for a larger organization.

Responsibility for technical functions of policy implementation and program operations was delegated to the various Directorates General, who have an interest in preserving control over development decisions concerning resources under their respective jurisdictions. Their primary mission is to manage development of these resources, and their responsibilities are most similar to line functions provided by an organization to the external public.

There was in 1996 some disagreement over the respective roles of the AMDAL Secretariat and the technical teams from the Directorates General in technical review of ANDAL. The BLT Director and staff of the AMDAL Secretariat viewed their role as active participants in technical review of ANDAL, asserting they must have a capability to "check the work" of the technical teams before presenting recommendations for approval to the AMDAL Commission. AMDAL Secretariat staff in BLT have expertise in technical subjects appropriate for such review, in excess of the expertise required for purely clerical tasks. In essence, their position was that each approval of an ANDAL was a policy decision made by the Secretary General, for which AMDAL Secretariat staff input was appropriate.

The Director General of Mining expressed a different view of the role of BLT, that would restrict the AMDAL Secretariat to a merely clerical function of coordinating reviews performed by others and processing paper, completely reserving the substance of technical review of ANDAL to the respective technical teams (comprised of staff from DTPU in the Directorate of Mines) which had responsibility for safety and environmental inspections of mining operations, and is housed within the Directorate General of Mining. In essence, this position was that review of ANDAL is a technical, policy implementing decision, for which the Secretary General should accept the advice of the technical teams without question. Perhaps coincidentally, this position would preclude a decision being made on environmental grounds that might be contrary to a development decision made by the Directorate General of Mining.

This position was apparently the basis for demands by the Director General of Mining to the Secretary General in 1996 that the TIPS workstation be moved to offices of the Directorate General of Mining, where it presumably would be used for technical review of ANDAL and other things. The Secretary General appeared to concur in this view of the role of the AMDAL Secretariat, and did not defend its need for a capability to "check the work" of the technical teams before approving ANDAL. It was not clear whether the Secretary General concurred in the view of BLT and the AMDAL Secretariat as a merely clerical function, but it appeared he might accept this view. In any event, the capabilities of BLT to perform technical reviews and solve environmental problems concerning any of the resources under the various Directorates General would have been reduced by relocation of the TIPS equipment and displacement of BLT staff in future TIPS training courses by DTPU staff.

The Directorate of Mines did not employ staff in Jakarta who were then capable of using the TIPS equipment. Two persons from the Directorate of Mines received very basic instruction in use of TIPS with BLT personnel during 1995, and it appeared the Director General of Mining wished these personnel to receive additional TIPS training. However they were not the most apt pupils. Moreover the only available person trained in UNIX system administration was employed by the Directorate General of Mines but was physically located at a research and development center in Bandung 3 hours away, which had no responsibility for review of ANDAL, no formal operating relationship with the Directorate of Mines, and therefore no budget justification for networking with offices in Jakarta.

Interviews conducted in June 1996 with the GIS and Remote Sensing Program Manager of the Mineral Technology Research and Development Center indicated there was a special section there with environmental expertise appropriate for review of ANDAL, but no near-term prospect for cable connections networking Jakarta with Bandung which would be adequate for interactive operation of TIPS. Such networking is commonplace in the U.S., and might be accomplished with expenditure of as little as US$8000 to purchase two microwave ground stations, but would require an ongoing operating budget of about US$2000 per month. Regardless of which organizational unit performed the technical review of ANDAL, it was essential that persons who would actually perform these reviews receive TIPS training.

In June 1996 discussions over the appropriate locus of technical review and the approval decision delayed selection of persons to be sent to the United States for TIPS training scheduled in August 1996 almost to the point of imperiling the training schedule. There was a need to resolve this issue before additional TIPS training was provided to Indonesian personnel, to ensure that persons who would ultimately use the TIPS capability actually received appropriate training. Concern about this situation eased somewhat during the next year. It had not been completely resolved in August 1997, but the TIPS equipment remained in offices of BLT, where training had taken place during the previous year.

Unmet Needs for Technology Training

A priority need remains for the Ministry to identify and engage local trainers for core software such as MapInfo or ARC/INFO, ARC/VIEW, AutoCAD, and Microsoft ACCESS Data Base Management System, to improve staff efficiency and performance, and provide a basis for more advanced TIPS training. It was not particularly efficient to use sophisticated Technology Team personnel for basic training in geospatial data, computer mapping, computer impact assessment, or basic coal geology, because this training was available from local vendors or from the Mineral Technology Research and Development Center in Bandung. There remains a need for opening routine staff level communication between DTPU and the Mineral Technology Research and Development Center to ensure a timely and continuous flow of ranked information for entry into each data base. There is also a need to provide access to this information for Kanwil offices and regional governments on other islands in Indonesia. With the shift in responsibility from national offices to regional government offices, the latter will increasingly become the source of much new data about mining operations. Dinas Pertambangan and BAPEDALDA need to understand the purpose for which data should be gathered and its value to their work. There are few indications they have this understanding currently.

Concerns for efficiency in use of scarce budgetary resources suggest there is a need for an inventory of software currently in use within the Ministry and some agreement about the utility of each software package, so some standardization could begin. There presently exists a large variety of software scattered among computers in the Ministry. Software that has wide applicability can be installed on a server and networked to personal computers within the Ministry less expensively than individual stand-alone versions. Units with special needs and/or applications can have stand-alone software or local area networks (LANS). Public domain software such as HC-GRAM and STORM could be installed more widely on machines of persons having responsibility for evaluating water quality, erosion and sediment control plans.

There was in 1998 a desire in some units of the Ministry to develop databases for tracking the processing of ANDAL (a management system) and evaluation of ANDAL environmental assessments (a technical database). There was little agreement about what each database would be used for so the proper data could be obtained and maintained. Priority needs which were not adequately addressed during the OSM project include development of a comprehensive database and management tracking system which includes integrated mine information and inspection findings. This database should provide historical inspection data about each mine and inspector as a tool for management to evaluate the inspection program. A computerized Applicant Violator System serves this function in the U.S. No schedule was developed for creation of this data base and management tracking system during the OSM project, leaving the Ministry with little ability to monitor and evaluate its inspection program.

Field Equipment for Mine Inspectors

Mine inspectors need to carry a small amount of highly specialized and portable testing equipment with them during inspections. This equipment allows mine inspectors to conduct field tests of soil and water quality to better support their observations of environmental problems in inspection reports. Mine inspection equipment used by OSM inspectors was purchased for 20 students in advance of the Training Team's presentation of technical courses on Principles of Inspection described above. This equipment included a compass and inclinometer, tape measure, water test kit and digital pH meter, a soil test meter, water sample bottles, and soil and rock sample bags for each student. As part of the Principles of Inspection Course, participants were taught how to use this equipment to measure environmental performance, and how to use the resulting information to support their findings when presented to company officials and their superiors. Subsequently, this equipment was distributed to mine inspectors in offices in Jakarta and six Kanwil offices.

Not all of the field equipment initially planned for acquisition was purchased. Discussions with mine inspectors during technical training courses in Indonesia produced revisions to the list of equipment planned for purchase. It was not necessary to purchase binoculars, hard hats, safety shoes or portable weirs because this equipment was already available to mine inspectors or else would not be useful. Video cameras were substituted for planned acquisition of 35 mm cameras, because video was used successfully by DTPU during an inspection of Kaltim Prima Coal in May 1996 and provides a more compelling record of violations. Interviews conducted in June 1996 with mine inspectors who used equipment provided indicate some concerns with it and possible needs for additional equipment. For example, one inspector indicated reluctance to use reagents provided with the water test kit for fear replacement supplies might not be available, and expressed concern for the effectiveness of the soil test kit in Indonesian circumstances. This suggested a need for more thorough training in use of the equipment, and that information about replacement supplies should be provided to inspectors. Additional equipment needed included dust samplers and cyanide analyzers for soils in gold mining areas.

Improving Mining Practices

Improvement of mining practices with respect to environmental management was an objective of the OSM project. The Assessment Team described a need to identify training for mine operators for which the Ministry has a responsibility, to insure a common understanding of environmental requirements and compliance technology. An effective way to provide training is often through seminars and demonstrations at exhibitions and conferences, in addition to regularly scheduled training courses. This has the advantage of reaching employees of mine operators with potential for improving mining practices, as well as compliance with mining environmental requirements.

Workshops and Presentations

During their visit to Indonesia to assist in delivery of technical training courses in August 1995, U.S. members of the Training Team presented a talk on "Reclamation in the United States" to over 300 faculty and students, and a seminar on "Reclamation and the Environment" to about 30 persons from offices of provincial governments and several mining operations at the Ministry's Manpower Development Center. U.S. members of the Policy Team presented papers at the First International Conference on Mining and the Environment held in the Ministry's Manpower Development Training Center, 6–7 March 1996 (Ministry of Mines and Energy, 1996b).[6] The 80–100 persons in attendance included personnel from the international mining community, senior officials of the Ministry's Jakarta and Kanwil offices, other national government personnel, and mining industry consultants.

A half-day workshop on prediction of acid mine drainage and water quality protection techniques was presented to a mixed class of 16–18 Ministry personnel and mining company technical staff at the Ministry's Manpower Development Center during March 1996. A seminar on "Mining and Reclamation Practices in the United States" was presented to about 60 persons from the Kanwil offices, the Directorate of Geology and Mineral Resources, Directorate of Oil and Gas, PLN, BAPEDAL, the Ministry of Forestry, and PT Timah during the BLT Annual Environmental Meeting in March 1996. During July 1996 the Bandung Institute of Technology sponsored a seminar on acid mine drainage to increase awareness within the Indonesian research and mining communities. A paper on "Overburden Analysis, Methods to Predict Acid Mine Drainage Potential" was presented to a mixed group of 36 university faculty, government officials and mine operators (Whitehouse, 1996c). The paper described a simple, consistent and relatively inexpensive procedure which helps mining companies and regulators develop strategies for preventing acid mine drainage during and after mining, and later appeared in a journal published by the Ministry (Whitehouse, 1998a). Identification of potential acid forming layers within the overburden allows mining plans to be prepared which may prevent or greatly reduce acid mine drainage.

A day-long seminar "Case Study on Mining in the United States" was presented to a mixed class of 25 Kanwil and provincial government officials and industry environmental managers at the Ministry's Manpower Development Center in August 1996 (Whitehouse, 1996a). Topics centered on erosion and sediment control, acid mine drainage prevention and control, top soil management, and revegetation. Technical papers and other information on specific topics were provided to several mining companies as a direct result of this outreach activity. A half-day seminar sponsored by the project was conducted by Dr R. Kelman Wieder, Professor of Biogeochemistry at Villanova University, Philadelphia, Pennsylvania on the use of wetlands for treatment of acid mine drainage during the last days of September 1996. The seminar was attended by about 20 persons from Indonesian universities, government and the mining industry.

In November 1996 at a plenary session on Strengthening National Capability in Mining Activities, a talk entitled "Economics and Environment Create a Role for Research" was presented by the OSM Project Director during the Ministry of Mines and Energy's Annual Colloquium and Exhibition on Mining in Bandung (Whitehouse, 1996b). The paper pointed out that society and government were gradually increasing demands on industry to "internalize" or pay the full price of the costs of coal production. In December 1996 a seminar was presented on "Mining Practices in the United States" to a training class of *kepala teknik* mining industry environmental managers in Bandung. In January 1997 the project sponsored a half-day seminar in Jakarta by Michael Harding, Technical Services Manager, Weyerhauser, Inc., Snoqualmie, Washington concerning management practices for erosion and sediment control. In March 1998 a seminar on Reclamation Guarantees was presented for a group of 25 mine industry *kepala teknik* at the Manpower Development Center in Bandung (Whitehouse, 1998b). Interviews conducted with the Director of the Manpower Development Center indicated that workshops and seminars offered there by OSM project participants were greatly appreciated, and energized both students and faculty. At the end of the project, there was continuing interest in sponsorship of similar future efforts and in adding technical training courses similar to those developed for the project to the curriculum of the Center. These occasions also proved useful in establishing contacts within the industry, greatly increasing the flow of technical information to the Indonesian mining industry during the OSM project.

Environmental Business Opportunities

The Assessment Team initiated and the Technology Team was charged with completing an evaluation of the quality and availability of existing equipment, laboratory facilities, and environmental services used to analyze and manage environmental impacts of mining operations. This review identified opportunities for improvement of environmental services and training which may be provided by the private sector to mine operators and government personnel.

In the United States, an environmental service industry has developed which may employ as many people as the mining industry itself. This service industry assists mining companies by preparing applications for mining permits, exploration and mine plans, and solving numerous environmental problems associated with mining, such as post-mining revegetation and water quality compliance. Service industry personnel work on the minesites alongside mining company personnel, helping them in areas where company employees are not expert. This expands the capabilities of mining company staff and expertise at a fraction of the cost of direct hires. These environmental service companies allow even small mining companies to meet the same environmental compliance standards as large, well capitalized companies.

With the exception of a few heavy equipment vendors, mining contractors, and a handful of engineering consulting firms seeking very large design projects, few environmental service providers or equipment suppliers are currently doing business in Indonesia supporting the mining industry. There is both a need and a market for this type of support. Miners are generally not adept at or interested in agriculture. Mine reclamation and revegetation is a specialized form of farming combining the skills of soil scientists, agronomists, biologists and foresters who restore mined land to productive future uses. Most mining engineers are not expert in these skills. Reclamation is often viewed grudgingly by mining engineers as a cost to be borne after mining rather than part of the active mining process. This view substantially increases reclamation costs because mining practices with short term efficiencies are employed at the expense of long term savings.

For example, mining companies in Indonesia often waste a great deal of money growing a small amount of nursery stock for revegetation efforts because of their inexperience in reclamation and lack of infrastructure. Reclamation on all Indonesian coal mines lags far behind mineral removal because regulations did not require and mine plans did not contemplate contemporaneous reclamation during mining operations. Consequently, mining reclamation nurseries are often too small to provide adequate stock, and reclamation appears to be a low priority. This is particularly ironic in Indonesia, where many individuals are adept at gardening and agricultural production. They know how to grow things, and vegetation is especially vigorous in Indonesia's tropical rain forest climate. Inexpensive agricultural labor is plentiful in Indonesia. Reclamation project management skills are not plentiful there.

The environmental business sector is only beginning to take shape in Indonesia, where regulatory programs are in early stages of development. Mine operators and government officials have little confidence in results provided by the few available soil and water quality testing laboratories. They often seek environmental engineering, monitoring equipment, and mine planning services overseas, in Singapore, Japan or Australia (Hamilton, 2000a). With over 210 million people and a rapidly expanding middle class, Indonesia is the fifth largest market in the world. Due to expanding coal production, the potential for increased trade is enormous. The market for mining equipment and services was estimated at $114-$145 million per year in 1994 (U.S. Department of

Commerce 1994), and Indonesian coal production has more than tripled since then.

There is a significant market for service industries to support the mining industry with laboratory services, contractors providing reclamation and revegetation services, and geological and engineering consultants to assist in mine design and preparation of environmental assessments. Vendors of equipment and business services used in environmental impact assessment, mine planning and engineering, land reclamation, extinguishing coal seam fires, inspecting and monitoring mining operations, processing coal and mine waste, and training services used by the Ministry and mining operations are in short supply in Indonesia, where most mines are located in frontier circumstances. For example, fine coal recovery systems hold great promise for increasing productivity and profits of existing and planned coal mining operations in Indonesia, as discussed in Chapter 5. Construction of preparation plants utilizing fine coal circuits may provide significant opportunities for development of new environmental businesses there. Increased royalties paid to the government on recovered fine coal would assist in securing environmentally sound sustainable development in Indonesian coal mines.

Monitoring of mining operations is only as effective as the equipment used to perform it. During the OSM project, several efforts were made to locate suppliers of equipment and services for the Ministry and the mining industry in Indonesia, with little success. This included suppliers of basic mine inspector equipment, such as inclinometers, compasses, soil probes, pH meters, and water test kits. Although there was some initial interest from local companies, either they were unable to obtain the equipment or prices were excessive. Consequently, most mine inspection and monitoring equipment provided during the OSM project was imported to Indonesia. Obviously there is a market for additional suppliers of such equipment in Indonesia. Other needs include service industries to support the mining industry with laboratory services, contractors providing reclamation and revegetation services, and geological and engineering consultants to assist in mine design and preparation of AMDAL documents. Although a great deal of coal mining is actually performed by contractors with heavy earth moving equipment, these types of suppliers and contractors are in short supply in the countryside outside Jakarta (Hamilton, 2000a). Adequate instruction on how to prepare environmental assessments and how to comply with environmental requirements is not currently available through the Ministry's Education and Training Center for Mineral and Coal Technology or in regional universities in coal mining provinces, but only in some private universities in Indonesia. The Education and Training Center for Mineral and Coal Technology provides instruction for mine operators in compliance with mine safety requirements, but not much on environmental requirements.

In the U.S. and other mining countries, there is a specialized service industry supporting the mining industry assisting in permit acquisition, mine plan design, reclamation, safety and environmental management training, and laboratory services for soil, water and overburden testing. In Indonesia, this is a fertile area for interested local and expatriate businesses to grow, especially as changes are made in mining environmental policy and the AMDAL process becomes more important in decision making. Equipment and business services used in environmental impact assessment and mining should be evaluated by government agencies of donor countries and the multinational banks with an interest in development of commercial enterprises in Indonesia. Opportunities should be identified for environmental services and training which may be provided by the private sector to mine operators and government organizations.

During the project, businesses providing environmental services or equipment were invited by BLT to demonstrate state-of-the-art mining environmental control technologies or training services, and some discussed opportunities for joint ventures between foreign and Indonesian firms. For example in January 1997 a representative of Weyerhauser, Inc., of Snoqualmie, Washington, presented a half-day seminar at Ministry offices in Jakarta concerning state-of-the-art technology and management practices for erosion and sediment control. In May 1997 a representative of Woodard and Curran Engineering from Portland, Maine attended Environmental Expo '97 in Jakarta to evaluate markets for environmental engineering services and explore opportunities for joint ventures with Indonesian firms. She met with BLT personnel to discuss environmental business opportunities in the mining sector. In June 1997 representatives of Tree Pro, Inc., West Lafayette, Indiana, a manufacturer of forest planting products, and Revegetation Consultants, Inc., Fort Collins, Colorado, a provider of training, reclamation and revegetation services, expressed interest and provided information to BLT concerning the capabilities of their firms and products.

More Industry Outreach is Needed

Training needs for mine operators identified by mine inspectors, AMDAL staff, mine operators, or observed needs during Assessment Team mine site visits include all phases of reclamation and revegetation of mined lands; stream channel diversions; surface and ground water hydrology; and water quality monitoring. Training provided to mine operators during the project has been described above. Additional training in these subjects is needed to improve performance on the ground at Indonesian mines. There is a continuing need for actions to stimulate development of environmental equipment vendors, business services and training opportunities in Indonesia. These efforts should identify businesses providing environmental services or equipment who may be invited to demonstrate state-of-the-art mining environmental control technologies or training services, and opportunities for joint ventures between foreign and Indonesian firms. In the United States, trade magazines such as *Land and Water* publish annual "buyer's guides" listing vendors of mining environmental services and equipment, and trade shows are held in conjunction with annual meetings of the National Coal Association and the American Society for Surface Mining and Reclamation. Efforts by the Indonesian Mining Association to further internationalize expositions associated with their annual meetings would be a useful step in this direction. Motivations of the multinational development banks in stimulating development of environmental business services are discussed in Chapter 9.

Notes

1 Kadar Wiryanto, Chief, Construction Services and Technology Analysis, Bureau of Environment and Technology, Witoro Soelarno and Ronald Tambunan, both Senior Mine Inspectors with the Directorate of Mines.
2 Rudiro Trisnardono from DTPU, Syamsudin Halik from BLT, Ilham Munandar from Kanwil SumBar, and Mohammed Thabrani from Kanwil SumSel.
3 Witoro Soelarno (DTPU), Ronald Tambunan (DTPU), and Kadar Wiryanto (BLT).
4 Witoro Soelarno, Ronald Tambunan, Otto Hasibuan, and Endri Erlangga (all from DTPU).
5 Lobo Balia, GIS and Remote Sensing Project Manager, Mineral Technology Research and Development Center, Bandung.
6 Papers were presented on: Water Quality Protection in America Under the Surface Mining Act (Whitehouse, 1996e); Permitting Process Overview: Surface Coal Mining and Reclamation (Clark, 1996); The Use of an Aerial Overflight Program for Monitoring Activity on Surface Coal Mining Operations (Blackburn, 1996); and Negotiated Rulemaking: Consensus Building and Issue Identification through Participatory Processes (Miller, 1996b).

Chapter 9

Origins and Motives for Assistance

Successes of diplomacy are not always reported through media stories about the signing of major international treaties but are often unheralded, based on small actions and relationships which build confidence, trust, and stability in bilateral relations. Numerous agencies of the U.S. government have for many years maintained science and technology or technical assistance relationships with counterpart agencies in foreign governments, with the approval of the U.S. State Department. In 2001 the various bureaus, services and offices of the U.S. Department of the Interior had over 100 formal agreements with 52 foreign governments and 15 multilateral agencies authorized by international treaties, program-specific legislation, general provisions of the Foreign Service Act of 1961 or the National Environmental Policy Act of 1969 (U.S.DOI, 2001). This number does not include similar agreements between the U.S. Environmental Protection Agency, the U.S. Department of Agriculture/Forest Service or other agencies and foreign governments or multilateral organizations. These are among the smaller instruments of foreign policy, scattered about numerous cabinet departments, independent agencies and their subunits. Yet these instruments are seldom examined and there is virtually no literature about their operation or efficacy.

More attention has been accorded to policies of multilateral development banks such as The World Bank. Phillipe Le Prestre described a usual process or "project cycle" of identification, preparation, appraisal, negotiation, implementation/supervision, and evaluation of major new projects financed by The World Bank (1989, pp. 47–59). He noted the volume of lending by the Bank was constrained by a lack of well-conceived and competent proposals. Consequently, although projects must be formally proposed by a country, they have long been as much the products of Bank initiative as of the borrowers' own determination of their needs.

Multilateral development banks such as The World Bank have been criticized since the early 1970s for supporting destruction of the environment in developing countries by financing major hydroelectric developments, mining ventures and forestry enterprises (Farvar and Milton, 1972; Le Prestre, 1989; Revkin, 1990; Rich, 1994; Watkins, 2000). They have also been praised for developing environmental awareness (Johnson and Stein, 1979). Between 1983 and 1986, the U.S. Congress held 17 hearings (Le Prestre, 1989, p. 33) and in 1987 alone held more than 20 hearings before six committees (Rich, 1994, p. 138) on allegations The World Bank failed to mitigate the environmental destructiveness of many development projects, supported ecological and human catastrophes in many countries, and failed to extend its lending to environmental projects.

Such criticisms, of course, are a matter of interpretation. Although an Office of Environmental and Health Affairs (now Office of Environmental and Scientific Affairs) was created in 1970 to review the environmental consequences of development projects it funded, The World Bank was slow to move towards financing environmental components of such projects, and even slower to embrace financing for projects designed to improve environmental conditions in developing countries (Le Prestre 1989, pp. 23, 28–29). Contrary to claims made by a series of World Bank presidents since 1970, Bruce Rich maintains the "greening" of The World Bank is a sham, suggesting the largest international development organization in the world has been slow – or unable – to learn how to finance environmental projects (1994, pp. 111–118, 148). In this chapter, some support is lent to this assertion by examining implementation of administrative policies and decision making during development of a project conducted by the U.S. Office of Surface Mining to develop institutional capacity for reformulating and implementing mining environmental policy in Indonesia. OSM assistance was requested by the Indonesian Ministry of Mines and Energy and financed by The World Bank in 1995.

Indonesia's Need for Technical Assistance

Indonesia's Basic Mining Law was enacted in 1967 and foreign investors were encouraged to develop mineral resources through Contracts of Work with the national government (Republic of Indonesia, 1967). As discussed in Chapter 2, although regulation of mining operations was decentralized to regional and local governments in 1999 (Republic of Indonesia, 1999b, 2000), this effectively changed the locus of administration but not the substance of mining policy. Changes to the mining law were under consideration in 2004 but its substance has not been significantly revised since 1967. The Ministry of Mines and Energy issued regulations governing environmental effects of mining activities in 1977 and several related decrees and circulars were subsequently issued, providing guidance for managing impacts of surface mining, underground mining, mineral treatment plants, and dredging activities (Ministry of Mines and Energy, 1992).

Separately the Act on Basic Provisions for Management of the Living Environment was enacted in 1982 and regulations were issued in 1986 (Republic of Indonesia, 1982; Ministry of Environment and Population, 1986) establishing an environmental impact assessment process. As discussed in Chapter 4, the Act and regulations were general in nature and did not specify regulations for particular sectors of development such as mining. A monitoring plan and reclamation plan were required, but generally were not prepared until after approval to mine was issued, reducing their value in determining conditions and requirements in the mining authorization. Thus regulations providing environmental protection from the effects of mining activities were issued in 1977, before the Act on Basic Provisions for Management of the Living Environment was enacted in 1982. At the time the OSM project was conceived, these mining regulations had not been but needed to be adjusted to bring them into conformance with the Act of 1982 and Government Regulation No. 29 of 1986 concerning environmental assessment (Kuntjoro, 1993).

Although steps were taken in the past 30 years to provide a basis for sustainable development, degradation of water quality and land increased as mining operations expanded. The Ministry of Mines and Energy made several efforts to improve administration of environmental protection policies and enforcement of its regulations. However insufficient numbers of technical staff and inspectors with knowledge and experience in the field of mining environmental management continued to be an obstacle to effective design of regulations and implementation of these policies. Additional training of personnel in regulatory program management, inspection procedures and guidelines, and use of modern equipment were needed (Ministry of Mines and Energy, 1992).

Development of Environmental Technical Assistance

A nation-state has two obligations to its people, which every nation-state pursues in its foreign policy: (1) national security, the survival and continued integrity of the nation-state; and (2) advancement of national interests in meeting the needs of its people, improving their human condition over time, and enhancing their public health, welfare and prosperity (Morgenthau, 1960). U.S. foreign aid to developing nations has been used as an instrument of foreign policy in resisting aggression from stronger states, attempting to minimize conflicts among developing states, and propagating congenial values and institutions to suit the conditions and traditions of poorer societies in the hope this will create a predisposition to community behavior in dealing with external problems (Kaplan, 1967, pp. 110–118). Bilateral technical assistance and science and technology exchanges which are not of a military nature have fallen mostly in the last category of foreign aid described above.

The Politics of Getting Foreign Assistance

The politics of getting international financial assistance may in part motivate pursuit of external funding for development projects by some recipient governments. Many years ago Norman Wengert advanced a theory of politics in which he maintained "American politicians will get as much as they can for their constituents, with only casual attention to the merits of the case and to the extent that they are not likely to be held directly accountable for costs" (1979, p. 17). Harold Lasswell's definition of politics as getting (1958) may find no better validation than in the history of attempts by less developed countries to obtain foreign aid. With apologies to Wengert for altering his conception and applying it in the international sphere, this history suggests strongly that officials in less developed countries will get as much as they can for their nation-states, to the extent they are not likely to be held accountable for the costs: that is, when they are not likely to be directly identified with raising domestic taxes or other adverse local effects. What better way for a government official to do this than securing a grant or loan for a technical assistance project that will be repaid (if repayment is necessary) by future generations or different agencies of government?

The search for local benefits at the expense of the global community has long been a goal in international relations. Often the product of imprudent borrowing (Cuddington, 1989, pp. 16–17), if foreign debt is any indicator of national priorities, government officials in many less developed countries have been quite successful in pursuing this goal at least since 1945, so much that their successes resulted in some repayment difficulties during the 1980s and 1990s (World Bank, 1994c). The politics of getting is a substantial component of international relations in the North-South conflict. Conceptions of international cooperation that advance the role of wealthier Northern states in funding compliance with international agreements by less wealthy Southern states using foreign aid dollars are useful political tools for officials of developing countries. For example, the concept of additionality, that existing development funds should not be diverted to environmental quality purposes, and *additional* funds should be provided for developing countries to carry out environmental provisions of international agreements (Caldwell, 1990, pp. 66, 199) has all the hallmarks of the politics of getting. Some bilateral negotiations exhibit similar attributes. That the politics of getting provided one motivation for Indonesian initiation of the OSM project does not denigrate the very real needs which also stimulated its development.

The Directorate of Mines, Ministry of Mines and Energy (now Ministry of Energy and Mineral Resources) has national permitting, inspection and enforcement authority over mining activities in Indonesia, similar in many respects to jurisdiction of the Office of Surface Mining in the United States. The Office of Surface Mining was approached in August 1991 by the Section Head for Mining Activities of this Directorate, who requested technical assistance to improve its regulatory program for controlling environmental effects of surface mining. Several proposals for similar technical assistance had previously been prepared by the Ministry and submitted to the Indonesian National Planning Agency, the Asian Development Bank, The World Bank and several offices of the United Nations but had not been approved. Apparently those proposals were short on specifics and failed to demonstrate the Ministry had the capability to implement them. Fearful previous criticism of multinational development banks might result in loss of development funding, and aware those banks in the 1990s had begun to support client country requests for environmental programs as part of major development projects, he wished to prepare a technical assistance proposal that might attract grant funding or be attached to an Indonesian request for a larger loan (Wiryanto, 1991). Repeated efforts to secure such funding suggest both the need for assistance in reforming mining environmental policy and the politics of getting may have motivated development of a technical assistance proposal and the search for international financial assistance to implement it. The creative energy of entrepreneurial bureaucracy may be rivaled only by its tenacity.

Figure 9.1   Finishing pond full of sediments discharging black water
Source: Alfred E. Whitehouse

OSM and Directorate staff worked closely over a period of three months to develop a draft proposal. When the Section Head for Mining Activities returned to Indonesia in November 1991, he carried a proposal describing a five-year US$7.2 million technical assistance project (Ministry of Mines and Energy, 1991). The proposal was submitted to the Indonesian National Planning Agency (BAPPENAS) in January 1992 (Ministry of Mines and Energy, 1992). BAPPENAS must approve proposals before they may be listed in the "Indonesia Blue Book" of projects eligible for external funding by multilateral development banks or foreign governments. After several revisions to the budget and scope of work, a three-year, US$3.2 million technical assistance project was approved in September 1992 by BAPPENAS (Kuntjoro, 1992). Thus identification and preparation of the technical assistance proposal required a bit more than one year from conception to approval by the originating government.

Thereafter, staff of the Directorate of Mines initiated contacts with The World Bank, the United Nations Development Program and the U.S. Agency for International Development concerning funding the proposed technical assistance. A delegation consisting of the Director of the Directorate of Mines and the Section Head for Mining Activities visited officials of these agencies in Washington, D.C. during October 1992, accompanied by OSM personnel. Meetings with U.N.D.P. and U.S.A.I.D. failed to generate interest in funding the proposal, but officials of The World Bank indicated a desire to accommodate a request from their client, Indonesia. Representations from a ranking Indonesian official that the proposed project was desired by the Ministry of Mines and Energy apparently made a favorable impression.

Reorganization of the Ministry of Mines and Energy

In December 1992 shortly after receiving the delegation from the Ministry, The World Bank expressed interest in financing the proposal as a technical assistance project to be added "to either the Cirata Hydropower Phase II Project, or the Outer Islands Power Project, tentatively scheduled for Board presentation in March 1993 and March 1994 respectively" (Scherer, 1992). The apparent lack of any rationale for attaching a technical assistance program for reform of mining environmental policy to financing for a hydroelectric project suggests a lack of concern on the part of The World Bank for the substance of the proposed assistance, but it began looking for a mechanism to support the project. Financing was contingent on establishment of institutional mechanisms to "cooperate and counterpart the international services to ensure successful institutional development and effective implementation of the work program" (Scherer, 1992). Unstated at that time, but evident later was a desire on the part of The World Bank to identify a specific office in the Ministry of Mines and Energy that would have few other responsibilities in addition to the proposed technical assistance project, to ensure funds proposed for computer equipment and salaries for two Indonesian "National Experts" would be spent as intended.

The Indonesian response was uncharacteristically swift: in January 1993 a new Bureau of Environment and Technology (BLT) was established in the General Secretariat of the Ministry of Mines and Energy (Gandataruna, 1993). As a staff arm of the Ministry, the principal function of the new Bureau was to review, develop, establish, and coordinate environmental policies and regulations within the Ministry of Mines and Energy. The Bureau had three divisions: Technology Analysis and Construction Services (headed by the previous Section Head for Mining Activities), Environmental Management and Spatial Zoning, and the Secretariat of the Central Commission for Environmental Impact Analysis (AMDAL). A separate Directorate of Technical Mining (previously the Directorate of Mines) retained the functions of mine inspection and enforcement of environmental regulations. After receiving news of the reorganization, The World Bank in April 1994 informed the Ministry it was prepared to finance a program of technical assistance and cooperation with OSM "as an environmental component in the proposed Outer Islands Power Project" (Scherer, 1994).

The project described here did not originate in the usual process described by Le Prestre (1989, pp. 47–59). Technical assistance for mining environmental policy was not created as an integral part of a proposal for an international development project loan by the Indonesian National Planning Agency or staff of The World Bank. Rather, the project originated in a regulatory subunit of the Ministry of Mines and Energy, was developed by that agency in cooperation with the expected provider of technical assistance, and was grafted onto a development loan identified by the Bank in response to a request from its client. Yet this technical assistance project did fit a pattern of environmental components being added to other larger development projects after they had been appraised (Le Prestre 1989, p. 30).

Authorization of a Technical Assistance Project

The technical assistance project was designed to improve the institutional capacity of the Ministry of Mines and Energy to regulate a rapidly expanding coal mining industry and reclaim mined lands for other productive uses. OSM provided technical assistance and advice to the Ministry in the following areas:

- regulatory policy development (e.g., relating general environmental policy and regulations specifically to mining operations);
- improvement of administrative organization and program management;
- technology cooperation concerning private sector environmental services and equipment;
- personnel training in mine inspection and other technical subjects;
- evaluation and improvement of data base adequacy for effective decision making;
- assessment and improvement of equipment and laboratory facilities for effective mine inspection and enforcement (U.S. DOI/OSM, 1995b).

This required preparation of agreements between OSM and the Ministry that would be approved by superiors in both governments and The World Bank.

The Office of Surface Mining Reclamation and Enforcement has national responsibility for regulating environmental effects of surface mining pursuant to the Surface Mining Control and Reclamation Act of 1977 (30 *U.S. Code* §§1202, 1211). In the National Environmental Policy Act (NEPA) of 1969, the U.S. Congress directed that:

> all agencies of the Federal government shall ... recognize the worldwide and long-range character of environmental problems and, where consistent with the foreign policy of the United States, lend appropriate support to initiatives, resolutions, and programs designed to maximize international cooperation in anticipating and preventing a decline in the quality of mankind's world environment.
>
> (42 *U.S. Code* §4332 (2)(F)).

Science and technology cooperation between U.S. agencies and other governments was authorized by the Mutual Educational and Cultural Exchange Act of 1961, as amended "to provide for interchanges and visits between the United States and other countries of scientists, scholars, leaders, and other experts in the fields of environmental science and environmental management" (22 *U.S. Code* §2452 (b) (11)). Cooperation and exchanges of technical information and personnel in the fields of natural resources, environment, and energy between agencies of the U.S. government and

Figure 9.2 Local villager fishing for family dinner between barge and coal covered shore of Mahakam River, East Kalimantan
Source: Michael S. Hamilton

Indonesia were authorized by bilateral international agreement in 1992 (U.S. Department of State, 1992).

Agreements between U.S. government agencies and foreign governments are authorized by the Foreign Assistance Act of 1961, for the purpose of assisting developing countries to exploit their energy resources and to achieve environmentally sound development (22 *U.S. Code* §§2151d (b) (1), 2151p). The Department of State in May 1993 determined that provision of technical assistance by OSM to Indonesia concerning improvement of mining environmental programs was consistent with U.S. foreign policy. Securing this determination required approvals by three different offices in the White House (Office of Management and Budget, Office of U.S. Trade Representative, Office of Science and Technology Policy), four U.S. cabinet departments (Interior, State, Defense and Commerce) and the U.S. Embassy in Jakarta. Similar reviews were conducted by the Indonesian government. OSM was authorized to negotiate and conclude a Memorandum of Understanding (MOU) concerning surface mining with the Indonesian Ministry of Mines and Energy (Bohlen, 1993). Despite the fact OSM had no direct mandate for international activities in its organic legislation, the National Environmental Policy Act and the Foreign Assistance Act provided legal authority for this agency to assist a foreign country in developing mining environmental policy.

In the 1990s, the United States pursued a foreign policy objective of welcoming newly industrializing countries like Indonesia to partnership in development projects and weaning them away from reliance on U.S. foreign aid funds. Due to their successes in previous economic development efforts, newly industrializing countries were considered to be less needy than developing countries. This change in status had significant repercussions for U.S. relations with Indonesia. In the early 1990s, the U.S. Agency for International Development mission in Indonesia was reduced substantially, and relations between the U.S. and Indonesia suffered. Consequently, development of a relationship between the Ministry of Mines and Energy and the U.S. Department of the Interior was viewed with favor by the U.S. Embassy in Jakarta.

In December 1994 an MOU was signed in Jakarta by the U.S. Ambassador and the Minister of Mines and Energy (U.S. DOI/OSM, 1994). The same month OSM received authority from the U.S. Trade and Development Agency pursuant to Section 607 of the Foreign Assistance Act to provide assistance "on an advance-of-funds or reimbursement basis" to the Indonesian Ministry of Mines and Energy, "to improve its regulatory program to control the environmental effects of surface mining and to restore mined land for other uses" (U.S. Trade and Development Agency, 1994). A technical assistance agreement was signed by OSM and BLT in February 1995 (U.S. DOI/OSM, 1995b) and accepted by The World Bank. Work began on the OSM project in April 1995 and was completed in August 1998.

The World Bank financed this $3.2 million technical assistance agreement as a small part of a $260.5 million loan to the Indonesian government for the Sumatera and Kalimantan Power Project (formerly the Outer Islands Power Project). Total cost of the Power Project was expected to run approximately $688.9 million, with the Indonesian government providing the balance of funds necessary, and the loan with interest to be repaid over 20 years from revenues charged for electric service (World Bank, 1993). Therefore all costs to OSM for providing this technical assistance were reimbursed by the Government of Indonesia.

The Sumatera and Kalimantan Power Project included several physical components related to rural electrification and a major policy component affecting reform of Persero (PLN), the government-owned national power company. Physical components are illustrated in Table 9.1. Technical assistance to PLN was provided by private consultants and intended to develop capability for environmental management and electricity demand management, reduce the growth rate in electricity demand, and curtail peak load in the interconnected grid system. Pilot projects were intended to transform operations in the respective regions into profit centers to improve the efficiency of electric system operations.

The policy component affecting reform of PLN entailed: (a) establishing a policy framework for private sector participation in provision of electric service in Indonesia; (b) restructuring PLN and establishing commercial operations as a limited liability company; (c) implementing a framework for regulatory oversight of the electric power sector; and (d) introducing formula-based power

Table 9.1   Physical Components of the Sumatera and Kalimantan Power Project

---

(a)   Construction of a 90 MWe hydroelectric plant at Besai, South Sumatera, with related transmission facilities;

(b)   Construction of a 130 MWe mine-mouth lignite-fired thermal power plant at Asam-Asam near Banjarmasin, South Kalimantan, with related transmission facilities;

(c)   Procurement of two barge-mounted power plants of 10 MWe (diesel) and 30 MWe (oil-fired turbines) generating capacity, and

(d)   Technical assistance comprising:
    (i)   strengthening the capability of the Ministry of Mines and Energy, Directorate of Mines for environmental management in the coal mining sector (provided by OSM);
    (ii)   engineering and construction supervision of Besai hydroelectric and Banjarmasin thermal power plants;
    (iii) implementing pilot projects for PLN's Regions IV and VI in South Sumatera and South Kalimantan to improve their operating efficiency, and
    (iv)   strengthening the environmental management capability of PLN (World Bank, 1993, 1994b).

---

tariffs, bulk power purchases and supply tariffs (World Bank, 1993). This policy component was designed to open the power generation sector to private investors, capitalize PLN and restructure it as a revenue generating operation, where previously it provided electric service as a subsidy to consumers. The policy component was designed to provide incentives for lowering costs and improving reliability of service, while the physical components implement government policies to extend electric service in remote areas of South Kalimantan and provide substitute resources to displace oil-fired capacity in South Sumatera.

The Asam-asam thermal power plant near Banjarmasin produces electricity at mine-mouth utilizing an inexpensive, low-quality source of fuel (lignite) that is not otherwise marketable and of no other commercial value. The West Asam-asam mine, operated by PT Arutmin Indonesia, will be capable of producing 2 million tonnes per year for the power plant. Preliminary engineering for the Besai hydro and Asam-Asam coal-fired plants was funded by The World Bank under previous loans. With the exception of technical assistance for environmental management provided by private consultants to PLN, and by OSM to the Ministry of Mines and Energy, this was a fairly typical World Bank development project designed to produce financial and structural reform of the electric generating sector in Indonesia.

Negotiations with The World Bank

Despite claims by a series of World Bank presidents to the effect every project it finances is reviewed by a special environmental unit (Rich, 1994, pp. 111–112), during negotiations between OSM and The World Bank over the terms of reference for this technical assistance project no meetings were held with or questions received from the Office of Environmental and Scientific Affairs. Technical and financial staff of the Bank posed few questions about the environmental content or substance of technical assistance to be provided, but focussed on three issues concerning how equipment would be purchased, payment of "honoraria" to Indonesian personnel, and audit procedures.

Procurement Issues

Staff of The World Bank initially were concerned that equipment be acquired through competitive bidding, and required a thorough explanation of the process used to place vendors and equipment on a general schedule and a detailed description of the purchasing procedures to be used. OSM proposed to use standard United States government procurement procedures to acquire necessary equipment from a general schedule of approved equipment and vendors maintained by the U.S. General Services Administration (GSA). The equipment would be purchased in the U.S. and shipped to Indonesia via the U.S. Embassy. When it became evident competitive bidding is used to place equipment on the GSA schedule at a "guaranteed lowest available price," staff of the Bank accepted the proposal, explaining they were principally concerned that purchasing of major items not be left in the hands of the Ministry of Mines and Energy.

Honoraria Issues

A question concerning payment of honoraria had a similar basis. Salaries for Indonesian government employees are generally paid in cash, do not provide a living wage, and must therefore be supplemented. Employment compensation includes a bewildering array of automatic and discretionary allowances and supplements, including a rice allowance and supplements for children and spouses. Following social practices predating the Dutch colonial period, a tradition of cash side-payments evolved throughout the government (and perhaps throughout the society), whereby small "honoraria" are paid by members of one subunit of government to members of other subunits who attend meetings of task forces, committees, and workshops of all sorts (Gray, 1979). Any activity receiving funds from the national development budget or foreign assistance pays project honoraria to employees, either monthly or per meeting. In some respects, honoraria appear analogous to the portion of a per diem paid for incidental expenses during official travel by U.S. government employees, except they are often paid for meetings in the same community, sometimes in the same building.

In other respects, the practice of paying honoraria is quite different from the per diem: the payment is often shared with members of the subunit represented by the receiving individual, apparently on the theory that subordinate staff must work harder to compensate for absent superiors.

Because attendance at extra meetings earns extra compensation, a superior who is able to get invited to many meetings is likely to have a devoted staff. Payments are for attendance, not necessarily for commitment to decisions discussed at meetings, but disagreements may be expected to be subsumed somewhat by a desire to be invited to future meetings. Amounts are usually small by American standards; an office head at middle management may garner the equivalent of US$20 to US$50 per meeting, and may divide shares among four to eight or more subordinates. Higher officials receive proportionately greater honoraria. For meetings where serious decisions are expected, amounts may be considerably greater. "An official inquiring after the fortunes of an acquaintance from another agency asks how many projects he has" (Gray, 1979, p. 88). In their more candid moments, some successful Indonesian officials estimate perhaps half of their annual income may be derived from distributed shares of honoraria; those in less successful units receive less.

Some Indonesian government officials may spend considerable time attending meetings, and little time supervising work of their own offices. Networks of reciprocal invitations develop, and influence is extended to broader cohort groups, who tend to follow each other in promotions and career paths. Because most of these funds probably originate as fees for agency services or budgeted revenues, accountability for government expenditures is limited by payment of honoraria. Consequently, officials at The World Bank tend to view this practice in an unfavorable light, perhaps as a form of corruption, but recognize it is a factor in effective implementation of development projects in Indonesia. Unable to abolish the practice, they seek to control what they perceive as a drain (e.g., "leakage") of project funds. As a practical matter, it is difficult to visualize any permanent solution to this dilemma other than providing adequate salaries for government employment and banning the practice of paying honoraria. There is a pressing need for a schedule of adequate compensation for government employees in Indonesia.

Because the tradition is well established, those who fail to pay honoraria may find it difficult to attract decision makers to their meetings. Staff of The World Bank were interested to learn how a U.S. government agency would handle honoraria. OSM proposed to stay within U.S. government requirements and pay per diem only, at U.S. government rates, for meetings that required travel to communities other than the principal duty station of the individual. Staff of the Bank seemed skeptical, but accepted this as necessary. Consequently, most meetings and training sessions were scheduled outside Jakarta, and OSM subsequently had difficulty attracting some decision makers to meetings in Jakarta held to discuss policy changes. Most such meetings were attended by subordinates, possibly slowing the pace of policy development. The pace and substance of policy development might have been improved considerably if the OSM project had been able to pay honoraria to decision makers attending meetings of the Joint Policy Team and Joint Program Management Team.

Audit Requirements

For many years, staff of The World Bank have suspected "leakage" of loan funds from Indonesian projects reduced the effectiveness of its economic development projects there. Staff of the Bank seemed pleased, perhaps even relieved when OSM accepted responsibility for all project expenditures and proposed to use standard accounting procedures, keep accurate and systematic accounts, permit inspection of its records, perform regular audits, and allow audits by persons appointed by the parties to the agreement (U.S. DOI/OSM, 1995b). Staff of the Ministry of Mines and Energy accepted OSM's position that OSM would be unable to provide technical assistance unless it was responsible for expenditures and accounting.

Greening or Clientelism?

All three of the foregoing issues concerned accountability for expenditures and are within the legitimate concerns of a lending institution. However the great amount of attention devoted to them, and lack of attention to the environmental program, suggest strongly that environmental issues are not yet on a par with lending issues as priorities with employees of The World Bank. It appeared no environmental assessment of the proposed technical assistance project was prepared by the Office of Environmental and Scientific Affairs. Throughout formulation and implementation of the technical assistance program described here, staff of The World Bank displayed little interest in the possibilities for improved environmental management of mining operations in Indonesia. They displayed more concern for issues of governance, transparency of decision making, and accountability for expenditures than they did about the substance of the environmental program. This experience provided little evidence that concern for environmental amenities and sustainable development have penetrated layers of management at multilateral development banks such as The World Bank who are responsible for identification, preparation, and review of major development projects.

Instead this experience lends support to Rich's hypothesis that "The Bank's environmental failure is systemic and rooted in deep institutional contradictions, not limited to individual problem projects" (Rich, 1994, p. 154). Because there are no institutional incentives and no organizational culture to support consideration of environmental amenities in design and approval of development projects, the pressure to finalize new loans and "move the money" may continue to override attempts to rationalize decision making procedures in The World Bank (Rich, 1994). Multilateral development banks like The World Bank are large organizations with cultures long focussed on national development and alleviation of poverty in less developed countries. It may require concerted, sustained effort and many years, perhaps generations, before their organizational cultures may be expected to display more than cosmetic evidence of change to embrace environmental amenities as part of their missions. A greater appreciation of the necessity for balance between the environmental costs and economic benefits of development as discussed in Chapter 1 might assist in this transition.

That this technical assistance program did not originate in the "usual" manner suggests that processes for development of World Bank environmental projects are more complex, variable and opportunistic than described by Le Prestre (1989, pp. 47–59). These processes may well be more diverse, less formal, and more entrepreneurial than previously believed. Environmental appraisal of this project appeared to be based on established relationships of clientelism: a patron responding to a request from an established borrower (Wilson, 1975, p. 88), rather than any greening of project review processes within The World Bank. Yet this is only one instance, and additional research is needed about implementation of policies of The World Bank concerning development of environmental projects and programs.

Chapter 10
Lessons Learned

The OSM project on Technical Assistance for Improvement of Mining Environmental Policy and Enforcement in Indonesia brought together government technical agencies in two countries in an effort to reduce the environmental impact of surface mining in Indonesia. Efforts initiated in 1991 produced a technical assistance project which was implemented beginning in 1995, and was followed by subsequent projects extending to the end of 2005. What lessons were learned from this attempt to develop institutional capacity for reformulation and implementation of mining environmental policy through bilateral technical assistance?

Transnational Policy Development is Possible

Technical assistance projects and science and technology exchanges are effective but little studied instruments of foreign policy which may be utilized to maintain or improve international relations between two or more countries. Successes of diplomacy are not limited to the signing of major international treaties, but are often unheralded, based on small actions and relationships which build confidence, trust, and stability in bilateral relations. Reformulation of environmental policy by a foreign government may be stimulated and assisted by the U.S. government through technical assistance and science and technology exchanges. Neither sustained diplomatic efforts of persuasion nor explicit threats are required. Policy change may often be accomplished through a more subtle, but deliberate exchange of ideas.

Yet in transnational policy development through technical assistance activities, a receptiveness to new ideas does not necessarily mean a willingness to change, and the resulting policies don't always turn out as expected. This was evident in discussion of the shortcomings of reclamation guarantee policy in Chapter 6, during the delivery of TIPS training described in Chapter 8, and after recommendations for decentralization of mining environmental regulatory functions to Kanwil regional offices of the national government were rejected in favor of decentralization to regional governments, as discussed in Chapter 7. In cross-cultural attempts at policy development and technical training, because recipient agencies view possibilities through their own cultural lenses and are free to make their own policy decisions, assisting agencies are often surprised at results and must always expect the unexpected.

Significant research questions remain about the efficacy and consistency of technical assistance projects in maintaining cordial relations between two countries. Of the many existing technical assistance agreements, what portion

Figure 10.1   Multiple seam mining with wet coal floor, East Kalimantan
Source: Alfred E. Whitehouse

involve active implementation at any given time, and what portion have merely symbolic significance? Is implementation a matter of convenience or whim on the part of one or both parties? Do technical assistance projects contribute to development and maintenance of stable relationships over time, or does their implementation rise and fall with the occurrence of bureaucratic entrepreneurship or shifting foreign policy priorities of different administrations? Is there any pattern in the manner in which new agreements are initiated, or old agreements terminated? How are they funded, and at what aggregate levels? Are they effectively monitored by the executive or legislative branches of government? Answers to these questions await future research.

Technology Diffusion Takes Time

Technical assistance projects and science and technology exchanges are effective but little understood means for diffusion of innovations and transfer of technologies between countries, whether sponsored by an assisting agency or fostered by international development organizations. Examination of the OSM project described in this book lends additional support to the observation that provision of technical assistance to developing countries through bilateral partnerships with similar but more mature organizations in other parts of the world has proved effective in transferring expertise, training personnel, improving management capabilities, and enhancing institutional capacities (Cooper, 1984). Government-to-government partnerships provide developing countries with access to expertise of a scope and depth that may not be available in a single consulting firm, especially concerning formulation and implementation of regulatory policy. Even the largest consulting firms in the United States would be hard pressed to demonstrate they maintain the staff experience or institutional capacity of organizations such as the U.S. Environmental Protection Agency or the U.S. Office of Surface Mining Reclamation and Enforcement in managing a permitting process or developing and applying environmental regulations. A government of a developing country which has already established commercial and regulatory relationships with industry and is aware of business views on many subjects may wish to acquire advice from a source known to have technical expertise but no appreciable stake in domestic regulatory outcomes. A government wary of the economic power and political influence of multinational corporations and concerned about the possibility of capture by business interests or exploitation of national resources on unfavorable terms – including trading short-term economic gains for a legacy of environmental degradation – may find cooperation with another government to be in its national interest.

Technology diffusion through technical assistance or science and technology exchanges is supported by the need of developing countries for foreign investment, which in turn encourages a politics of getting foreign assistance and willingness to consider imported ideas, as described in Chapter 1. Yet results are limited by language and cultural barriers, the usual vagaries of personnel management (e.g., divorce, illness, a tendency to bring family problems into the workplace), the instability of developing countries, the amount of time for accomplishment of objectives, and availability of adequate funding. Understanding the process of policy innovation requires a time perspective of a decade or more to determine whether new ideas and technologies have been adopted and successfully implemented (Sabatier, 1988, p. 131). That some significant policy changes were stimulated by the OSM project within a three year period provided unusually rapid evidence of success in planned diffusion of innovations. Rapid change was also evident in adoption of fine coal recovery technology at three mines discussed in Chapter 5, adoption of sediment and erosion control techniques at the PT Adaro mine discussed in Chapter 8, and development of the training capabilities of the Ministry of Energy and Mineral Resources described in Chapter 8. Some visible change may come sooner than expected; some changes are probably invisible; and some may never happen. Matters such as these often require a substantial period for discussion before sufficient agreement is reached for action to be taken. Because it is impossible to assess with accuracy how many "innovators" and "early adopters" (Rogers, 1995, pp. 263–264) of these innovations may have participated in the OSM project, the most significant results may not be evident for many years in the future, until Indonesian participants move into more influential decision making positions and make further efforts to implement what they have learned.

Figure 10.2   Wetland rice near Mahakam River at Samarinda, East Kalimantan
Source: Michael S. Hamilton

Entrepreneurial Bureaucracy Produces Results

The OSM project described here was the product of Indonesian bureaucratic entrepreneurs seeking funding for technical assistance from international development agencies, and American bureaucratic entrepreneurs seeking expansion of a domestic mission and opportunities for professional development for technical and professional staff in an agency that had largely accomplished its original mission. These efforts were encouraged by the dynamics of the politics of getting foreign aid in the contemporary international political system, which in turn was supported by the development needs of a transitional society. Moreover these efforts found support in the clientele orientation of a multilateral development bank which responded positively to a request for funding from its client Indonesia. The relationships between these factors are ripe for future research.

Interagency Cooperation Requires Constant Effort

Increasing the effectiveness of cooperation between related government agencies concerned with mining and environmental policy in Indonesia was an objective of the OSM project. Where mining operations are concerned, the most frequent interagency interactions of concern to the Ministry of Mines and

Energy are probably with the Ministry of Forestry concerning forest resources, the Ministry of Environment concerning environmental disruption and, at least potentially, with local communities concerning post-mining land uses.

The Ministry Central AMDAL Commission was historically the principal forum for interagency cooperation on environmental issues associated with mining activities in Indonesia. Several Ministries were represented on this Commission, which reviewed ANDAL environmental assessments for proposed mining operations. Decentralization abolished sectoral AMDAL Commissions and moved the locus of environmental reviews to regional governments, where representation by different agencies has been obscured and perhaps slow to occur. BAPEDAL and its regional government counterparts BAPEDALDA have assumed greater prominence in environmental reviews, but BAPEDALDA have not been created in every regional jurisdiction where mining occurs. Interagency cooperation between levels of government has been reduced under decentralization of functions, although this result may be reversed in future with some effort on the part of regional governments. The greatest barrier to this happening is lack of staff expertise, as discussed in Chapter 8. Development of an oversight role by the Ministry will be necessary to coordinate actions by regional governments nationwide during mining environmental policy implementation in Indonesia.

Use of Joint Teams is Effective

Because institutional and human resource development were major goals of the OSM technical assistance, it was deemed essential that Indonesian personnel who would implement any recommended changes acquire a sense of ownership of the recommendations. Consequently the project was conducted through a series of training activities, technical consultations and problem-solving seminars by teams comprised of both American and Indonesian personnel. To emphasize a sense of partnership in deliberations, and minimize any sense of superior/subordinate relationships, each team had Indonesian and American co-chairs, and involved presentations by both U.S. and Indonesian team members. This approach helped to reduce the impact of language and cultural barriers on team deliberations and products.

A phased seminar format was designed to present general principles, methods and techniques in initial meetings, using discussion to identify alternative courses of action and obtain feedback, before holding later meetings to further discuss and prepare recommendations. Postponing development of recommendations until later meetings allowed consultations between Indonesian team members and their respective superiors and organizational units to influence them. This approach was designed to maximize transfer of concepts and skills to Indonesian personnel, while allowing them to provide cultural and organizational filters for team recommendations. This was intended to increase the probability work products would achieve an appropriate fit with Indonesian cultural circumstances and political will.

Use of joint-teams produced the policy changes and recommendations for program reorganization and management described in Chapters 6 and 7. Technical training and, after an initial adjustment, TIPS training were provided using a similar three-step "training the trainers" method developed for this project. The World Bank or another multilateral development organization should consider preparing a publication describing this approach for the benefit of contractors designing future technical assistance projects for its clients. The utility of this approach for development of human resources, institutional capability, and transfer of technical knowledge has much to commend it to future development projects of all types.

National Staff Support Requires Management

Provision of offices for the OSM Project Director within the Ministry of Mines and Energy was intended to associate the program of work for the project as closely as possible with daily operations of the Ministry. This was a departure from the more common practice of housing technical assistance staff in embassy compounds or office buildings separate from Ministry offices. Staffing the project with two full-time positions for Indonesian personnel from the Ministry, an Expert Program Manager and an Expert Technical Manager, was further intended to overcome difficulties of language and coordination while developing human resources and institutional capability of the Ministry. The Expert Program Manager in particular was expected to contribute greatly, because he initiated the project and had been involved with it since inception. However during the first 15 months the OSM Project Director rarely received full-time support from either of the assigned Indonesian nationals, due to the press of their continuing responsibilities as Division Chiefs within the Bureau of Environment and Technology. Moreover on separate occasions these individuals each received assignments from the Ministry for mandatory management training that required their absence from the office for periods of three months each, without replacement staff being provided.

Although this might reasonably be attributed to a severe shortage of personnel with appropriate expertise within the Ministry, it did not allow the project to function as efficiently as planned. Use of Indonesian personnel for project staff was less valuable than it might have been during the first 15 months of the project. Problems of coordination and execution became apparent, and there were lapses of administrative support. On several occasions, routine housing, travel and meeting arrangements for joint team discussions were postponed until the last minute or were not timely completed. Required appointments were not made in advance or were changed at the last minute. On other occasions Indonesian members of joint teams either were not given enough advance notice of meetings with their OSM counterparts visiting from the U.S., or were not briefed or otherwise prepared sufficiently in advance of important working meetings. On some occasions they may even have been surprised by the nature of the proceedings. The situation described in Chapter 8 concerning the TIPS demonstration and location of equipment is one example in point: it probably would not have happened if all parties had been adequately informed. Although it is impossible to determine with certainty, lack of coordination, inadequate preparation of materials for joint team members, and insufficient follow-up on discussions after initial meetings may have slowed development of mining environmental policy during the first 15 months of the project. To resolve this situation and improve coordination with key decision makers and personnel with line functions, personnel from the Directorate General of Mines were temporarily assigned to the project as full-time counterpart staff on a rotational basis.

During the second year of the project, adjustments were necessary in the positions of Expert Program Manager and Expert Technical Manager to accommodate the realities of workload and career development of Ministry personnel. On 1 October 1996, a Senior Mine Inspector from the Directorate of Technical Mining (Directorat Teknik Pertambangan Umum-DTPU) was substituted for the BLT Division Chief of Environmental Management and Spatial Zoning as Expert Technical Manager. Full-time participation by the BLT Division Chief for Construction Services and Technology Analysis was reduced to half-time, and the duties of Expert Program Manager were divided equally between these two BLT Division Chiefs. This was accomplished without increasing project costs. Thereafter, line personnel from DTPU were assigned to the project as full-time or half-time counterpart staff on a rotation of approximately 3–6 months each. Six months was the maximum period allowed by regulations of the Ministry of Mines and Energy for temporary duty assignments without loss of one's permanent position. The Ministry preferred its personnel to share these developmental assignments rather than appointing one person full-time and vacating a permanent position in DTPU.

This change in staffing expanded the OSM Project Director's daily interaction with the line operations organization within the Ministry responsible for mine inspection, approval of Contracts of Work, mine feasibility studies, and technical review of ANDAL. It also provided more balance in project participation between the administrative and operating arms of the Ministry, allowed new people with fresh ideas and energy to rotate through the project office, and provided a greater number of professional development opportunities for Indonesian technical staff. This resulted in a more realistic use of Ministry permanent staff and one which greatly benefited the project. Interviews conducted in July 1998 in Jakarta indicated the Project Director, BLT Director and the five persons assigned on rotation to the position of Expert Technical Manager during the project were satisfied with this modification of the staffing plan. Liaison between the OSM Project Director, DTPU, and the BLT Director improved dramatically and participation in the project by Ministry personnel outside BLT increased substantially during the second and third years of the project.

In future it will be important not to place excessive reliance on a limited number of national personnel to staff technical assistance project implementation, but instead continually cultivate additional personnel for this purpose. Plans for rotating full-time assignments to positions held by national personnel for at least three months per person, and if possible for six months per person, are recommended for future projects in Indonesia. Regular rotation of national personnel through a limited number of positions reduces dependence on particular individuals and develops knowledge and expertise of a greater number of persons in the substance of the project and skills needed for project implementation.

Audit, Leakage and Policy-oriented Learning

Audits of OSM project expenditures were performed each year and a final audit completed after the end of the project by the Indonesian Ministry of Finance and the U.S. Department of the Interior Office of the Inspector General. These efforts produced clean audits with no unanswered questions. Nonetheless rent-seeking behavior by Indonesian officials at various levels of government was observed at several stages of the OSM project, including auditors from the Ministry of Finance who were at least theoretically performing a final audit in the U.S. to "keep the project honest." Inability of the OSM project to attract key decision makers to some meetings and seminars without advance offers of honoraria probably impeded the achievement of results, especially in the area of policy development discussed in Chapter 6.

Using honoraria as a form of compensation for government employees may impede environmentally sound sustainable development. A lack of account-ability within the Indonesian government during payment of honoraria to technical teams and decision makers may help explain the historical approval of some ANDAL, Mining Authorizations and Contracts of Work which contain insufficient information to provide an adequate basis for inspection or enforcement of their environmental requirements. These documents continue to allow the mineral resources of the nation to be wasted, and the land and water of its human environment to be degraded, raising fundamental questions about the sustainability of Indonesian economic development. Officials who approved such documents in previous years may have received substantial honoraria for attending meetings at which their approval was expressed. Under what circumstances might it be considered unpatriotic for an Indonesian government official to place their own private interest in additional compensation above the interests of the nation in the principles of social justice and nationalism articulated in the Pancasila, and the right to a good and healthy environment advanced by the 1945 Constitution?

In the Indonesian circumstance, honoraria benefiting an individual often benefits a social group, assists it to function more productively within a larger organization, and may therefore be rationalized as a benefit to the organization and the nation. Government and business organizations in both the United States and Indonesia often encourage loyalty and identification of employees with the organization as a means to more effective implementation of their respective missions. Honoraria in Indonesia appear to be a social artifact of a transitional culture which are related to status and low compensation rates more than any widespread evidence of corruption or greed. For reasons outlined in Chapter 1 and discussed throughout this book, in a prismatic society like Indonesia, overlapping social and official functions of government administration are conducted by a bureaucratic elite in which the norms of *bapakism* are deeply embedded, social status is more important than wealth, and compensation is a routine way of conducting business. Consequently the legal forms expressed in public policy are often more prominent than substantive results during implementation, and the rule of law is sometimes set aside in favor of other social goals.

The issue of honoraria is particularly difficult to manage in technical assistance projects where assisting agencies are restricted by their own domestic policies from even the appearance of making side-payments. Nonetheless this issue can probably can be handled appropriately if the assisting agency acknowledges in advance that honoraria are considered to be a legitimate portion of government employee compensation for national personnel working on development projects. Successful policy transfer and policy-oriented learning (Sabatier, 1988) may be facilitated or accellerated by careful management of honoraria. Both the potential for significant leakage of project funds and the possibility of undermining project effort must be carefully managed. Often the amounts sought appear to be small, may be satisfied with inexpensive token gestures such as souvenirs or personal items like T-shirts, and may be minimized with proper management. A learned assisting agency needs to expect and budget for honoraria when working in Indonesia and attentively manage it to effectively implement a project while keeping costs under control. Specific budget items should be prepared for honoraria, and expenditures accounted for in a systematic manner similar to those used ordinarily for office petty cash accounts.

Technical Assistance is Needed

There continues to be a great need for additional technical assistance for policy and human resources development in Indonesia, especially in regional governments. The national government has a significant role to play in establishing environmental performance standards which are sector-specific and require adherence and elaboration by regional governments. Uniform national performance standards implemented by regional governments are the only way to ensure a stable marketplace for foreign investment without encouraging competition between regional governments tending towards lax regulation. To be effective, regulatory standards must provide an even playing field for all members of the same industry in all regions of the country. Development of an oversight role by national government agencies will be essential to this effort. Allocation of a portion of mineral royalties to regional governments might be made contingent on environmental performance, with any funds withheld placed in a trust account for repairing environmental damage attributable to lax regulation. Bilateral or multilateral technical assistance in these areas will facilitate continued improvement of mining environmental policy in a decentralized governance structure.

What Does this Mean for Sustainable Development?

Sustainable development and coal mining can be compatible if mineral development is conducted in a manner that: (1) effectively controls environmental impacts and costs during mining, (2) reclaims mined land for other productive uses after mining, and (3) provides revenues from mineral extraction for investments in infrastructure and enterprises which sustain economic and social development and promote diversification as mineral reserves are depleted. Laying a sound foundation for sustainable development of mineral resources requires successful implementation of effective policies to minimize both the short-term and long-term environmental costs of mining, while ensuring the future productivity of mined land for non-mining uses. Indonesian mining environmental policy has not yet been successful in meeting the first two conditions described here. There is some evidence Indonesian investment policies have been at least partially effective in promoting diversification of the national economy and expansion of infrastructure and other non-mining enterprises which might sustain development (Hamilton 1997a). The Republic of Indonesia has displayed leadership among the newly industrializing countries of the world by making substantial investments of revenues from mineral extraction in infrastructure which have assisted the development and diversification of other sectors of the economy. In any event, fulfilling the third condition will not be sufficient if the first two more fundamental conditions are not attained. There is a potential in Indonesia for creating a large deficit of unpaid costs of development in the form of environmental degradation which will threaten human health, public safety, and biodiversity. Natural resources currently are being wasted, and development revenues lost through the use of unsustainable mining practices.

It bears repeating that for economic development to be sustainable over time, total benefits of development must accumulate more rapidly than total costs. If unpaid environmental costs of development – such as disturbed and unproductive land, toxic waste dumps, or polluted rivers – are allowed to exceed the benefits, demands for payment of these development costs may eventually drain capital and labor resources: Development will not be sustainable in the long term. When expenditures necessary to protect the health and welfare of people exceed income, development may stall. The costs of development must always be paid, if not today, then tomorrow. It is the nature of development costs that they will eventually demand payment, if not in money, then in reduced health and welfare of people who live near wasted land, breathe polluted air, or consume contaminated fish, fruit, grain or vegetables for food. Destruction of rain forest biodiversity may cause catastrophic flood events due to excessive deforestation, loss of opportunities for medical discoveries and ecotourism that could provide sustainable economies.

There is some evidence that training provided through this project to personnel of mining operations has produced changes on the ground in the manner in which operations are conducted, discussed in Chapters 5 and 8. That Indonesian mining operations increasingly viewed the OSM project and the

Ministry as a source of expertise and information concerning improved mining environmental practices is evident in requests it received for publications, technical advice, and sources of information to solve specific problems such as control and recovery of coal fines, prevention of acid mine drainage, and extinguishment of coal seam fires. OSM lived up to and enhanced its international reputation as a storehouse of expertise and a pre-eminent provider of technical assistance and training in the field of mining environmental protection and policy development.

The success of this project depended heavily on the capabilities, cultural sensitivity and group dynamics of the various joint teams. U.S. participants were carefully briefed on cultural differences and encouraged to consider how they might affect meetings and recommendations (U.S.DOI/OSM, 1995c). The tendency of Indonesian culture to place high value on social harmony and long-term relationships had a great effect on the pace of development, nature, and tone of recommendations concerning the role and methods of enforcement, embracing a regulatory approach which is significantly different from that in use in the U.S. Similarly, direct training of mine operators in methods of compliance, provision of economic incentives, and waivers or reduction of penalties for compliance after notice of violation may be more consonant with Indonesian culture than administrative enforcement of criminal sanctions.

In Indonesia, improved environmental management of mining operations will reduce degradation of water, air and land. Introduction of new ideas and technologies can dramatically reduce the long-term impacts of mining in Indonesia with acceptable financial costs, some of which can be offset by improvements in the efficiency of mining operations. The largest essential investments in equipment and infrastructure have already been made at most operating mines in Indonesia. Necessary earth moving equipment is in place to make significant improvements in sediment control and water quality, spoil dump stability, and site preparation for reclamation. The size of the "footprint" of mining operations, the amount of water pollution from storm water drainage and erosion, and the resulting destruction of biological diversity can all be reduced substantially by improving mining practices and training in mine planning, overburden handling, and reclamation. Disturbed wildlife habitat can be reclaimed instead of abandoned after mining. Mined land can be reclaimed for productive postmining land uses which can make additional contributions to sustainable development.

References

Allison, Graham (1971), *Essence of Decision: Explaining the Cuban Missile Crisis*, Boston: Little, Brown & Co.

Anderson, Fred R. (1973), *NEPA in the Courts: A Legal Analysis of the National Environmental Policy Act*, Baltimore: Johns Hopkins University Press.

Almond, Gabriel A. and G. Bingham Powell, Jr. (1978), *Comparative Politics: System, Process, and Policy*, 2nd, Boston: Little, Brown & Co.

Anderson, Fred R. (1973), *NEPA in the Courts: A Legal Analysis of the National Environmental Policy Act*, Baltimore: Johns Hopkins University Press.

——— Daniel Mandelker, and A. Dan Tarlock (1999), *Environmental Protection: Law and Policy*, 3d, Aspen: Aspen Law & Business.

Antal, Ariane Berthoin, Meinolf Dierkes and Hans N. Weiler (1987), 'Cross-National Policy Research: Traditions, Achievements and Challenges,' in Dierkes, M., H.N. Weiler, and A.B. Antal (eds), *Comparative Policy Research: Learning from Experience*, New York: St. Martin's Press, pp. 13–30.

Aspinall, Clive (2000), 'Implications of Indonesia's New General Mining Law (Still in Draft),' *Indonesian Mining Risk Analysis*, Jakarta: Clive Aspinall, 18 May.

Baas, Pieter, Scientific Director, Rijksherbarium/Hortus Botanicus, Leiden University (1996), Correspondence with the author from Leiden, The Netherlands, 2 July.

BAPEDAL [Badan Pengendalian Dampak Lingkungan-Environmental Impact Management Agency] (2000a), *Community Involvement and Information Openness in the Process of Environmental Impact Assessment* Decree No. 08/2000, Jakarta: BAPEDAL, Republic of Indonesia.

——— (2000b), *Guidelines for Preparation of Environmental Impact Assessment* Decree No. 09/2000, Jakarta: BAPEDAL, Republic of Indonesia.

——— (1994), *Guidelines for Determination of Significant Impact*, Decree No. Kep-056/ 1994, Jakarta: BAPEDAL, Republic of Indonesia.

Bass, Ron and Scott Warner (1979), 'Streamlining NEPA: a Look at the Council on Environmental Quality's new EIS regulations,' *Environmental Comments*, August, pp. 14–19.

'The Battle for Kalimantan's Coal' (2002), *Tempo* 17 June, 47–52.

Bauzó, Edwin B. Fernández (1987), 'Cross-Cultural Training: A Multicultural, Pluralistic Approach,' in Reddy, W. Brendan and Clenard C. Henderson, Jr. (eds.), *Training Theory and Practice*, Arlington, VA: NTL Institute for Applied Behavioral Science, pp. 249–258.

Baxter, F.P., L.D. Bradfield, R.G. Edwards, L. Erickson and N.S. Fischman (1983), *A National Inventory of AML Problems: An Emphasis on Health, Safety, and General Welfare Impacts*, NTIS No. DE84-016741, Oak Ridge, TN: Oak Ridge National Laboratory.

Bear, Dinah (1989), 'NEPA at 19: A Primer on an "Old" Law with Solutions to New Problems,' *Environmental Law Reporter* 19 (February): 10060–10069.

Benveniste, Guy (1977), *Bureaucracy*, San Francisco: Boyd and Fraser.

Berry, Frances Stokes (1994), 'Sizing Up State Policy Innovation Research,' *Policy Studies Journal* 22: 442–456.

Bierregaard, R.O., Jr., T.E. Lovejoy, V. Kapos, A. Augusto dos Santos, and R.W. Hutchings (1992), 'The Biological Dynamics of Tropical Rainforest Fragments,' *BioScience* 42: 849–866.

Blackburn, Joseph (1996), 'The Use of an Aerial Overflight Program for Monitoring Mining Activity on Surface Coal Mining Operations,' in *Proceedings of the First International Conference on Mining and the Environment, March 6–8, Bandung, Indonesia*, Jakarta: Ministry of Mines and Energy, Republic of Indonesia.

Bohlen, C. (1993), 'Request for Circular 175 Authority to Negotiate and Conclude a Memorandum of Understanding between the Department of the Interior of the United States of America and the Ministry of Mines and Energy of the Republic of Indonesia on Cooperation and Exchange in the Field of Surface Mining,' Washington, D.C.: U.S. Department of State, May 19.

Braybrooke, David, and Charles Lindblom (1963), *A Strategy of Decision*, New York: Free Press.

Buckley, Roger, and Jim Caple (2000), *Theory and Practice of Training*, 4th, London: Kogan Page, Ltd.

Caldwell, Lynton Keith (1990), *International Environmental Policy*, 2d. Durham, NC: Duke University Press.

—— (1989), 'A Constitutional Law for the Environment,' *Environment* 31: 6–28.

Calvert Cliffs Coordinating Committee v. Atomic Energy Commission (1972), 449 F. 2d 1109 (D.C. Cir. 1971), cert. denied, 404 U.S. 942.

Carter, Russell A. (1995a), Careful Planning Solves the Logistics Puzzle, *Coal* August, 30–31.

—— (1995b), Kaltim Prima: Energy from the Equator, *Coal*, August, 33–35.

Carvounis, Chris C. (1986), *The Foreign Debt/National Development Conflict*, New York: Quorum Books.

Caulfield, Henry P. (1984), 'U.S. Water Resources Development Policy and Intergovernmental Relations,' in Francis, John G. and Richard Ganzel (eds.), *Western Public Lands: Management of Natural Resources in a Time of Declining Federalism* (Totowa, NJ: Rowman and Allanheld, pp. 215–231.

Clark, Billie (1996), 'Permitting Process Overview: Surface Coal Mining and Reclamation,' in *Proceedings of the First International Conference on Mining and the Environment, March 6–8, Bandung, Indonesia*, Jakarta: Ministry of Mines and Energy, Republic of Indonesia.

Clark, Ray, and Larry Canter (1997), *Environmental Policy and NEPA: Past, Present and Future*, Boca Raton: St. Lucie Press.

Coddington, J.A., C.E. Griswold, D.S. Dávila, E. Peñaranda, and S.F. Larcher (1991), 'Designing and Testing Sampling Protocols to Estimate Biodiversity in Tropical Ecosystems,' in Dudley, E.C. (ed.), *The Unity of Evolutionary Biology: Proceedings of the Fourth International Congress of Systematic and Evolutionary Change*, Portland, OR: Dioscorides Press.

Colfer, Carol J. Pierce, Nancy Peluso and Chin See Shung (1997), *Beyond Slash and Burn: Building on Indigenous Management of Borneo's Tropical Rain Forests*, Bronx: New York Botanical Garden.

Congressional Quarterly, Inc. (1977), 'Strip Mining Control Bill Signed,' in *Congressional Quarterly Almanac-1977*, Washington, D.C.: Congressional Quarterly, Inc., pp. 617–626.

—— (1969), 'Environmental Quality Council,' in *Congressional Quarterly Almanac-1969*, 525–527, Washington, D.C.: Congressional Quarterly, Inc.

Conservation International, Inc. (2005), *Rapid Assessment Program*, Internet address:

www.biodiversityscience.org/xp/CABS/research/rap/terrestrial_rap/terrestrap.xml [accessed 12 January].

—————— (1995), *Rapid Assessment Program: Global Conservation Priority Setting*, Washington, D.C.: Conservation International, Inc.

Cooper, Lauren (1984), *The Twinning of Institutions – Its Use as a Technical Assistance Delivery System*, World Bank Technical Paper No. 23. Washington, D.C.: The World Bank.

Cuddington, John T. (1989), 'The Extent and Causes of the Debt Crisis of the 1980s,' in Husain, Ishrat, and Ishac Diwan (eds), *Dealing with the Debt Crisis*, Washington, D.C.: The World Bank.

Cyert, R., and J. March (1963), *A Behavioral Theory of the Firm*, Englewood Cliffs, NJ: Prentice-Hall.

Desai, Uday (ed.) (2002), *Environmental Politics and Policy in Industrialized Countries*, Cambridge: MIT Press.

—————— (1993), *Moving the Earth: Cooperative Federalism and Implementation of the Surface Mining Act*, Westport: Greenwood Press.

DeSombre, Elizabeth R. (2000), *Domestic Sources of International Environmental Policy*. Cambridge: MIT Press.

Dick, J. and L. Bailey (1992), *Indonesia's Environmental Assessment Process (AMDAL): Progress, Problems and a Suggested Blueprint for Improvement*, Jakarta & Halifax: Environmental Management Development in Indonesia Project.

Dror, Y. (1964), 'Muddling Through – "Science" or Inertia?' *Public Administration Review*, 24: 153–57.

Dodge Building Cost Services (1987), *Dodge Guide to Public Works and Heavy Construction Costs*, New York: McGraw-Hill Information Systems Co.

Duckett, Gerald P., President, Duckett Resources, Inc. (2002), Interview by author, 26 September, Mt. Lake Park, Maryland.

—————— (2001a), Interview by author, 15 March, Mt. Lake Park, Maryland.

—————— (2001b), Interview by author, 8 August, Mt. Lake Park, Maryland.

—————— (2001c), *Equipment Costs: Clean Coal Circuit*, Workshop for Provincial and Local Governments on Fine Coal Recovery Systems, sponsored by the Ministry of Energy and Mineral Resources, Republic of Indonesia, Samarinda, East Kalimantan, June.

—————— (1989), Buffalo Coal Accelerates Output, *Coal Magazine* 56 (September): 19–21.

Eichbaum, William M. and Hope M. Babcock (1982), 'A Question of Delegation: The Surface Mining Control and Reclamation Act of 1977 and State-Federal Relations. An Inquiry Into the Success with which Congress May Provide Detailed Guidance for Executive Agency Action,' *Dickinson Law Review* 86 (Summer): 615–646.

Emmerson, Donald K. (1978), 'Bureaucracy in Political Context: Weakness in Strength,' in Jackson. Karl D., and Lucian W. Pye (eds), *Political Power and Communications in Indonesia*, Berkeley: University of California Press, pp. 82–136.

Engelsman, Coert (1985), *Heavy Construction Cost File*, New York: Van Nostrand Reinhold Co.

Etzioni, Amatai (1968), *The Active Society*, New York: Free Press.

—————— (1967), 'Mixed-Scanning: a "Third" Approach to Decision Making,' *Public Administration Review*, 27: 385–92.

Farrel, Tom (2000), *Discussion Paper: Mining Environmental Management in Indonesia, Opportunities for Collaboration in Science and Technology*, Canberra City, Australia: Industry Sciences & Resources.

Farvar, Mitaghi and John P. Milton (1972), *The Careless Technology: Ecology and International Development*, Garden City, NY: Natural History Press.

Fields, Timothy (2002), 'Former Assistant Administrator of EPA Talks to GW Students About EPA's Superfund Program,' *Environmental & Energy Management Newsletter*, 3 (Spring), George Washington University Engineering Management & Systems Engineering Department, Internet address: http://www.gwu.edu/~eemnews/ spring2002-superfund.htm [accessed 21 November].

Galloway, L. Thomas and J. Davitt McAteer (1980), 'Surface Mining Regulation in the Federal Republic of Germany, Great Britain, Australia, and the United States: A Comparative Study,' *Harvard Environmental Law Review* 4: 261–309.

Gandataruna, Kosim (1993), Letter from the Director General, Directorate General of Mines, Indonesian Ministry of Mines and Energy, to Peter Scherer, Chief, Industry and Energy Operations Division, East Asia Country Department 3, The World Bank, Jakarta, February 8.

George, Alexander (1974), 'Adaptation to Stress in Political Decisionmaking: The Individual, Small Group and Organizational Contexts,' in Coelho, G.V., D.A. Hamburg and J.E. Adams (eds) *Coping and Adaptation*, New York: Basic Books, pp. 176–245.

'Government Prepares Provinces for Autonomy in Mining' (2000), *Jakarta Post*, 15 May, 1.

'Government Urged to take Strong Action on Polluting Industries' (1992), *Jakarta Post*, 25 August, 1.

Gray, Clive (1979), 'Civil Service Compensation in Indonesia,' *Bulletin of Indonesian Economic Studies* 15: 85–113.

Gray, Virginia (1973), 'Innovation in the States: A Diffusion Study,' *American Political Science Review* 67 (December): 1174–1185.

Greenberg, Michael R. (1987), *Public Health and the Environment*, New York: Guilford Press.

Halperin, Morton (1974), *Bureaucratic Politics and Foreign Policy*, Washington, D.C.: Brookings Institution.

Hamilton, Michael S (2004), 'Natural Resource Policy and Underinvestment in Indonesian Mining Operations,' *Natural Resource Management* 7 (September): 25–34.

—— (2001), 'Prospects for Increasing Profits by Improving Water Quality at Indonesian Coal Mines,' *Minerals & Energy* 16 (February): 3–13.

—— (2000a), 'Developing Mining Environmental Policy and Opportunities for Environmental Business in Indonesia,' in *Overcoming Barriers to Environmental Improvement: Proceedings of the 25th Annual Meeting of the National Association of Environmental Professionals*, Cary, NC: National Association of Environmental Professionals.

—— (2000b), 'Policy Drivers: Estimating Capital Recovery Periods for Investments in Fine Coal Circuits at Indonesian Coal Preparation Plants,' *Indonesian Mining Journal* 6: 69–77.

—— (1998a), 'Lost Profits, Lost Royalties: Formulating Policy to Recover the Value of Lost Coal Fines from Indonesian Mining Operations,' *Indonesian Mining Journal*, 4: 71–78.

—— (1998b), *Mining Environmental Policy in Indonesia: Program Evaluation of the Joint BLT-OSM Mining Environmental Project, Third Year of Effort*, Portland, ME: Prepared in fulfillment of a contract with the Office of Surface Mining Reclamation and Enforcement, U.S. Department of the Interior.

—— (1997a), 'Coal Mining Can be Compatible with Sustainable Development,' *Buletin Informasi Lingkungan* [*Environmental Data Bulletin*] 1 (3): 31–34.

—— (1997b), *Development of Mining Environmental Policy in Indonesia: Program Evaluation of the Joint BLT-OSM Mining Environmental Project, Second Year of*

Effort, Portland, ME: Prepared in fulfillment of a contract with the Office of Surface Mining Reclamation and Enforcement, U.S. Department of the Interior.

––––– (1996), *Program Evaluation of the Joint BLT-OSM Mining Environmental Project: First Year of Effort*, Portland, ME: Prepared in fulfillment of a contract with the Office of Surface Mining Reclamation and Enforcement, U.S. Department of the Interior.

––––– (1990), 'Introduction: An Intergovernmental Model of Policy Implementation,' in Hamilton, M. (ed.), *Regulatory Federalism, Natural Resources, and Environmental Management*, Washington, D.C.: American Society for Public Administration, pp. 1–24.

––––– and Norman Wengert (1980), *Environmental, Legal and Political Constraints on Power Plant Siting in the Southwestern United States*, A Report to the Los Alamos Scientific Laboratory, Fort Collins, CO: Colorado State University Experiment Station.

––––– Alfred E. Whitehouse, and W. Hord Tipton (1992), *Report on Trip to Indonesia, March 22–April 1, 1992*, Washington, D.C.: U.S. Department of the Interior, Office of Surface Mining Reclamation and Enforcement.

––––– Kadar Wiryanto, and Alfred E. Whitehouse (1996), 'Building Institutional Capacity for Policy Development Through Bilateral Science and Technology Cooperation Using Joint-Teams: Mining Environmental Policy in Indonesia,' *Indonesian Mining Journal* 2 (October): 71–83.

Harris, Richard A. (1985), *Coal Firms under the New Social Regulation*, Durham, NC: Duke University Press.

Heady, Ferrel (1996), *Public Administration: A Comparative Perspective*, 5th, New York: Marcel Dekker, Inc.

Heidenheimer, Arnold J., Hugh Heclo and Carolyn Teich Adams (1990), *Comparative Public Policy: The Politics of Social Choice in Europe and America*, 3d, New York: St. Martin's Press.

Hildebrand, Stephen G. (ed.) (1993), *Environmental Analysis: The NEPA Experience*, Boca Raton: Lewis Publishers.

Hirsch, Philip, and Carol Warren (1998), *The Politics of Environment in Southeast Asia*, New York: Routledge.

Honadle, George (1999), *How Context Matters: Linking Environmental Policy to People and Place*, West Hartford: Kumarian Press.

Huntington, Samuel (1968), *Political Order in Changing Societies*, New Haven, CT: Yale University Press.

Imhoff, Edgar, Thomas O. Fritz and James R. LeFevers (1976), *A Guide to State Programs for the Reclamation of Surface Mined Areas*, Washington, D.C.: U.S. Department of the Interior, Geological Survey.

Ingram, Helen, and Dean Mann (1992), *Why Policies Succeed or Fail*, Beverly Hills: Sage Publications.

Irving, Alan, Senior Environmental Engineer, PT Kaltim Prima Coal (1992), Interview with the author, Sangatta Baru, East Kalimantan, Indonesia, 28 March.

Jackson, Karl D. (1980), *Traditional Authority, Islam and Rebellion*, Berkeley: University of California Press.

––––– (1978a), 'Bureaucratic Polity: A Theoretical Framework for the Analysis of Power and Communications in Indonesia,' in Jackson, Karl D. and Lucian W. Pye (eds.), *Political Power and Communications in Indonesia*, Berkeley: University of California Press, pp. 3–22.

––––– (1978b), 'The Political Implications of Structure and Culture in Indonesia,' in

Jackson, Karl D. and Lucian W. Pye (eds), *Political Power and Communications in Indonesia*, Berkeley: University of California Press, pp. 23–44.

———— (1978c), 'Urbanization and the Rise of Patron-Client Relations: The Changing Quality of Interpersonal Communications in the Neighborhoods of Bandung and the Villages of West Java,' in Jackson, Karl D. and Lucian W. Pye (eds.), *Political Power and Communications in Indonesia*, Berkeley: University of California Press, pp. 343–394.

Janis, Irving L. and Leon Mann (1977), *Decision Making*, New York: Free Press.

Jay, R. (1969), *Javanese Villagers: Social Relations in Rural Modjokuto*, Southeast Asia Studies Monograph, New Haven, CT: Yale University Press.

Johnson, Brian, and Robert Stein (1979), *Banking on the Biosphere? Environmental Procedures and Practices of Nine Multilateral Development Agencies*, Washington, D.C.: International Institute for Environment and Development.

Johnson, Wilton and George C. Miller (1979), *Abandoned Coal-Mined Lands: Nature, Extent, and Cost of Reclamation*, Washington, D.C.: U.S. Government Printing Office.

Kaplan, Jacob J. (1967), *The Challenge of Foreign Aid*, New York: Frederick A. Praeger.

Kartasasmita, Ginandjar (1991), *Sumber Daya Energi Dan Mineral Indonesia*, Jakarta: Ministry of Mines and Energy, Republic of Indonesia.

Kessler, Paul, Research Group on Tropical Phanerograms, Rijksherbarium/Hortus Botanicus, Leiden University (1996), Interview with the author, Leiden, The Netherlands, 11 July.

King, Dwight Y. (1995), 'Bureaucracy and Implementation of Complex Tasks in Rapidly Developing States: Evidence from Indonesia,' *Studies in Comparative International Development* 30 (Winter): 78–92.

Kunanayagam, Ramanie and Ken Young (1998), 'Mining, Environmental Impact and Dependent Communities: The View from Below in East Kalimantan,' in Hirsch, Philip, and Carol Warren (eds), *The Politics of Environment in Southeast Asia*, New York: Routledge, pp. 139–158.

Kuntjoro, Dibyo (1994), 'Environmental Policy Planning and Strategy as an Effort to Achieve Sustainable Development in the Mining Industry Towards the Second Long Term Development Program in Indonesia,' in Marangin Simatupang and Beni N. Wahju (eds), *Environmental Aspects of Mining in Indonesia*, Jakarta: Indonesian Mining Association.

———— (1993), 'Minerals Management and Environmental Issues in Indonesia,' paper presented at the American Mining Congress Exhibition and Conference, Las Vegas, NV, 21 October.

———— (1992), Letter from the Director, Directorate of Mines, Ministry of Mines and Energy to the International Programs Coordinator, U.S. Office of Surface Mining Reclamation and Enforcement, Washington, D.C., October 15.

Lachica, Eddie (2004), 'Conclusions: Failure of the State?' Presentation to the U.S.-Indonesia Society by a former *Asia Wall Street Journal* reporter, Conference on the Future of US-Indonesia Relations, November 17, 2004, Washington, D.C.

Lasswell, Harold (1958), *Politics: Who Gets What, When, How?* New York: Meridian Press.

Le Prestre, Phillipe (1989), *The World Bank and the Environmental Challenge*, Selinsgrove: Susquehanna University Press.

Lerner, Daniel and Harold D. Lasswell (eds) (1951), *The Policy Sciences*, Stanford: Stanford University Press.

Lindblom, Charles (1965), *The Intelligence of Democracy*, New York: Free Press.

——— (1959), 'The Science of "Muddling Through",' *Public Administration Review*, 19: 79–88.

Liroff, Richard S. (1976), *A National Policy for the Environment: NEPA and Its Aftermath*, Bloomington, IN: Indiana University Press.

Lucas, Anton (1998), 'River Pollution and Political Action in Indonesia,' in Hirsch, Philip, and Carol Warren (eds), *The Politics of Environment in Southeast Asia*, New York: Routledge.

MacAndrews, Colin (1994), 'The Indonesian Environmental Impact Management Agency (BAPEDAL): Its Role, Development and Future,' *Bulletin of Indonesian Economic Studies* 30(April): 85–103.

MacKinnon, Kathy (1997), *The Ecology of Kalimantan*, Oxford: Oxford University Press/Periplus.

Mahmud, Ridwan (1996), 'Indonesia's Strategy in Developing Coal Resources,' *IPTEK*, August, 13–17.

Mangkusubroto, Kuntoro (1996a), 'Indonesian Coal Policy and Development Strategy,' *Indonesian Mining Journal* 2 (June): 45–52.

——— (1996b), 'Modernizing the Indonesian Mineral Industry,' *Indonesian Mining Journal* 2(June): 62–67.

Mazmanian, Daniel S. and Michael E. Kraft (2001), *Toward Sustainable Communities: Transition and Transformations in Environmental Policy*, Cambridge: MIT Press.

Maynard, Bernard (1998a), *Soils, Erosion, Sediment Control, and Revegetation: Integrated Concepts*, Part I: *Soils, I-A Soil Properties, I-B Soil Water*, Jakarta: Ministry of Mines and Energy, BLT-OSM Mining Environmental Project.

——— (1998b), *Soils, Erosion, Sediment Control, and Revegetation: Integrated Concepts, Part II: Erosion and Sediment Control*, Jakarta: Ministry of Mines and Energy, BLT-OSM Mining Environmental Project.

——— (1998c), *Soils, Erosion, Sediment Control, and Revegetation: Integrated Concepts, Part III: Revegetation*, Jakarta: Ministry of Mines and Energy, BLT-OSM Mining Environmental Project.

Mayo, G. Douglas, and Philip H. DuBois (1987), *The Complete Book of Training: Theory, Principles, and Techniques*, San Diego, CA: University Associates, Inc.

McMahon, Gary, Elly R. Subdibyo, Jean Aden, Aziz Bouzaher, Giovanna Dore and Ramanie Kunanayagam (2000), *Mining and the Environment in Indonesia: Long-Term Trends and Repercussions of the Asian Economic Crisis*, Washington, D.C.: The World Bank.

Miller, D.W. and M.K. Starr (1967), *The Structure of Human Decisions*, Englewood Cliffs, NJ: Prentice-Hall.

Miller, Richard O. (1999), 'Mining, Environmental Protection, and Sustainable Development in Indonesia,' in Vig, Norman, and Regina Axelrod (eds), *The Global Environment: Institutions, Law, and Policy*, Washington, D.C.: Congressional Quarterly, Inc.

——— (1996a), 'The Design and Implementation of Regulatory Standards,' Seminar presented to the Ministry of Mines and Energy, Jakarta, Indonesia, 12 March.

——— (1996b), 'Negotiated Rulemaking: Consensus Building and Issue Identification Through Participatory Processes,' in *Proceedings of the First International Conference on Mining and the Environment, March 6–8, Bandung, Indonesia*, Jakarta: Ministry of Mines and Energy, Republic of Indonesia.

——— (1993), 'The Surface Mining Control and Reclamation Act: Policy Structure, Policy Choices, and the Legacy of Legislation,' in Desai, Uday (ed.), *Moving the Earth: Cooperative Federalism and Implementation of the Surface Mining Act*, Westport: Greenwood Press, pp. 17–30.

Ministry of Energy and Mineral Resources (2004), *Indonesia Mineral & Coal Statistics 2004*, Jakarta: Ministry of Energy and Mineral Resources, Republic of Indonesia.

—— (2003), *Indonesia Mineral & Coal Statistics 2003*, Jakarta: Ministry of Energy and Mineral Resources, Republic of Indonesia.

—— (2001), *Mining and Energy Yearbook of Indonesia, 2000*, Jakarta: Ministry of Energy and Mineral Resources, Republic of Indonesia.

—— (2000), *Technical Guidelines of the Government Authority on General Mining*, Decree 1453 K/29/MEM/2000, Jakarta: Ministry of Energy and Mineral Resources, Republic of Indonesia.

Ministry of Environment (2003), *Effluent Standards of Wastewater from Coal Mining Activities*, Decree No. 113/2003, Jakarta: Ministry of Environment, Republic of Indonesia.

—— (2001), *Types of Business and/or Activity Plans that are Required to be Completed with the Environmental Impact Assessment*, Decree No. 17/2001, Jakarta: Ministry of Environment, Republic of Indonesia.

—— (2000a), *Guidelines for AMDAL Document Evaluation*, Decree No. 2/2000, Jakarta: Ministry of Environment, Republic of Indonesia.

—— (2000b), *Guidelines for Work System of Evaluator Committee for Environmental Impact Assessment*, Decree No. 40/2000, Jakarta: Ministry of Environment, Republic of Indonesia.

—— (2000c), *Guidelines for Establishment of Regencial/Municipal Evaluator Committee for Environmental Impact Assessment*, Decree No. 41/2000, Jakarta: Ministry of Environment, Republic of Indonesia.

—— (2000d), *Membership Composition of Central Evaluator Committee and Technical Team for Environmental Impact Assessment*, Decree No. 42/2000, Jakarta: Ministry of Environment, Republic of Indonesia.

—— (1995), *Water Quality Control for Industrial Activity,* Decree No. 51/MENLH/10/95, Jakarta: Ministry of Environment, Republic of Indonesia.

—— (1994a), *Cancellation of Decrees No. KEP-49/MENKLH/6/1987 to KEP-53/MENKLH/6/1987*, Decree No. Kep-10/MENKLH/3/1994, Jakarta: Ministry of Environment, Republic of Indonesia.

—— (1994b), *Types of Business or Activities Required to Prepare an Environmental Impact Assessment*, Decree No. Kep-11/MENKLH/3/1994, Jakarta: Ministry of Environment, Republic of Indonesia.

—— (1994c), *Guidelines for Environmental Management Procedures and Environmental Monitoring Procedures*, Decree No. Kep-12/MENKLH/3/1994, Jakarta: Ministry of Environment, Republic of Indonesia.

—— (1994d), *Guidelines for AMDAL Commission Membership and Working Procedures*, Decree No. Kep-13/MENKLH/3/1994, Jakarta: Ministry of Environment, Republic of Indonesia.

—— (1994e), *General Guidelines for Preparation of Environmental Impact Assessments*, Decree No. Kep-14/MENKLH/3/1994, Jakarta: Ministry of Environment, Republic of Indonesia.

—— (1994f), *Establishment of an Environmental Impact Assessment Commission for Integrated/Multisectoral Activities*, Decree No. Kep-15/MENKLH/3/1994, Jakarta: Ministry of Environment, Republic of Indonesia.

Ministry of Environment and Population (1987a), *Guidelines for Determination of Significant Impact*, Decree No. Kep-49/MENKLH/6/1987, Jakarta: Ministry of Environment and Population, Republic of Indonesia.

—— (1987b), *Guidelines for Environmental Impact Analysis of Proposed Projects*,

Decree No. Kep-50/MENKLH/6/1987, Jakarta: Ministry of Environment and Population, Republic of Indonesia.

—— (1987c), *Guidelines for Environmental Impact Analysis of Existing Projects*, Decree No. Kep-51/MENKLH/6/1987, Jakarta: Ministry of Environment and Population, Republic of Indonesia.

—— (1987d), *Time Limits for Carrying Out Environmental Impact Analysis of Existing Projects*, Decree No. Kep-52/MENKLH/6/1987, Jakarta: Ministry of Environment and Population, Republic of Indonesia.

—— (1987e), *Guidelines for AMDAL Commission Membership, Composition and Working Procedures*, Decree No. Kep-53/MENKLH/6/1987, Jakarta: Ministry of Environment and Population, Republic of Indonesia.

—— (1986), *Government Regulation No 29 on Analysis of Environmental Impacts*, Jakarta: Ministry of Environment and Population, Republic of Indonesia.

Ministry of Health (1977), *Regulations on Water Pollution Control for Relevant Utilization Related to Public Health*, No. 173/MENKES/PER/VIII/77, Jakarta: Ministry of Health, Republic of Indonesia.

Ministry of Mines (1977), *Prevention and Measures for Disturbance and Pollution Caused by Mining Activities*, Decree No. 04/P/M/1977, Jakarta: Ministry of Mines, Republic of Indonesia.

Ministry of Mines and Energy (1999a), *Indonesian Coal Statistics*, Jakarta: Ministry of Mines and Energy, Republic of Indonesia.

—— (1999b), *Mining and Energy Yearbook of Indonesia, 1998*, Jakarta: Ministry of Mines and Energy, Republic of Indonesia.

—— (1997), *Coal Production by Company, 1990–1996*, [Table: Prodcom-1], Jakarta: Ministry of Mines and Energy, Republic of Indonesia.

—— (1996a), *Guidelines for Environmental Impact Analysis for Mining and Energy Activities*, Decree No. 1256/K/008/M.PE/1996, Ministry of Mines and Energy, Republic of Indonesia.

—— (1996b), *Proceedings of the First International Conference on Mining and the Environment, March 6–8, Bandung, Indonesia*, Jakarta: Ministry of Mines and Energy, Republic of Indonesia.

—— (1996c), *Reclamation Guarantee*, Decision 336.K/271/DDJP/1996, Jakarta: Director General of General Mining, Ministry of Mines and Energy, Republic of Indonesia.

—— (1996d), *Technical Guidelines for Erosion and Sediment Control for Mining Activity*, Decree No. 693.K/008/DDJP/1996, Jakarta: Ministry of Mines and Energy, Republic of Indonesia.

—— (1995a), *1994 Mining and Energy Yearbook*, Jakarta: Ministry of Mines and Energy, Republic of Indonesia.

—— (1995b), *Prevention and Mitigation of Environmental Damage and Pollution in General Mining Operations*, Decree No. 1211.K/008/M.PE/1995, Jakarta: Ministry of Mines and Energy, Republic of Indonesia.

—— (1993), *Mine Inspectors*, Decree No. 2555.K/201/M.PE/1993, Jakarta: Ministry of Mines and Energy, Republic of Indonesia.

—— (1992), *Project Proposal for Development and Improvement of Mining Environmental Policy and Regulatory Program to Strengthen Enforcement Strategy and Capability*, Jakarta: Ministry of Mines and Energy, Republic of Indonesia.

—— (1991), *Draft Project Proposal for Development and Improvement of Mining Environmental Policy and Regulatory Program to Strengthen Enforcement Strategy and Capability*, Jakarta: Ministry of Mines and Energy, Republic of Indonesia.

—— (1989), *Guidelines Procedure for Environmental Impact Assessment for Mining and Energy Development Projects*, Decree No. 0158/K/008/MPE/1989, Jakarta: Ministry of Mines and Energy, Republic of Indonesia.

———— (1988a), *Environmental Impact Assessment Implementation Procedure for General Mining, Oil and Gas and Geothermal Resources Sectors*, Circular No. 02E/008/M.PE/ 1988, Jakarta: Ministry of Mines and Energy, Republic of Indonesia.

———— (1988b), *Technical Guidelines for Initial Environmental Evaluation Report and Environmental Impact Assessment Report Presentation for General Mining, Oil and Gas and Geothermal Resources Sectors*, Decree No. 0185K/008/M.PE/1988, Jakarta: Ministry of Mines and Energy, Republic of Indonesia.

———— (1988c), *Technical Guidelines for the Preparation of Initial Environmental Evaluation Report and Environmental Impact Assessment*, Decree No. 0158.K/008/ M.PE/1988, Jakarta: Ministry of Mines and Energy, Republic of Indonesia.

———— (1978a), *Prevention and Measures for Disturbance and Pollution as a Result of Mineral Processing and Refining*, Decree No. 09/DU/1978, Jakarta: Ministry of Mines and Energy, Republic of Indonesia.

———— (1978b), *Prevention and Measures for Disturbance and Pollution Affected by Open Cut Mining*, Decree No. 07/DU/1978, Jakarta: Ministry of Mines and Energy, Republic of Indonesia.

———— (1977), *Prevention and Measures for Disturbance and Pollution Affected by Mining Activities*, Decree No. 04/P/M/1977, Jakarta: Ministry of Mines and Energy, Republic of Indonesia.

Mintrom, Michael (1997), 'Policy Entrepreneurs and the Diffusion of Innovation,' *American Journal of Political Science* 41(July): 738–770.

Mittermeier, Russell A. (1988), 'Primate Diversity and the Tropical Forest: Case Studies from Brazil and Madagascar and the Importance of the Megadiversity Countries,' in Wilson, E.O. (ed.), *Biodiversity*, Washington, D.C.: National Academy Press, pp. 145–154.

Modak, Prasad and Asit K. Biswas (1999), *Conducting Environmental Impact Assessment in Developing Countries*, New York: United Nations University Press.

Morgenthau, Hans J. (1960), *Politics Among Nations*, New York: Alfred A. Knopf.

Nagel, Stuart (ed.) (1994), *Encyclopedia of Policy Studies*, 2d ed, New York: Marcel Dekker.

National Mining Association (2003), *Trends in U.S. Coal Mining 1923–2002*, Washington, D.C.: National Mining Association, Internet address: www.nma.org/ pdf/trends_us_coal_mining2001.pdf

O'Leary, Rosemary (1993), *Environmental Change: Federal Courts and the EPA*, Philadelphia: Temple University Press.

O'Leary, Rosemary, Robert F. Durant, Daniel J. Fiorino, and Paul S. Weiland (1999), *Managing for the Environment: Understanding the Legal, Organizational, and Policy Challenges*, San Francisco: Jossey-Bass Publishers.

Peluso, Nancy Lee (2000), 'The Role of Government Intervention in Creating Forest Landscapes and Resource Tenure in Indonesia,' in Gaul, Karen K. and Jackie Hiltz (eds), *Landscapes and Communities on the Pacific Rim*, Armonk, New York: M.E. Sharpe, pp. 147–166.

Petrich, Carl H. (1994), 'The Applicability of the U.S. NEPA Experience to Indonesia's Environmental Management Challenge,' paper presented at the DOEME Workshop II Sponsored by the Ministry of Mines and Energy, Directorate General of Electric Power and New Energy, Jakarta, 13–14-October.

Prasodjo, Edi (1996), 'Minimizing Environmental Impact of Coal Production and Utilization in Indonesia,' *PETROMINER* 15 March, 67–73.

PT Arutmin Indonesia (1994), *Environmental Impact Assessment of the West Asam-asam*

Coal Mining Project in Block 6, South Kalimantan Indonesia, Jakarta: PT Arutmin Indonesia.

Ramage, Douglas E. (1995), *Politics in Indonesia*, New York: Routledge.

Republic of Indonesia (2001), *Government Regulation No. 82/2001 on the Management of Water Quality and Pollution Control*, Jakarta: The Ministry/State Secretary of the Republic of Indonesia.

—— (2000), *Government Regulation No. 25/2000 on the Authority of the Government and the Autonomy of a Province as an Autonomous Region*, State Gazette No. 54, Jakarta: The Ministry/State Secretary of the Republic of Indonesia.

—— (1999a), *Act on Financial Equilibrium between the Central and Regional Government No. 25/1999*, Statute Book of the Republic of Indonesia No. 72, Jakarta: The Ministry/State Secretary of the Republic of Indonesia.

—— (1999b), *Act on Regional Administrations No. 22/1999*, State Gazette No. 60, Jakarta: The Ministry/State Secretary of the Republic of Indonesia.

—— (1999c), *Government Regulation No. 27/1999 on Analysis of Environmental Impacts*, State Gazette No. 59, Jakarta: The Ministry/State Secretary of the Republic of Indonesia.

—— (1998), *Contract of Work between the Government of the Republic of Indonesia and PT Sorikmas Mining*, Jakarta: Presiden Republik Indonesia.

—— (1997), *Act on Environmental Management No. 23/1997*, State Gazette No. 68, Jakarta: The Ministry/State Secretary of the Republic of Indonesia.

—— (1993), *Government Regulation No. 51/1993 on Analysis of Environmental Impacts*, State Gazette No. 84, Jakarta: The Ministry/State Secretary of the Republic of Indonesia.

—— (1992), *Government Regulation No. 67/1992 on Organization of the Secretariate General, Ministry of Mines and Energy*, Jakarta: The Ministry/State Secretary of the Republic of Indonesia.

—— (1990a), *Act on Conservation of Bio Natural Resources and its Ecosystem 5/1990*, State Gazette No. 49, Jakarta: The Ministry/State Secretary of the Republic of Indonesia.

—— (1990b), *Government Regulation No. 23/1990 on Agency for Environmental Impact Control*, Jakarta: The Ministry/State Secretary of the Republic of Indonesia.

—— (1990c), *Government Regulation No. 32/1990 on Management of Protected Areas*, Jakarta: The Ministry/State Secretary of the Republic of Indonesia.

—— (1986), *Government Regulation No. 29/1986 on Analysis of Environmental Impacts*, State Gazette No. 42, Jakarta: The Ministry/State Secretary of the Republic of Indonesia.

—— (1982), *Act on Basic Provisions for the Management of the Living Environment*, UU No. 4, State Gazette No. 12, Jakarta: The Ministry/State Secretary of the Republic of Indonesia.

—— (1969), *Government Regulation of the Republic of Indonesia Number 32 of 1969*, State Gazette No. 60, Jakarta: The Ministry/State Secretary of the Republic of Indonesia.

—— (1967), *Basic Mining Law*, UU No. 11, State Gazette No. 22, Jakarta: The Ministry/State Secretary of the Republic of Indonesia.

Revkin, Andrew (1990), *The Burning Season: The Murder of Chico Mendes and the Fight for the Amazon Rain Forest*, Boston: Houghton Mifflin.

Rich, Bruce (1994), *Mortgaging the Earth: The World Bank, Environmental Impoverishment, and the Crisis of Development*, Boston: Beacon Press.

Rieffel, Lex (2004), 'Indonesia's Quiet Revolution,' *Foreign Affairs* 83 (September/October): 98–110.

Riggs, Fred W. (1966), *Thailand: The Modernization of a Bureaucratic Polity*, Honolulu: East-West Center Press.

———— (1964), *Administration in Developing Countries: The Theory of Prismatic Society*, Boston: Houghton Mifflin.

Roberts, J. Scott (1998), *Acid Mine Drainage: Prediction, Prevention, and Treatment*, Jakarta: Ministry of Mines and Energy, BLT-OSM Mining Environmental Project.

Rochow, K. W. James (1979), 'The Far Side of Paradox: State Regulation of the Environmental Effects of Coal Mining,' 81 *West Virginia Law Review* 559.

Rogers, Everett M. (1995), *Diffusion of Innovations*, 4th ed, New York: Free Press.

Roos, M. C. Leader, Research Group on Tropical Phanerograms, Rijksherbarium/ Hortus Botanicus, Leiden University (1996), Interview with the author, Leiden, The Netherlands, 11 July.

'Ryas Tenders his Resignation from Cabinet' (2001), *Jakarta Post*, 3 January, 1.

Sabatier, Paul (1988), 'An Advocacy Coalition Framework of Policy Change and the Role of Policy-Oriented Learning Therein,' *Policy Sciences* 21: 129–168.

Sablatura, Bob (1995), 'Superfund's Super Mess,' *Houston Chronicle*, 20 October, 1.

Savage, Robert L. (1985), 'Diffusion Research Traditions and the Spread of Policy Innovations in a Federal System,' *Publius* 15 (Fall): 1–27.

Scheberle, Denise (1997), *Federalism and Environmental Policy: Trust and the Politics of Implementation*, Washington, D.C.: Georgetown University Press.

Scherer, Peter (1994), Letter from the Chief, Industry and Energy Operations Division, East Asia Country Department 3, The World Bank, to Kosim Gandataruna, Director General, Directorate General of Mines, Indonesian Ministry of Mines and Energy, Washington, D.C., April 27.

———— (1992), Letter from the Chief, Industry and Energy Operations Division, East Asia Country Department 3, The World Bank, to Kosim Gandataruna, Director General, Directorate General of Mines, Indonesian Ministry of Mines and Energy, Washington, D.C., December 8.

Shover, Neal, Donald A. Clelland and John Lynzwiler (1986), *Enforcement or Negotiation: Constructing a Regulatory Bureaucracy*, Albany: State University of New York Press.

Sigit, Soetaryo (1988), *Collection of Papers on Indonesia Mining Review and Mineral Development Policy*, Jakarta: Ministry of Mines and Energy, Directorate General of Mines.

Simatupang, Marangin, and Beni N. Wahju (eds) (1994), *Environmental Aspects of Mining in Indonesia*, Jakarta: Indonesian Mining Association.

Simon, Herbert (1976), *Administrative Behavior*, 3d ed, New York: Free Press.

———— (1960), *New Science of Management Decision*, Englewood Cliffs, NJ: Prentice-Hall.

Slovic, P. (1971), 'Limitations of the Mind of Man: Implications for Decision Making in the Nuclear Age,' *Oregon Research Institute Bulletin*, 11: 41–49.

Soemarwoto, Otto (1996), 'Environmental Impact of Mining,' *PETROMINER*, 15 March, 78–80.

Soesilo, Indroyono (ed.) (1993), *Remote Sensing Technology in Indonesia*, Jakarta: Ministry of Technology.

Standard Laboratories, Inc. (2001), Certificate of Analysis No. 1997–9307, Gormania, WV: 16 March.

Stone, David (1994), *Tanah Air: Indonesia's Biodiversity*, Singapore: Archipelago Press.

Subagyo, Iskandar (1994), 'Environmental Enforcement for Mining in Indonesia,' paper presented at the UNCTAD Seminar on Capacity Building for Environmental Management in Asian/Pacific Mining, Jakarta, Indonesia, September 6–8.

Sunardi, R.A. (1997), 'Indonesia's Coal Production and the Challenges Facing the Industry in the 21st Century,' *Indonesian Mining Journal* 3 (February): 56–66.

Taufiqurohman, M., SetriYasra, Ali Nur Yasin and Thomas Hadiwinata (2003), 'Back to Square One,' *Tempo*, 4 August, 56–58.

Terborgh, J. (1992), 'Maintenance of Diversity in Tropical Forests,' *Biotropica* 24: 283–292.

Turner, Mark, Owen Podger, Maria Sumardjono, and Wayan K. Tirthayasa (2003), *Decentralisation in Indonesia: Redesigning the State*, Canberra: Asia Pacific Press/ Australian National University.

Udall, Morris K. (1979), 'Enactment of the Surface Mining Control and Reclamation Act of 1977 in Retrospect,' 81 *West Virginia Law Review* 553.

U.N. Conference on Trade and Development (UNCTAD) (1994), *Proceedings of the Seminar on Capacity Building for Environmental Management in Asian/Pacific Mining, Jakarta, Indonesia, 6–8 September*, New York: UNCTAD.

U.S. Congress (1969), *National Environmental Policy Act of 1969*, 91st Cong., 1st sess, House Report 765 to accompany S. 1075, Conference Report.

U.S. Congress, Senate Committee on Interior and Insular Affairs (1969), *National Environmental Policy Act of 1969*, 91st Cong., 1st sess, Senate Report 296 to accompany S. 1075.

U.S. Council on Environmental Quality (U.S.CEQ) (1978), 'Council on Environmental Quality,' *U.S. Code of Federal Regulations*, Vol. 40, §1500.

—— (1976), *Environmental Impact Statements: An Analysis of Six Years Experience by Seventy Federal Agencies*, Washington, D.C.: Government Printing Office.

—— (1973), 'Preparation of Environmental Impact Statements: Guidelines,' *Federal Register*, 38: 20550–62.

U.S. Department of Commerce (1994), *Indonesia – Coal Mining Profile*, Washington, D.C.: U.S. Department of Commerce, International Trade Administration, June 10.

U.S. Department of Energy (U.S. DOE) (2002), *Annual Energy Review, 2002*, Washington, D.C.: U.S. Government Printing Office.

U.S. Department of the Interior (U.S.DOI) (2001), *Overview of DOI International Activities*, Washington, D.C.: U.S. Department of the Interior, Office of Policy Analysis.

U.S. Department of the Interior, Office of Surface Mining Reclamation and Enforcement (U.S.DOI/OSM) (2004), *Abandoned Mine Land Fund Status*, Internet address: http://www.osmre.gov/fundstat.htm [accessed 31 December].

—— (2002), *2002 Annual Report*, Washington, D.C.: U.S. Department of the Interior, Office of Surface Mining Reclamation and Enforcement.

—— (2001), *National Technical Training Catalog FY 2002*, Washington, D.C.: Office of Surface Mining Reclamation and Enforcement.

—— (2000), *Governance, Decentralization and Sustainable Development: Technical Assistance to the Ministry of Energy and Mineral Resources, Republic of Indonesia*, Washington, D.C.: Office of Surface Mining Reclamation and Enforcement.

—— (1998a), *Phase III Extension Proposal: Technical Assistance for Improvement of Mining Environmental Policy and Enforcement in Indonesia*, Washington, D.C.: U.S. Department of the Interior, Office of Surface Mining Reclamation and Enforcement.

—— (1998b), *Technical Assistance for Improvement of Mining Environmental Policy and Enforcement in Indonesia: Quarterly Report, January 1, 1998–March 31, 1998*, Washington, D.C.: U.S. Department of the Interior, Office of Surface Mining Reclamation and Enforcement.

—— (1996a), *Technical Assistance for Improvement of Mining Environmental Policy and Enforcement in Indonesia: Quarterly Report, January 1, 1996–March 31, 1996*,

Washington, D.C.: U.S. Department of the Interior, Office of Surface Mining Reclamation and Enforcement.

——— (1996b), *A Review of the Effectiveness and Efficiency of Existing Organizational Arrangements Within the Ministry of Mines and Energy for Implementing Environmental Management and Protection Under Conditions of Industry Expansion: Report of the Program Management Team*, Washington, D.C.: U.S. Department of the Interior, Office of Surface Mining Reclamation and Enforcement.

——— (1995a), *Assessment Trip Report*, Washington, D.C.: U.S. Office of Surface Mining Reclamation & Enforcement, June 10.

——— (1995b), *Technical Assistance Agreement in the Area of Mining Environmental Policy and Enforcement*, Washington, D.C.: U.S. Department of the Interior, Office of Surface Mining Reclamation and Enforcement.

——— (1995c), *Technical Assistance for Improvement of Mining Environmental Policy and Enforcement in Indonesia: Briefing Book*, Washington, D.C.: U.S. Department of the Interior, Office of Surface Mining Reclamation and Enforcement.

——— (1995d), *Technical Assistance for Improvement of Mining Environmental Policy and Enforcement in Indonesia: Phase I Report*, Washington, D.C.: U.S. Department of the Interior, Office of Surface Mining Reclamation and Enforcement.

——— (1994), 'Memorandum of Understanding Between the Department of Mines and Energy of the Republic of Indonesia and the Department of the Interior of the United States of America on Cooperation and Exchange in the Fields of Surface Mining,' Washington, D.C.: U.S. Department of the Interior, Office of Surface Mining Reclamation and Enforcement.

——— (1991), *Basic Inspection Workbook*, Washington, D.C.: U.S. Department of the Interior, Office of Surface Mining Reclamation and Enforcement.

——— (1988), *Application for a Permit to Conduct Surface Coal Mining and Reclamation Operations under the Federal Program for California*, Washington, D.C.: U.S. Department of the Interior, Office of Surface Mining Reclamation and Enforcement.

——— (1979), *Permanent Regulatory Program Implementing Section 501(b) of the Surface Mining Control and Reclamation Act of 1977*, Washington, D.C.: U.S. Department of the Interior, Office of Surface Mining Reclamation and Enforcement.

U.S. Department of State (1992), 'Agreement Between the Government of the United States of America and the Government of the Republic of Indonesia for Cooperation in Scientific Research and Technological Development,' 15 January, Washington, D.C.: U.S. Department of State.

U.S. Energy Information Administration (2003), *Monthly Energy Review-June*, Washington, D.C.: U.S. Government Printing Office. Internet address: http://www.eia.doe.gov/

——— (2001), *International Energy Annual-2001*, Washington, D.C.: U.S. Government Printing Office. Internet address: http://www.eia.doe.gov/

U.S. Environmental Protection Agency (U.S.EPA) (2004), *Cleaning up the Nation's Waste Sites: Markets and Technology Trends*, Washington, D.C.: U.S. Environmental Protection Agency, Office of Solid Waste and Emergency Preparedness.

——— (2003a), *Superfund Accomplishment Figures, Summary Fiscal Year (FY) 2003*, Internet address: http://www.epa.gov/superfund/action/process/numbers.htm [accessed 21 November].

——— (2003b), *Superfund Budget History*, Internet address: http://www.epa.gov/superfund/action/process/budgethistory.htm [accessed 22 November].

U.S. Library of Congress (1973), *The Constitution of the United States of America, Analysis and Interpretation*, Washington, D.C.: U.S. Government Printing Office.

U.S. Trade and Development Agency (1994), 'Reimbursable Technical Assistance by the United States Department of the Interior, Office of Surface Mining Reclamation and Enforcement to the Government of Indonesia,' Washington, D.C.: U.S. Trade and Development Agency.

Waier, Phillip R., and Barbara Balboni (2003), *Building Construction Cost Data*, Kingston, MA: R.S. Means Co.

Walker, Jack L. (1969), 'The Diffusion of Innovations Among the American States,' *American Political Science Review* 63 (September): 880–899.

Warren, Carol, and Kylie Elston (1994), *Environmental Regulation in Indonesia*, Nedlands, Western Australia: University of Western Australia Press.

Watkins, Neil (2000), 'Campaign Launched to Boycott World Bank Bonds,' *World Rainforest Report* 27 (Summer): 6. See also: www.worldbankboycott.org

Watson, William, and Richard Bernknopf (1979), *Economic Analysis of Maximum Economic Recovery of Federal Coal*, Reston, VA: U.S. Geological Survey.

Welles, Holly (1995), 'EIA Capacity-Strengthening in Asia: The USAID/WRI Model,' *Environmental Professional* 17: 103–116.

Wengert, Norman (1979), 'The Energy Boom Town: An Analysis of the Politics of Getting,' in Lawrence, Robert M. (ed.) *New Dimensions to Energy Policy*, Lexington, MA: Lexington Books pp. 17–24.

Whitehouse, Alfred E., Project Director, Governance, Decentralization and Sustainable Development in Indonesia Project (2005). Interview by author, 28 March, Jakarta, Indonesia.

—— (2003), Interviews by author, 30 June and 22 August, Jakarta, Indonesia.

—— (2002), Interview by author, 17 April, Jakarta, Indonesia.

—— (2001), Interview by author, 9 July, Jakarta, Indonesia.

—— (1999), 'The Indonesian Mining Sector: The Regulatory Process, Results on the Ground, and Opportunities for Improvement,' *Indonesian Mining Journal* 5 (June): 61–79.

—— (1998a), 'Acid Mine Drainage: Prediction of Risk and Management of Impacts,' *Buletin Informasi Lingkungan* 2: 37–44.

—— (1998b), 'Reclamation Guarantees,' Paper presented to a Seminar for Mining Industry Environmental Managers, Manpower Development Center, Bandung, March 11.

—— (1996a), 'Case Study on Mining in the United States,' paper presented at the Ministry of Mines and Energy, Manpower Development Center, Bandung, August.

—— (1996b), 'Economics and Environment Create a Role for Research,' paper presented at a Plenary Session of 'Strengthening National Capability in Mining Activities,' Annual Colloquium and Exhibition on Mining of the Ministry of Mines and Energy, Bandung, Indonesia, November.

—— (1996c), 'Overburden Analysis, Methods to Predict Acid Mine Drainage Potential,' paper presented at the Bandung Institute of Technology, Bandung, Indonesia, July.

—— (1996d), 'Reclamation Guarantees or Bonds' *Buletin Informasi Lingkungan* 1: 22–24.

—— (1996e), 'Water Quality Protection in America Under the Surface Mining Act,' in *Proceedings of the First International Conference on Mining and the Environment, March 6–8, Bandung, Indonesia*, Jakarta: Ministry of Mines and Energy, Republic of Indonesia.

—— and Asep A.S. Mulyana (2004), 'Coal Fires in Indonesia,' *International Journal of Coal Geology* 59: 91–97.

Whitten, Tony (2000), *The Ecology of Sumatra*, Oxford: Oxford University Press/ Periplus.

Wijayanta, Hanibal W.Y., Wahyu Muryadi, Imron Rosyid and Adi Mawardi (2004), 'A Threat from Kramat Raya,' *Tempo*, 7–13 December, pp. 12–15.

Wilson, James Q. (1975), 'The Rise of the Bureaucratic State,' *The Public Interest* 41 (Fall): 77–103.

Wiryanto, Kadar, Section Head for Mining Activities, Directorate of Mines, Ministry of Mines and Energy, Republic of Indonesia (1991), Interview with the author, Washington, D.C., 12 September.

World Bank (1995), *Industrial Pollution Prevention & Abatement Handbook*, Washington, D.C.: The World Bank.

——— (1994a), *Indonesia: Environment and Development*, Washington, D.C.: The World Bank.

——— (1994b), *Loan Agreement: Sumatera and Kalimantan Power Project*, Loan Number 3761 IND, Washington, D.C.: The World Bank.

——— (1994c), *World Debt Tables 1994–95*, Vol. I, Washington, D.C.: The World Bank.

——— (1993), *Aide-Memoire, Outer Island Power Project*, Washington, D.C.: The World Bank, 5 May.

——— (1985), *Handbook on Consulting Services*, Washington, D.C.: The World Bank.

World Commission on Environment and Development (Brundtland Commission) (1987), *Our Common Future*, London: Oxford University Press.

Wright, Andrew (1984), 'Plant Designed with Flexibility in Mind,' *Coal Age* (December): 78–83.

Young, S. (1966), *Management: A Systems Analysis*, Glenview, Il: Scott Foresmen.

Yusgiantoro, Purnomo (2003), 'Energy and Mining Sector Update,' Presentation to the U.S.-Indonesia Society by the Minister of Energy and Mineral Resources, Republic of Indonesia, December 19, 2003, Washington, D.C.

Appendix:
Persons Interviewed

Ahmad Thabri Akma, Director
  Manpower Development Centre for Mines
  Ministry of Mines and Energy
  Bandung

Lobo Balia, GIS and Remote Sensing Project Manager
  Research and Development Center for Mineral Technology
  Ministry of Mines and Energy, Bandung

Chitta R. Bhattacharya, Senior Procurement Specialist
  The World Bank
  Jakarta

Joseph L. Blackburn, Deputy Field Office Director
  Office of Surface Mining Reclamation and Enforcement
  Lexington Field Office, Lexington, KY

Victoria Bryan, Reclamation Bond Specialist
  Western Support Center
  Office of Surface Mining Reclamation and Enforcement
  Denver, Colorado

Dr. Edward F. Carey, Physical Scientist
  State Program Support Branch
  Eastern Support Center
  Office of Surface Mining Reclamation and Enforcement
  Pittsburgh, Pennsylvania

Billie E. Clark, Jr., Supervisory Physical Scientist
  Branch of Federal and Indian Permits
  Western Support Center
  Office of Surface Mining Reclamation and Enforcement
  Denver, Colorado

Dr. Bukin Daulay, Coal Specialist
  Mineral Technolgy Research and Development Center
  Ministry of Mines and Energy, Bandung

Mike Dunn, Geologist, Computer Applications
  Branch of Geology and Applied Technology
  Office of Surface Mining Reclamation and Enforcement
  Pittsburgh, Pennsylvania

Bambang Edi, Staff Petroleum Engineer
  Bureau of Environment and Technology
  Ministry of Mines and Energy

Ir. Mulyono Hadiprayitno, Program Regular
  Manpower Development Centre for Mines
  Ministry of Mines and Energy, Bandung

Otto Hasibuan, Mine Inspector
  Directorate of Technical Mining/DTPU
  Ministry of Mines and Energy

Ir. Dadang Kadarisman, Director
  Manpower Development Centre for Mines
  Ministry of Mines and Energy, Bandung

Dibyo Kuntjoro, Director
  Bureau of Environment and Technology
  Ministry of Mines and Energy

Hikman Manaf, Director
  Manpower Development Center
  Ministry of Mines and Energy, Bandung

Kuntoro Mangkusubroto,
  Minister of Mines and Energy, 1998
  Director General, Directorate of General Mining, 1996
  Ministry of Mines and Energy

Richard McNeer, Attorney
  Division of Surface Mining
  Office of the Solicitor
  U.S. Department of the Interior
  Washington, D.C.

Dr. Richard O. Miller, Chief
  Planning and Analysis Staff
  Office of Surface Mining, Reclamation and Enforcement
  Washington, D.C.

Nining Sudini Ningrum
  Mineral Technolgy Research and Development Center
  Ministry of Mines and Energy, Bandung

Eko Priyatno, Site General Manager
   PT Bukit Baiduri Enterprises
   Merandi, East Kalimantan

Gene Robinson, Surface Mining Reclamation Specialist
   Branch of Regulatory Programs
   Office of Surface Mining Reclamation and Enforcement
   Wyoming Field Office
   Casper, Wyoming

Peter L. Rutledge, Supervisory Program Manager
   Western Support Center
   Office of Surface Mining Reclamation and Enforcement
   Denver, Colorado

Umar Said, Secretary General
   Ministry of Mines and Energy

Charles E. Sandburg, Acting Assistant Director
   Western Support Center
   Office of Surface Mining Reclamation and Enforcement
   Denver, Colorado

M. Ridha Sanusi
   Manpower Development Centre for Mines
   Ministry of Mines and Energy, Bandung

Kurt Schenk, Power Specialist
   Energy and Mining Sector Unit
   East Asia and Pacific Region
   The World Bank
   Washington. D.C.

Dr. Thamrin Sihite, Chief
   Division of Environmental Managment and Spatial Zoning
   Bureau of Environment and Technology
   Ministry of Mines and Energy

P.H. Silitonga, Assistant Minister for Environment
   Ministry of Mines and Energy

Ir. Supriatna Suhala, Chief Mine Inspector, DTPU, Jakarta and Director,
Mineral Technology Research and Development Center
   Ministry of Mines and Energy, Bandung

E. Supriydi, Mining Engineer
   PT Tanito Harum
   Tenggarong, East Kalimantan

Sensus Suyanto, Quality Control and Environmental Department Head
  PT Tanito Harum
  Tenggarong, East Kalimantan

Ronald Tambunan, Senior Mine Inspector
  Directorate of Mines/DTPU
  Ministry of Mines and Energy

Rudiro Trisnandono, Mine Inspector
  Directorate of Mines/DTPU
  Ministry of Mines and Energy

Alfred Whitehouse, Project Director
  BLT-OSM Mining Environmental Project
  Ministry of Mines and Energy
  Jakarta
  [Formerly Chief, Program Support Division
  Eastern Support Center
  Office of Surface Mining Reclamation and Enforcement
  Pittsburgh, Pennsylvania]

B.S. Wydianto, Branch Manager
  PT Tanito Harum
  Samarinda, East Kalimantan

Kadar Wiryanto, Chief
  Division of Construction Services and Technology Assessment
  Bureau of Environment and Technology
  Ministry of Mines and Energy

Ir. Soemarno Witoro, Senior Mine Inspector
  Directorate of Technical Mining/DTPU
  Ministry of Mines and Energy

Index

Printed in the United States
by Baker & Taylor Publisher Services